Yang-Baxter Equation and Quantum Groups: An Algebraic Approach

Yang-Baxter Equation and Quantum Groups: An Algebraic Approach

Editor: Danny Hunt

NY RESEARCH PRESS
New York

Published by NY Research Press
118-35 Queens Blvd., Suite 400,
Forest Hills, NY 11375, USA
www.nyresearchpress.com

Yang-Baxter Equation and Quantum Groups: An Algebraic Approach
Edited by Danny Hunt

International Standard Book Number: 978-1-64725-441-4 (Hardback)

Cataloging-in-Publication Data

Yang-Baxter equation and quantum groups : an algebraic approach / edited by Danny Hunt.
p. cm.
Includes bibliographical references and index.
ISBN 978-1-64725-441-4
1. Yang-Baxter equation. 2. Quantum groups. 3. Algebra. 4. Mathematical physics.
5. Quantum field theory. I. Hunt, Danny.
QC174.52.Y36 Y36 2023
530.143--dc23

Contents

Preface VII

Chapter 1 **An Independent Set of Axioms of MV-Algebras and Solutions of the Set-Theoretical Yang–Baxter Equation** 1
Tahsin Oner, Ibrahim Senturk and Gulsah Oner

Chapter 2 **On Solutions to the Set-Theoretical Yang-Baxter Equation in Wajsberg-Algebras** 10
Tahsin Oner and Tugce Katican

Chapter 3 **On Transcendental Numbers: New Results and a Little History** 23
Solomon Marcus and Florin F. Nichita

Chapter 4 **The Hecke Bicategory** 30
Alexander E. Hoffnung

Chapter 5 **Valued Graphs and the Representation Theory of Lie Algebras** 58
Joel Lemay

Chapter 6 **Frobenius–Schur Indicator for Categories with Duality** 91
Kenichi Shimizu

Chapter 7 **Hasse-Schmidt Derivations and the Hopf Algebra of Non-Commutative Symmetric Functions** 127
Michiel Hazewinkel

Chapter 8 **Hopf Algebra Symmetries of an Integrable Hamiltonian for Anyonic Pairing** 133
Jon Links

Chapter 9 **Quasitriangular Structure of Myhill–Nerode Bialgebras** 144
Robert G. Underwood

Chapter 10 **The Duality between Corings and Ring Extensions** 160
Florin F. Nichita and Bartosz Zieliński

Chapter 11 **Bundles over Quantum Real Weighted Projective Spaces** 172
Tomasz Brzeziński and Simon A. Fairfax

Chapter 12 **The Sum of a Finite Group of Weights of a Hopf Algebra** 194
Apoorva Khare

Chapter 13 **From Coalgebra to Bialgebra for the Six-Vertex Model: The Star-Triangle Relation as a Necessary Condition for Commuting Transfer Matrices** 221
Jeffrey R. Schmidt

Permissions

List of Contributors

Index

Preface

The Yang–Baxter equation refers to a consistency equation which is based on the concept that particles may preserve their momentum while changing their quantum internal states in some scattering situations. It plays a significant role in theoretical physics and has numerous uses in various areas, ranging from condensed matter to string theory. The Yang-Baxter equation is linked to the universal gates from quantum computing and realizes a unification of some non-associative structures. The quantum Yang-Baxter equation led to the development of the theory of quantum groups. The theory was proposed as the language of quantum groups which is the suitable algebraic language for the solutions of quantum Yang-Baxter equation. This book aims to shed light on some of the unexplored aspects of Yang-Baxter equation and quantum groups. It presents researches and studies performed by experts across the globe. This book will serve as a reference to a broad spectrum of readers.

Various studies have approached the subject by analyzing it with a single perspective, but the present book provides diverse methodologies and techniques to address this field. This book contains theories and applications needed for understanding the subject from different perspectives. The aim is to keep the readers informed about the progresses in the field; therefore, the contributions were carefully examined to compile novel researches by specialists from across the globe.

Indeed, the job of the editor is the most crucial and challenging in compiling all chapters into a single book. In the end, I would extend my sincere thanks to the chapter authors for their profound work. I am also thankful for the support provided by my family and colleagues during the compilation of this book.

Editor

1

An Independent Set of Axioms of MV-Algebras and Solutions of the Set-Theoretical Yang–Baxter Equation

Tahsin Oner [1], Ibrahim Senturk [1,*] and Gulsah Oner [2]

1 Department of Mathematics, Ege University, İzmir 35100, Turkey; tahsin.oner@ege.edu.tr
2 Multidisciplinary Vocational School Technical Programme, Dokuz Eylul University, İzmir 35210, Turkey; gulsah.darilmaz@deu.edu.tr
* Correspondence: ibrahim.senturk@ege.edu.tr.

Academic Editor: Florin Felix Nichita

Abstract: The aim of this paper is to give a new equivalent set of axioms for *MV*-algebras, and to show that the axioms are independent. In addition to this, we handle Yang–Baxter equation problem. In conclusion, we construct a new set-theoretical solution for the Yang–Baxter equation by using *MV*-algebras.

Keywords: *MV*-algebras; equivalence of set of axioms; independence; Yang–Baxter equation

---

1. Introduction

An axiomatic system is composed of a certain undefined or primitive term (or terms) together with a set of statements which are called axioms. Axioms are presupposed to be true. A theorem is any statement that can be deduced from the axioms by using inference rules. So, when an axiomatic system is given, a natural question arises: are its axioms independent? For instance, independence of the Axiom of Choice from Zermelo–Fraenkel set theory's axioms took many decades to be confirmed.

An independent axiomatization of any set of formulas from propositional or predicate logic was difficult to settle down. Many logicians have tried to undertake a solution, but they failed. The first partial success for countable sets was obtained by Tarski [1]. Kreisel [2] worked through research on independent recursive axiomatization.

A set of sentences Γ is *independent* if for all $\varphi \in \Gamma$, φ is not a logical consequence of $\Gamma - \{\varphi\}$, or equivalently, if there is a model of $(\Gamma - \{\varphi\}) \cup \{\neg\varphi\}$, where $\neg$ is the negation operation.

To show that a given finite set of formulas $\{F_1, F_2, ..., F_n\}$ is independent, it is sufficient to find an assignment of truth values that satisfies all the formulas F_j for each $1 \leq i \leq n$, where $j \neq i$, and that does not satisfy F_i. On the contrary, to demonstrate that it is not independent, one shows that one of the formulas F_i is a consequence of the other formulas.

Two sets of formulas are said to be *equivalent* if any formula of one set is a consequence of the other set, and conversely.

Oner and Terziler [3] have exemplified the situation by considering axioms for Boolean algebras; the proof of independence is obtained by using model forming. Additionally, Chajda and Kolařík [4] have proved that the axioms of basic algebras given in Chajda and Emanovský [5] are not independent.

Here, we consider MV-algebras which were originally introduced by Chang [6] and further developed by Chang [7]. A simplified axiomatization of MV-algebras in use today can be found in the monograph Algebraic Foundations of Many-valued Reasoning [8].

Definition 1. *[8] An MV-algebra is an algebra $\mathcal{A} = (A, \oplus, \neg, 0)$ satisfying the following axioms (where A is a nonempty set, $\oplus$ is a binary operation on A, $\neg$ is a unary operation on A, and 0 is a constant element of A):*
$(MV1)$ $x \oplus 0 = x$
$(MV2)$ $\neg\neg x = x$
$(MV3)$ $x \oplus y = y \oplus x$
$(MV4)$ $(x \oplus y) \oplus z = x \oplus (y \oplus z)$
$(MV5)$ $\neg(\neg x \oplus y) \oplus y = \neg(\neg y \oplus x) \oplus x$
$(MV6)$ $x \oplus \neg 0 = \neg 0$.

In [9,10], this axiomatic system for $MV-$algebras was shown not to be independent. Moreover, M. Kolařík [10] proved that if the commutativity axiom is omitted in the above set of axiomatic system, the remaining axioms are independent.

On the other hand, firstly used in theoretical physics [11] and statistical mechanics [12–14], the Yang–Baxter equation has received more and more attention from researchers in a wide range of disciplines in the last years. Intending to apply the Yang–Baxter equation to individual works with different aspects from theoretical to practical in almost all sciences, technology, and industry, researchers have found uncountable and varied applications of this equation in fields such as link invariants, quantum computing, quantum groups, braided categories, knot theory, the analysis of integrable systems, quantum mechanics, etc. ([15,16]). In addition to these, one of the areas which the Yang–Baxter equation was practised is pure mathematics, and there have been many studies in this field, as in other areas. Several researchers have benefited from the axioms of various algebraic structures to solve this equation. Since it is well known that $MV-$algebras which provide an algebraic proof of the completeness theorem of infinite-valued Logics (or Łukasiewicz logics) are important algebraic tools to study in quantum structures, set-theoretical solutions for the Yang–Baxter equation could be examined in $MV-$algebras.

In this paper, we give some theorems in $MV-$algebras and obtain an equivalent set of axioms for $MV-$algebras in Section 2. In Section 3, we prove the independence of this equivalent set of axioms. In the last section, we find some solutions of the set-theoretical Yang–Baxter equation in $MV-$algebras.

2. An Equivalent Set of Axioms for *MV*-Algebras

In this section, we introduce a modified set of axioms for $MV-$algebras. To this end, we add a new axiom and remove axioms $(MV1)$ and $(MV6)$ from the above axiomatization of $MV-$algebras, and we prove that these two sets of axioms are equivalent to each other.

Theorem 1. *Let $\mathcal{A} = (A, \oplus, \neg, 0)$ be an $MV-$algebra. Then, there exists a unique element $y = 0 \in A$ such that $y \oplus x = x$ for all $x \in A$.*

Proof. By using $(MV1)$ and $(MV3)$, there exists a $0 \in A$ such that $0 \oplus x = x$ for all $x \in A$. For the uniqueness of this element, we take $x = 0$. Then, by the hypothesis:

$$y \oplus 0 = 0,$$

and since $(MV1)$ implies that $y \oplus 0 = y$, we have

$$y = 0.$$

For $x = 0$, the solution set of y contains only 0. In that case, the intersection sets of y solution values which verify this equation also contains 0. Thus, $y = 0$ is the unique element of A which holds $y \oplus x = x$ for all $x \in A$. □

Theorem 2. *Let* $\mathcal{A} = (A, \oplus, \neg, 0)$ *be an* $MV-$*algebra. Then*

$$\forall x \in A, x \oplus y = x \Leftrightarrow y = 0.$$

Proof. *Necessity*: It is clear by $(MV3)$ and Theorem 1.
Sufficiency: It follows from $(MV1)$. □

Theorem 3. $\mathcal{A} = (A, \oplus, \neg, 0)$ *is an* $MV-$*algebra if and only if it satisfies the following axioms:*
$(MV1')$ $x \oplus y = y \Leftrightarrow x = 0$
$(MV2)$ $\neg\neg x = x$
$(MV3)$ $x \oplus y = y \oplus x$
$(MV4)$ $(x \oplus y) \oplus z = x \oplus (y \oplus z)$
$(MV5)$ $\neg(\neg x \oplus y) \oplus y = \neg(\neg y \oplus x) \oplus x.$

Proof. Necessity follows from Theorem 1 and [10]. For sufficiency, we have to show that the axioms $(MV1)$ and $(MV6)$ follow from the axioms in Theorem 3. We have $0 \oplus x = x$ by $(MV1')$. For $(MV1)$, it follows from $(MV3)$ that $0 \oplus x = x = x \oplus 0$. Now, we give a deduction of $(MV6)$ as follows:

$$\begin{array}{lcll}
\neg x \oplus 0 & = & \neg x & \text{(by } (MV1)) \\
& \Rightarrow & \neg(\neg x \oplus 0) = \neg\neg x & \\
& \Rightarrow & \neg(\neg x \oplus 0) = x & \text{(by } (MV2)) \\
& \Rightarrow & \neg(\neg x \oplus 0) \oplus 0 = x \oplus 0 = x & \text{(by } (MV1)) \\
& \Rightarrow & \neg(\neg 0 \oplus x) \oplus x = x & \text{(by } (MV5)) \\
& \Rightarrow & \neg(\neg 0 \oplus x) = 0 & \text{(by } (MV1')) \\
& \Rightarrow & \neg\neg(\neg 0 \oplus x) = \neg 0 & \\
& \Rightarrow & \neg 0 \oplus x = \neg 0 & \text{(by } (MV2)) \\
& \Rightarrow & x \oplus \neg 0 = \neg 0 & \text{(by } (MV3)).
\end{array}$$

Therefore, the set of axioms $(MV1') - (MV5)$ is an equivalent set of axioms for MV-algebras. □

3. Proof of the Independence of the Equivalent Set of Axioms

Theorem 4. *The axioms* $(MV1')$, $(MV2)$, $(MV3)$, $(MV4)$, *and* $(MV5)$ *are independent.*

Proof. To prove this claim, we construct a model for each axiom in which this axiom is false while the others are true. Let $\mathcal{M}$ be our model with the universe $U = \{0, 1, 2\}$. The symbols $\oplus^{\mathcal{M}}$, $\neg^{\mathcal{M}}$, and $0^{\mathcal{M}}$ are interpreted as a binary operation, a unary operation, and a constant, respectively. Let $\mathcal{A} = (\{0, 1, 2\}, \oplus, \neg, 0)$ be an algebra whose operations are defined as in the following tables:

(1) **Independence of** $(MV1')$:
We can define the operations $\oplus^{\mathcal{M}}$ and $\neg^{\mathcal{M}}$ as in the Table 1:

Table 1. Operations tables for the independence of $(MV1')$.

$\oplus$	0	1	2
0	0	0	0
1	0	1	1
2	0	1	1

x	0	1	2
$\neg x$	0	1	2

The algebra $\mathcal{A}$ is not a model for $(MV1')$ because, if we choose $x = 0$ and $y = 1$, we obtain $x \oplus y = 0 \oplus 1 = 0$, which is a contradiction with $(MV1')$. Clearly, the algebra $\mathcal{A}$ satisfies $(MV2)$ and $(MV3)$. To see that $(MV4)$ and $(MV5)$ are satisfied by the algebra $\mathcal{A}$, it is sufficient to consider the following:

For $(MV4)$:

(i) If $x = 0$, then $0 \oplus (y \oplus z) = 0 = (0 \oplus y) \oplus z$.
(ii) If $x = 1$, then $1 \oplus (y \oplus z) = (1 \oplus y) \oplus z$ is true.
(iii) If $x = 2$, then $2 \oplus (y \oplus z) = (2 \oplus y) \oplus z$ is also true.

As for $(MV5)$:

(i) If $y = 0$, then $\neg(\neg x \oplus 0) \oplus 0 = 0 = \neg(\neg 0 \oplus x) \oplus x$.
(ii) If $y = 1$, then $\neg(\neg x \oplus 1) \oplus 1 = \neg(\neg 1 \oplus x) \oplus x$. It can be checked from the operations in Table 1 that the equation holds for $y = 1$.
(iii) If $y = 2$, then $\neg(\neg x \oplus 2) \oplus 2 = \neg(\neg 2 \oplus x) \oplus x$. It can be also controlled from the operations in Table 1 that the equation holds for $y = 2$.

(2) **Independence of** $(MV2)$**:**
The operations $\oplus^{\mathcal{M}}$ and $\neg^{\mathcal{M}}$ are defined as in the Table 2:

Table 2. Operations Tables for Independence of $(MV2)$.

$\oplus$	0	1	2
0	0	1	2
1	1	2	2
2	2	2	2

x	0	1	2
$\neg x$	2	2	2

The algebra $\mathcal{A}$ satisfies $(MV1')$, $(MV3)$, $(MV4)$, $(MV5)$ but not $(MV2)$ because $\neg\neg 0 \neq 0$.

(3) **Independence of** $(MV3)$**:**
The operations $\oplus^{\mathcal{M}}$ and $\neg^{\mathcal{M}}$ are defined as in the Table 3:

Table 3. Operations Tables for Independence of $(MV3)$.

$\oplus$	0	1	2
0	0	1	2
1	1	1	1
2	1	1	1

x	0	1	2
$\neg x$	0	1	2

The algebra $\mathcal{A}$ satisfies $(MV1')$, $(MV2)$, $(MV4)$, $(MV5)$, but not $(MV3)$ because $0 \oplus 2 = 2$ but $2 \oplus 0 = 1$.

(4) **Independence of** $(MV4)$**:**
The operations $\oplus^{\mathcal{M}}$ and $\neg^{\mathcal{M}}$ are defined as in the Table 4:

Table 4. Operations tables for the independence of $(MV4)$.

$\oplus$	0	1	2
0	0	1	2
1	1	2	1
2	2	1	1

x	0	1	2
$\neg x$	1	0	2

The algebra $\mathcal{A}$ satisfies $(MV1')$, $(MV2)$, $(MV3)$, $(MV4)$, but not $(MV5)$ because $(1 \oplus 1) \oplus 2 = 1$ but $(1 \oplus 1) \oplus 2 = 2$. So, $(1 \oplus 1) \oplus 2 \neq 1 \oplus (1 \oplus 2)$.

(5) **Independence of $(MV5)$:**
The operations $\oplus^{\mathcal{M}}$ and $\neg^{\mathcal{M}}$ are defined as in the Table 5:

Table 5. Operations tables for the independence of $(MV5)$.

$\oplus$	0	1	2
0	0	1	2
1	1	1	1
2	2	1	1

x	0	1	2
$\neg x$	2	1	0

The algebra $\mathcal{A}$ satisfies $(MV1')$, $(MV2)$, $(MV3)$, $(MV5)$, but not $(MV4)$, because $\neg(\neg 0 \oplus 2) \oplus 2 = 1$ but $\neg(\neg 2 \oplus 0) \oplus 0 = 2$. So, $\neg(\neg 0 \oplus 2) \oplus 2 \neq \neg(\neg 2 \oplus 0) \oplus 0 = 2$.

□

4. Solutions of the Yang–Baxter Equation in *MV*-Algebras

Firstly used in theoretical physics by C.N. Yang and in statistical mechanics by R.J. Baxter almost fifty years ago, the Yang–Baxter equation has been studied as the master equation in integrable models in statistical mechanics and quantum field theory. Recent progress in other fields such as C^*- algebras, Hopf algebras, simple Lie algebras, representation theory, conformal field theory, etc. shed light on the significance of the equation, and has drawn attention among many researchers which have used the axioms of these algebraic structures in order to solve this equation.

In this section, we present solutions of the set-theoretical Yang–Baxter equation in $MV-$algebras. Let k be a field and tensor products be defined over the field k. For V a k-space, we denote by $\tau : V \otimes V \to V \otimes V$ the twist map defined by $\tau(v \otimes w) = w \otimes v$ and by $I : V \to V$ the identity map of the space V; for $R : V \otimes V \to V \otimes V$ a $k-$linear map, let $R^{12} = R \otimes I$, $R^{23} = I \otimes R$, and $R^{13} = (I \otimes \tau)(R \otimes I)(\tau \otimes I)$.

Definition 2. *[16] A Yang–Baxter operator is $k-$linear map $R : V \otimes V \to V \otimes V$, which is invertible, and it satisfies the braid condition (called the Yang–Baxter equation):*

$$R^{12} \circ R^{23} \circ R^{12} = R^{23} \circ R^{12} \circ R^{23}. \quad (1)$$

If R satisfies Equation (1), then both $R \circ \tau$ and $\tau \circ R$ satisfy the quantum Yang–Baxter equation (QYBE):

$$R^{12} \circ R^{13} \circ R^{23} = R^{23} \circ R^{13} \circ R^{12}. \quad (2)$$

Lemma 1. *[16] Equations (1) and (2) are equivalent to each other.*

To establish a relation between the set-theoretical Yang–Baxter equation and $MV-$algebras, we need the following definition.

Definition 3. *[16] Let$X$ be a set and $S : X \times X \to X \times X$, $S(u,v) = (u',v')$ be a map. The map S is a solution for the set-theoretical Yang–Baxter equation if it satisfies the following equation:*

$$S^{12} \circ S^{23} \circ S^{12} = S^{23} \circ S^{12} \circ S^{23}, \tag{3}$$

which is also equivalent

$$S^{12} \circ S^{13} \circ S^{23} = S^{23} \circ S^{13} \circ S^{12}, \tag{4}$$

where

$$\begin{aligned} S^{12} &: X \times X \times X \to X \times X \times X, \;\; S^{12}(u,v,w) = (u',v',w), \\ S^{23} &: X \times X \times X \to X \times X \times X, \;\; S^{23}(u,v,w) = (u,v',w'), \\ S^{13} &: X \times X \times X \to X \times X \times X, \;\; S^{13}(u,v,w) = (u',v,w'). \end{aligned}$$

Now, we give a new method to construct solutions of the set theoretical Yang–Baxter equation by using $MV-$algebras.

Lemma 2. *Let $(A, \oplus, \neg, 0)$ be an $MV-$algebra. Then, $S(x,y) = (x \oplus y, 0)$ is a solution of the set-theoretical Yang–Baxter equation.*

Proof. Let S^{12} and S^{23} be defined as follows:

$$\begin{aligned} S^{12}(x,y,z) &= (x \oplus y, 0, z), \\ S^{23}(x,y,z) &= (x, y \oplus z, 0). \end{aligned}$$

We verify the equality $S^{12} \circ S^{23} \circ S^{12} = S^{23} \circ S^{12} \circ S^{23}$ for all $(x,y,z) \in A \times A \times A$. By using $(MV1')$ and $(MV4)$, we get

$$\begin{aligned} (S^{12} \circ S^{23} \circ S^{12})(x,y,z) &= S^{12}(S^{23}(S^{12}(x,y,z))) \\ &= S^{12}(S^{23}(x \oplus y, 0, z) \\ &= S^{12}(x \oplus y, 0 \oplus z, 0) \\ &= S^{12}(x \oplus y, z, 0)) \\ &= ((x \oplus y) \oplus z, 0, 0) \\ &= (x \oplus (y \oplus z), 0, 0) \\ &= (x \oplus (y \oplus z), 0 \oplus 0, 0) \\ &= S^{23}(x \oplus (y \oplus z), 0, 0) \\ &= S^{12}(S^{23}(x, y \oplus z, 0)) \\ &= (S^{23} \circ S^{12} \circ S^{23})(x,y,z). \end{aligned}$$

Hence $S(x,y) = (x \oplus y, 0)$ is a solution of the set-theoretical Yang–Baxter equation in $MV-$algebras. □

Lemma 3. *Let $(A, \oplus, \neg, 0)$ be an $MV-$algebra. Then, $S(x,y) = (\neg y, \neg x)$ is a solution of the set-theoretical Yang–Baxter equation.*

Proof. We define S^{12} and S^{23} as follows:

$$S^{12}(x,y,z) = (\neg y, \neg x, z),$$
$$S^{23}(x,y,z) = (x, \neg z, \neg y).$$

We show that the equality $S^{12} \circ S^{23} \circ S^{12} = S^{23} \circ S^{12} \circ S^{23}$ holds for all $(x,y,z) \in A \times A \times A$. By using $(MV2)$, we attain

$$\begin{aligned}(S^{12} \circ S^{23} \circ S^{12})(x,y,z) &= S^{12}(S^{23}(S^{12}(x,y,z))) \\ &= S^{12}(S^{23}(\neg y, \neg x, z) \\ &= S^{12}(\neg y, \neg z, \neg\neg x) \\ &= S^{12}(\neg y, \neg z, x)) \\ &= (\neg\neg z, \neg\neg y, x) \\ &= (z, y, x)\end{aligned}$$

and

$$\begin{aligned}(S^{23} \circ S^{12} \circ S^{23})(x,y,z) &= S^{23}(S^{12}(S^{23}(x,y,z))) \\ &= S^{23}(S^{12}(x, \neg z, \neg y) \\ &= S^{23}(\neg\neg z, \neg x, \neg y) \\ &= S^{23}(z, \neg x, \neg y)) \\ &= (z, \neg\neg y, \neg\neg x) \\ &= (z, y, x).\end{aligned}$$

Thus, $S(x,y) = (\neg y, \neg x)$ is a solution of the set-theoretical Yang–Baxter equation in $MV-$algebras. □

Definition 4. *[8] Let A be an $MV-$algebra.The binary relation $\leq$ defined on A as below*

$$x \leq y \text{ if and only if } \neg x \oplus y = 1$$

is a partial order on A. It is called the natural order of A.

Proposition 1. *[8] Let A be an $MV-$algebra with the natural order. If the join and the meet operators are defined as below:*

$$x \vee y = \neg(\neg x \oplus y) \oplus y \text{ and } x \wedge y = \neg(\neg x \oplus \neg(\neg x \oplus y)),$$

then the natural order determines a lattice structure on A.

Lemma 4. *[8] Any $MV-$algebra A is a Boolean algebra if and only if the operation $\oplus$ is idempotent on A.*

Lemma 5. *[17] Let $(A, \vee, \wedge, 0, 1, \neg)$ be a Boolean algebra. Then $S(x,y) = (x \vee y, x \wedge y)$ is a solution of the set-theoretical Yang–Baxter equation.*

Proposition 2. *Let $(A, \oplus, \neg, 0)$ be an MV−algebra. If the identities $\neg(\neg(x \oplus y) \oplus \neg z) = \neg(\neg x \oplus \neg z) \oplus \neg(\neg y \oplus \neg z)$ or $\neg(\neg x \oplus \neg y) \oplus z = \neg(\neg(x \oplus z) \oplus \neg(y \oplus z))$ are satisfied for all $x, y, z \in A$, then*

$$S(x,y) = (\neg(\neg x \oplus y) \oplus y), \neg(\neg x \oplus \neg(\neg x \oplus y)) \text{ and } S(x,y) = (x \oplus y, \neg(\neg x \oplus \neg y))$$

are solutions for the set-theoretical Yang–Baxter equation in MV−algebra A.

Proof. Assume that $\neg(\neg(x \oplus y) \oplus \neg z) = \neg(\neg x \oplus \neg z) \oplus \neg(\neg y \oplus \neg z)$ or $\neg(\neg x \oplus \neg y) \oplus z = \neg(\neg(x \oplus z) \oplus \neg(y \oplus z))$ holds for all $x, y, z \in A$. Now substituting $[x := 1, y := 1]$ or $[x := 0, y := 0]$ in these identities, respectively, we get $z = z \oplus z$ or $\neg z = \neg z \oplus \neg z$ for all $z \in A$. Then, the operation $\oplus$ is idempotent on A and so by Lemma 4, A is a Boolean algebra. From the definitions of meet and join operators, we obtain $S(x,y) = (\neg(\neg x \oplus y) \oplus y, \neg(\neg x \oplus \neg(\neg x \oplus y))) = (x \vee y, x \wedge y)$. Therefore, $S(x,y) = (\neg(\neg x \oplus y) \oplus y, \neg(\neg x \oplus \neg(\neg x \oplus y)))$ is a solution of the Yang–Baxter equation in $MV-$algebra A from Lemma 5. In addition, since the operation is idempotent, we have $(\neg(\neg x \oplus y) \oplus y, \neg(\neg x \oplus \neg(\neg x \oplus y))) = (x \oplus y, \neg(\neg x \oplus \neg y))$. Hence, $S(x,y) = (x \oplus y, \neg(\neg x \oplus \neg y))$ is also a solution of Yang–Baxter equation in $MV-$algebra A. □

Lemma 6. *[8] The below equations hold in every MV−algebra A:*

(i) $x \oplus y \oplus \neg(\neg x \oplus \neg y) = x \oplus y$,
(ii) $\neg(\neg x \oplus y) \oplus \neg(\neg(x \oplus \neg y) \oplus \neg y) = x$,
(iii) $\neg(\neg x \oplus \neg y) \oplus \neg(\neg(x \oplus y) \oplus \neg z) = \neg(\neg x \oplus \neg z) \oplus \neg(\neg(x \oplus z) \oplus \neg y)$.

Munduci et al. [8] needed the following proposition for the proof of the above lemma.

Proposition 3. *[8] The following identity*

$$\neg(\neg x \oplus \neg y) \oplus \neg(\neg(x \oplus y) \oplus \neg z) = \neg(\neg(x \oplus y) \oplus \neg(\neg(\neg x \oplus \neg y) \oplus z))$$

holds for each MV−algebra.

Theorem 5. *Let $(A, \oplus, \neg, 0)$ be an MV−algebra. Then, $S(x,y) = (x \oplus y, \neg(\neg x \oplus \neg y))$ is a solution of the set-theoretical Yang–Baxter equation in MV−algebra A.*

Proof. Through the instrumentality of equalities in Lemma 6 and Proposition 3, we obtain the following equalities:

$$\begin{aligned}
(S^{12} \circ S^{23} \circ S^{12})(x,y,z) &= S^{12}(S^{23}(S^{12}(x,y,z))) \\
&= S^{12}(S^{23}(x \oplus y, \neg(\neg x \oplus \neg y), z)) \\
&= S^{12}(x \oplus y, \neg(\neg x \oplus \neg y) \oplus z, \neg(\neg\neg(\neg x \oplus \neg y) \oplus \neg z)) \\
&= S^{12}(x \oplus y, \neg(\neg x \oplus \neg y) \oplus z, \neg((\neg x \oplus \neg y) \oplus \neg z)) \\
&= ((x \oplus y) \oplus (\neg(\neg x \oplus \neg y) \oplus z), \neg(x \oplus y) \oplus \neg(\neg(\neg x \oplus \neg y) \oplus z), \\
&\quad \neg((\neg x \oplus \neg y) \oplus \neg z)) \\
&= (((x \oplus y) \oplus \neg(\neg x \oplus \neg y)) \oplus z, \neg(\neg(x \oplus y) \oplus \neg(\neg(\neg x \oplus \neg y) \oplus z)), \\
&\quad \neg((\neg x \oplus \neg y) \oplus \neg z)) \\
&= (x \oplus y \oplus z, \neg(\neg(x \oplus y) \oplus \neg(\neg(\neg x \oplus \neg y) \oplus z)), \neg(\neg x \oplus \neg y \oplus \neg z)),
\end{aligned}$$

and for the other direction of the set-theoretical Yang baxter equation, we attain

$$
\begin{aligned}
(S^{23} \circ S^{12} \circ S^{23})(x,y,z) &= S^{23}(S^{12}(S^{23}(x,y,z))) \\
&= S^{23}(S^{12}(x, y \oplus z, \neg(\neg y \oplus \neg z))) \\
&= S^{23}(x \oplus (y \oplus z), \neg(\neg x \oplus \neg(y \oplus z)), \neg(\neg y \oplus \neg z)) \\
&= (x \oplus (y \oplus z), \neg(\neg x \oplus \neg(y \oplus z)) \oplus \neg(\neg y \oplus \neg z), \neg(\neg\neg(\neg x \oplus \neg y \oplus z) \\
&\quad \oplus \neg\neg(\neg y \oplus \neg z))) \\
&= (x \oplus y \oplus z, \neg(\neg x \oplus \neg(y \oplus z)) \oplus \neg(\neg y \oplus \neg z), \neg((\neg x \oplus \neg(y \oplus z)) \\
&\quad \oplus(\neg y \oplus \neg z))) \\
&= (x \oplus y \oplus z, \neg(\neg x \oplus \neg(y \oplus z)) \oplus \neg(\neg y \oplus \neg z), \neg(\neg x \oplus (\neg(y \oplus z) \\
&\quad \oplus(\neg y \oplus \neg z)))) \\
&= (x \oplus y \oplus z, \neg(\neg x \oplus \neg y) \oplus \neg(\neg(x \oplus y) \oplus \neg z), \neg(\neg x \oplus \neg y \oplus \neg z).
\end{aligned}
$$

Therefore, $S(x,y) = (x \oplus y, \neg(\neg x \oplus \neg y))$ is a solution of the set-theoretical Yang–Baxter equation in $MV-$algebra A. □

Acknowledgments: The authors thank the academic editor for his valuable comments and suggestions and the anonymous referees for his/her remarks which helped them to improve the presentation of the paper.

Author Contributions: The coauthors, Tahsin Oner, Ibrahim Senturk and Gulsah Oner wrote the paper jointly and contributed equally to this work.

References

1. Tarski, A. *Logic, Semantics, Metamathematics*; The Clarendon Press: Oxford, UK, 1956.
2. Kreisel, G. Independent recursive axiomatization. *J. Symb. Log.* **1957**, *22*, 109.
3. Oner, T.; Terziler, M. Independent of countable sets of formulas of the propositional logic. *ARS Comb.* **2013**, *12*, 73–80.
4. Chajda, I.; Kolařík, M. Independence of the axiomatic system of basic algebras. *Soft Comput.* **2009**, *13*, 41–43.
5. Chajda I.; Emanovský, P. Bounded lattices with antitone involutions and properties of MV-Algebras. *Discuss. Math. Gen. Algebra Appll.* **2008**, *24*, 31–42.
6. Chang, C.C. Algebraic analysis of many-valued logics. *Trans. Am. Soc.* **1958**, *88*, 467–490.
7. Chang, C.C. A new proof of the completeness of the Łukasiewicz axioms. *Trans. Am. Soc.* **1958**, *93*, 74–80.
8. Cignoli, R.; D'Ottaviano, I.; Mundici, D. *Algebraic Foundations of Many-Valued Reasoning*; Kluwer Academic Publishers: Dordrech, The Netherlands, 2000.
9. Cattaneo, G. Independent axiomatization of MV-Algebras. *Tatra Mt. Math. Publ.* **1998**, *15*, 227–232.
10. Kolařík, M. Independence of the axiomatic system for MV-Algebras. *Math. Slov.* **2013**, *63*, 1–4.
11. Perk, J.H.H.; Au, Y.H. Yang-Baxter Equations. In *Encyclopedia of Mathematical Physics*; Françoise, J.-P., Naber, G.L., Tsou, S.T., Eds.; Elseiver: Oxford, UK, 2006; Volume 5, pp. 465–473.
12. Baxter, R.J. *Exactly Solved Models in Statistical Mechanics*; Academy Press: London, UK, 1982.
13. Baxter, R.J. Partition function of the eight-vertex lattice model. *Ann. Phys.* **1972**, *70*, 193–228.
14. Yang, C.N. Some exact results for the many-body problem in one dimension with repulsive delta-function interaction. *Phys. Rev. Lett.* **1967**, *19*, 1312–1315.
15. Nichita, F.F. Special Issue "Hopf Algebras, Quantum Groups and Yang–Baxter Equations, 2014". Available online: http://www.mdpi.com/journal/axioms/special_issues/hopf_algebras_2014 (accessed on 22 June 2017).
16. Nichita, F.F. Yang-Baxter Equations, Computational Methods and Applications. *Axioms* **2015**, *4*, 423–435.
17. Nichita, F.F. On the set-theoretical Yang-Baxter equation. *Acta Univ. Apulensis Math.-Inf.* **2003**, *5*, 97–100.

2

On Solutions to the Set-Theoretical Yang-Baxter Equation in Wajsberg-Algebras

Tahsin Oner * and **Tugce Katican**

Department of Mathematics, Ege University, 35100 Izmir, Turkey; tugcektcn@gmail.com
* Correspondence: tahsin.oner@ege.edu.tr

Abstract: In this work, we introduce Wajsberg algebras which are equivalent structures to MV-algebras in their implicational version, and then we define new notions and give new solutions to the set-theoretical Yang-Baxter equation by using Wajsberg algebras.

Keywords: Yang-Baxter equation; Wajsberg-algebra

1. Introduction

The Yang-Baxter equation which was initially used in theoretical physics [1] and statical mechanics [2–4] has gradually attracted the attention of researchers from various areas of science. In particular, this equation is considered in areas such as link invariant, C^*-algebras, conformal field theory, quantum computing, quantum groups, quantum mechanics, knot theory, intregrable systems, non-commutative geometry, etc. (see, for example, [5–11]).

The problem to find and study (set-theoretical) solutions of the Yang-Baxter equation has attracted many authors.

The Yang-Baxter equation involves a linear operator $R : V \otimes V \longrightarrow V \otimes V$, where V is a vector space and has the form

$$R^{12} \circ R^{23} \circ R^{12} = R^{23} \circ R^{12} \circ R^{23} \text{ in } End(V \otimes V \otimes V)$$

where $1 \leq n,\ m \leq 3$ and R^{nm} means R acting in the n-th and m-th components. In the last years, many set-theoretical solutions of this equation have given rise to the connection with various mathematical structures, such as quantum binomial algebras [12,13], semigroups of I-type and Bieberbach groups [14,15], bijective 1-cocyles [16], semisimple minimal triangular Hopf algebras [17], dynamical systems [18], and geometric crystals [19].

Since Wajsberg has shown that ∞-valued Lukasiewicz logics were complete with respect to the axioms postulated by Lukasiewicz, these logics were shown to play an important role in the study of quantum physics. We then wish to investigate the Yang-Baxter equation rather in relation with Wajsberg-algebras than quantum physics.

The set-theoretical solutions to the Yang-Baxter equation using MV-algeras were given by [11].

In this paper, we give some solutions to the set-theoretical Yang–Baxter equation in Wajsberg algebras.

2. Preliminaries

In this section, we present some definitions and properties of Wajsberg-algebras.

Definition 1. [20] *A Wajsberg algebra (briefly, a W-algebra) is a structure $\langle A, \longrightarrow, \neg, 1\rangle$ satisfying the following equations, where A is a nonempty set, $\neg$ is a unary operation on A, $\longrightarrow$ is a binary operation on A, and 1 is a distinguished element of A:*

(W1) $1 \longrightarrow a = a$

(W2) $(a \longrightarrow b) \longrightarrow ((b \longrightarrow c) \longrightarrow (a \longrightarrow c)) = 1$

(W3) $(a \longrightarrow b) \longrightarrow b = (b \longrightarrow a) \longrightarrow a$

(W4) $(\neg a \longrightarrow \neg b) \longrightarrow (b \longrightarrow a) = 1.$

Moreover, bounded commutative BCI/BCK-algebras (BCK-algebras are special cases of BCI-algebras, for example, see [21]) , in their implicational notation, are known as Wajsberg algebras (see also [22]).

Lemma 1. [20] *Let $\langle A, \longrightarrow, 1\rangle$ be a system satisfying (W1), (W2) and (W3). Then the following properties hold for every a, b and c in A:*

(W5) $a \longrightarrow a = 1$

(W6) *If* $a \longrightarrow b = b \longrightarrow a = 1$, *then* $a = b$.

(W7) $a \longrightarrow 1 = 1$

(W8) $a \longrightarrow (b \longrightarrow a) = 1$

(W9) *If* $a \longrightarrow b = b \longrightarrow c = 1$, *then* $a \longrightarrow c = 1$.

(W10) $(a \longrightarrow b) \longrightarrow ((c \longrightarrow a) \longrightarrow (c \longrightarrow b)) = 1$

(W11) $a \longrightarrow (b \longrightarrow c) = b \longrightarrow (a \longrightarrow c)$.

Lemma 2. [20] *The following equations hold in every W-algebra:*

(i) $\neg 1 \longrightarrow a = 1,$

(ii) $\neg a = a \longrightarrow \neg 1,$

(iii) $\neg\neg a = a,$

(iv) $a \longrightarrow b = \neg b \longrightarrow \neg a.$

3. Solutions to the Yang-Baxter Equation in W-Algebras

In this section, we provide solutions to the set-theoretical Yang-Baxter equation in W-algebras. Let V be a vector space over a field F. We denote by $\tau : V \otimes V \longrightarrow V \otimes V$ the twist map defined by $\tau(v \otimes w) = w \otimes v$ and by $I : V \longrightarrow V$ the identity map over the space V; for a F-*linear* map $R : V \otimes V \longrightarrow V \otimes V$, let $R^{12} = R \otimes I$, $R^{23} = I \otimes R$, and $R^{13} = (I \otimes \tau)(R \otimes I)(\tau \otimes I)$.

Definition 2. [10] *A Yang-Baxter operator is an invertible F-linear map $R : V \otimes V \longrightarrow V \otimes V$, and it satisfies the braid condition (also called the Yang-Baxter equation):*

$$R^{12} \circ R^{23} \circ R^{12} = R^{23} \circ R^{12} \circ R^{23}. \tag{1}$$

If R satisfies Equation (1), then both $R \circ \tau$ and $\tau \circ R$ satisfy the quantum Yang-Baxter equation:

$$R^{12} \circ R^{13} \circ R^{23} = R^{23} \circ R^{13} \circ R^{12}. \tag{QYBE}$$

The following definition enables us to constitute a relationship between the set-theoretical Yang-Baxter equation and W-algebras.

Definition 3. [10] *Let X be a set and $S : X^2 \longrightarrow X^2$, $S(p_1, p_2) = (p'_1, p'_2)$ be a map. The map S is a solution to the set-theoretical Yang-Baxter equation if it satisfies the following identity:*

$$S^{12} \circ S^{23} \circ S^{12} = S^{23} \circ S^{12} \circ S^{23}, \tag{2}$$

where

$$S^{12} : X^3 \longrightarrow X^3, \quad S^{12}(p_1, p_2, p_3) = (p'_1, p'_2, p_3),$$

$$S^{23} : X^3 \longrightarrow X^3, \quad S^{23}(p_1, p_2, p_3) = (p_1, p_2', p_3'),$$
$$S^{13} : X^3 \longrightarrow X^3, \quad S^{13}(p_1, p_2, p_3) = (p_1', p_2, p_3').$$

Now, we provide solutions to the set-theoretical Yang-Baxter equation by using W-algebras.

Theorem 1. *Let $\langle A, \longrightarrow, \neg, 1\rangle$ be a W-algebra. Then $S(a,b) = (\neg a \longrightarrow b, \neg 1)$ is a solution to the set-theoretical Yang-Baxter equation.*

Proof. S^{12} and S^{23} are defined in the following forms:

$$S^{12}(a,b,c) = (\neg a \longrightarrow b, \neg 1, c), \quad S^{23}(a,b,c) = (a, \neg b \longrightarrow c, \neg 1).$$

For all $(a,b,c) \in A^3$, we get

$$\begin{aligned}
(S^{12} \circ S^{23} \circ S^{12})(a,b,c) &= (S^{12} \circ S^{23})(S^{12}(a,b,c)) \\
&= (S^{12} \circ S^{23})(\neg a \longrightarrow b, \neg 1, c) \\
&= S^{12}(S^{23}(\neg a \longrightarrow b, \neg 1, c)) \\
&= S^{12}(\neg a \longrightarrow b, \neg\neg 1 \longrightarrow c, \neg 1) \\
&= S^{12}(\neg a \longrightarrow b, 1 \longrightarrow c, \neg 1) && (\textit{Lemma } 2\ (iii)) \\
&= S^{12}(\neg a \longrightarrow b, c, \neg 1) && (W1) \\
&= (\neg(\neg a \longrightarrow b) \longrightarrow c, \neg 1, \neg 1) \\
&= (\neg c \longrightarrow (\neg b \longrightarrow a), \neg 1, \neg 1) && (\textit{Lemma } 2\ (iii)\text{–}(iv)) \\
&= (\neg b \longrightarrow (\neg c \longrightarrow a), \neg 1, \neg 1) && (W11) \\
&= (\neg b \longrightarrow (\neg a \longrightarrow c), \neg 1, \neg 1) && (\textit{Lemma } 2\ (iii)\text{–}(iv)) \\
&= (\neg a \longrightarrow (\neg b \longrightarrow c), \neg 1, \neg 1) && (W11) \\
&= (\neg a \longrightarrow (\neg b \longrightarrow c), \neg\neg 1 \longrightarrow \neg 1, \neg 1) && ((W1) \textit{ and Lemma } 2\ (iii)) \\
&= S^{23}(\neg a \longrightarrow (\neg b \longrightarrow c), \neg 1, \neg 1) \\
&= S^{23}(S^{12}(a, \neg b \longrightarrow c, \neg 1)) \\
&= (S^{23} \circ S^{12})(a, \neg b \longrightarrow c, \neg 1) \\
&= (S^{23} \circ S^{12})(S^{23}(a,b,c)) \\
&= (S^{23} \circ S^{12} \circ S^{23})(a,b,c)
\end{aligned}$$

Then, $S(a,b) = (\neg a \longrightarrow b, \neg 1)$ is a solution to the set-theoretical Yang-Baxter equation in the W-algebra A. □

Lemma 3. *Let $\langle A, \longrightarrow, \neg, 1\rangle$ be a W-algebra. Then $S(a,b) = (\neg b, \neg a)$ is a solution to the set-theoretical Yang-Baxter equation.*

Theorem 2. *Let $\langle A, \longrightarrow, \neg, 1\rangle$ be a W-algebra. Then $S(a,b) = ((a \longrightarrow b) \longrightarrow b, b)$ is a solution to the set-theoretical Yang-Baxter equation.*

Proof. S^{12} and S^{23} are defined in the following forms:

$$S^{12}(a,b,c) = ((a \longrightarrow b) \longrightarrow b, b, c), \quad S^{23}(a,b,c) = (a, (b \longrightarrow c) \longrightarrow c, c).$$

For all $(a,b,c) \in A^3$, we have

$$\begin{aligned}
(S^{12} \circ S^{23} \circ S^{12})(a,b,c) &= S^{12}(S^{23}(S^{12}(a,b,c))) \\
&= S^{12}(S^{23}((a \longrightarrow b) \longrightarrow b,b,c)) \\
&= S^{12}((a \longrightarrow b) \longrightarrow b,(b \longrightarrow c) \longrightarrow c,c) \\
&= ((((a \longrightarrow b) \longrightarrow b) \longrightarrow \\
&\quad ((b \longrightarrow c) \longrightarrow c)) \longrightarrow \\
&\quad ((b \longrightarrow c) \longrightarrow c), \\
&\quad (b \longrightarrow c) \longrightarrow c,c) \\
&= ((((a \longrightarrow b) \longrightarrow b) \longrightarrow \\
&\quad ((c \longrightarrow b) \longrightarrow b)) \longrightarrow \\
&\quad ((c \longrightarrow b) \longrightarrow b), \\
&\quad (b \longrightarrow c) \longrightarrow c,c) && \text{(W3)} \\
&= (((c \longrightarrow b) \longrightarrow (((a \longrightarrow b) \\
&\quad \longrightarrow b) \longrightarrow b)) \longrightarrow ((c \longrightarrow \\
&\quad b) \longrightarrow b),(b \longrightarrow c) \longrightarrow c,c) && \text{(W11)} \\
&= (((c \longrightarrow b) \longrightarrow ((b \longrightarrow \\
&\quad (a \longrightarrow b)) \longrightarrow (a \longrightarrow b))) \\
&\quad \longrightarrow ((c \longrightarrow b) \longrightarrow b), \\
&\quad (b \longrightarrow c) \longrightarrow c,c) && \text{(W3)} \\
&= (((c \longrightarrow b) \longrightarrow (a \longrightarrow b)) \\
&\quad \longrightarrow ((c \longrightarrow b) \longrightarrow b), \\
&\quad (b \longrightarrow c) \longrightarrow c,c) && ((W8)\ and\ (W1)) \\
&= ((a \longrightarrow ((b \longrightarrow c) \longrightarrow c)) \\
&\quad \longrightarrow ((b \longrightarrow c) \longrightarrow c), \\
&\quad (b \longrightarrow c) \longrightarrow c,c) && ((W11)\ and\ (W3))
\end{aligned}$$

and

$$\begin{aligned}
(S^{23} \circ S^{12} \circ S^{23})(a,b,c) &= S^{23}(S^{12}(S^{23}(a,b,c))) \\
&= S^{23}(S^{12}(a,(b \longrightarrow c) \longrightarrow c,c)) \\
&= S^{23}((a \longrightarrow ((b \longrightarrow c) \longrightarrow c)) \\
&\quad \longrightarrow ((b \longrightarrow c) \longrightarrow c), \\
&\quad (b \longrightarrow c) \longrightarrow c,c) \\
&= ((a \longrightarrow ((b \longrightarrow c) \longrightarrow c)) \longrightarrow \\
&\quad ((b \longrightarrow c) \longrightarrow c),(((b \longrightarrow \\
&\quad c) \longrightarrow c) \longrightarrow c) \longrightarrow c,c) \\
&= ((a \longrightarrow ((b \longrightarrow c) \longrightarrow c)) \\
&\quad \longrightarrow ((b \longrightarrow c) \longrightarrow c), \\
&\quad (b \longrightarrow c) \longrightarrow c,c) && ((W3),\ (W8)\ and\ (W1),\ respectively).
\end{aligned}$$

Then, $S(a,b) = ((a \longrightarrow b) \longrightarrow b,b)$ is a solution to the set-theoretical Yang-Baxter equation in the W-algebra A. □

Corollary 1. *Let $\langle A, \longrightarrow, \neg, 1\rangle$ be a W-algebra. Then $S(a,b) = ((b \longrightarrow a) \longrightarrow a,b)$, $S(a,b) = ((b \longrightarrow a) \longrightarrow a,a)$ and $S(a,b) = ((a \longrightarrow b) \longrightarrow b,a)$ are solutions to the set-theoretical Yang-Baxter equation.*

Proof. The proof is completed from (W3) and Theorem 2. □

Proposition 1. *[23] Let A be a W-algebra. The binary relation $\leq$ defined on A as follows*

$$a \leq b \text{ if and only if } a \longrightarrow b = 1$$

is a partial order on A.

Proposition 2. *[23] Let A be a W-algebra and $\leq$ be a partial order on A. If the join and the meet operators are defined by*

$$p \vee q = (p \longrightarrow q) \longrightarrow q$$

and

$$p \wedge q = \neg(p \longrightarrow \neg(p \longrightarrow q)),$$

then this partial order determines a lattice on A.

Proposition 3. *[23] Any W-algebra satisfies the following implications and equations*

(1) $\neg(a \vee b) = \neg a \wedge \neg b$
(2) $\neg(a \wedge b) = \neg a \vee \neg b$
(3) $(a \vee b) \longrightarrow c = (a \longrightarrow c) \wedge (b \longrightarrow c)$
(4) $a \longrightarrow (b \vee c) = (a \longrightarrow b) \vee (a \longrightarrow c)$
(5) $(a \wedge b) \vee c = (a \vee c) \wedge (b \vee c)$
(6) $a \longrightarrow (b \wedge c) = (a \longrightarrow b) \wedge (a \longrightarrow c)$
(7) $(a \longrightarrow b) \vee (b \longrightarrow a) = 1$

Lemma 4. *Any W-algebra A is a Boolean algebra if and only if $\neg a \longrightarrow a = a$ for all $a \in A$.*

Proof. ($\Rightarrow$) Assume that a W-algebra A is a Boolean algebra. By using properties of Boolean algebras and proposition 2, we obtain

$$1 = \neg a \vee a = (\neg a \longrightarrow a) \longrightarrow a \qquad (*_1)$$

for all $a \in A$. Since we already have that $a = 1 \longrightarrow a$ for all $a \in A$ from (W1), we get

$$\begin{aligned} a &= 1 \longrightarrow a && \text{(W1)} \\ &= ((\neg a \longrightarrow a) \longrightarrow a) \longrightarrow a && (*_1) \\ &= (a \longrightarrow (\neg a \longrightarrow a)) \longrightarrow (\neg a \longrightarrow a) && \text{(W3)} \\ &= 1 \longrightarrow (\neg a \longrightarrow a) && \text{(W8)} \\ &= (\neg a \longrightarrow a). && \text{(W1)} \end{aligned}$$

Consequently, $\neg a \longrightarrow a = a$ for all $a \in A$.
($\Leftarrow$) Suppose that A is a W-algebra and let $\neg a \longrightarrow a = a$ for all $a \in A$. Since

$$\begin{aligned} \neg a \vee a &= ((\neg a \longrightarrow a) \longrightarrow a) && (\textit{Proposition } 2) \\ &= a \longrightarrow a && (\textit{Hypothesis}) \\ &= 1 && \text{(W5)} \end{aligned}$$

and

$$\begin{aligned} \neg a \wedge a &= \neg(\neg a \longrightarrow \neg(\neg a \longrightarrow a)) && (\textit{Proposition } 2) \\ &= \neg(\neg a \longrightarrow \neg a) && (\textit{Hypothesis}) \\ &= \neg 1 && \text{(W5)} \\ &= 0, \end{aligned}$$

we have that the negation $\neg$ is a complementation, and the least element of A is 0 defined as $\neg 1 = 0$ and 1 is the greatest element in A. Thus, A is a bounded lattice with a complementation.

It remains to see that A is distributive. By Proposition 3 (5), we already have that $(a \wedge b) \vee c = (a \vee c) \wedge (b \vee c)$ for all $a, b, c \in A$. Besides, we obtain that

$$\begin{array}{rcll}
(a \vee b) \wedge c & = & \neg((a \vee b) \longrightarrow \neg((a \vee b) \longrightarrow c)) & (\textit{Proposition } 2) \\
& = & \neg((\neg c \longrightarrow \neg(a \vee b)) \longrightarrow \neg(a \vee b)) & (\textit{Lemma } 2\ (iii)\text{–}(iv)) \\
& = & \neg((\neg(a \vee b) \longrightarrow \neg c) \longrightarrow \neg c) & (\mathrm{W3}) \\
& = & \neg((c \longrightarrow (a \vee b)) \longrightarrow \neg c) & (\textit{Lemma } 2\ (iv)) \\
& = & \neg(((c \longrightarrow a) \vee (c \longrightarrow b)) \longrightarrow \neg c) & (\textit{Proposition } 3\ (4)) \\
& = & \neg(((c \longrightarrow a) \longrightarrow \neg c) \wedge ((c \longrightarrow b) \longrightarrow \neg c)) & (\textit{Proposition } 3\ (3)) \\
& = & \neg((c \longrightarrow a) \longrightarrow \neg c) \vee \neg((c \longrightarrow b) \longrightarrow \neg c) & (\textit{Proposition } 3\ (2)) \\
& = & \neg((\neg a \longrightarrow \neg c) \longrightarrow \neg c) \vee \neg((\neg b \longrightarrow \neg c) \longrightarrow \neg c) & (\textit{Lemma } 2\ (iii)\text{–}(iv)) \\
& = & \neg(\neg a \vee \neg c) \vee \neg(\neg b \vee \neg c) & (\textit{Proposition } 2) \\
& = & (\neg\neg a \wedge \neg\neg c) \vee (\neg\neg b \wedge \neg\neg c) & (\textit{Proposition } 3\ (1)) \\
& = & (a \wedge c) \vee (b \wedge c). & (\textit{Lemma } 2\ (iii))
\end{array}$$

Hence, A is a bounded distributive lattice with a complementation, that is, a W-algebra A is a Boolean algebra. □

The proof of the following theorem is given by F.F. Nichita.

Theorem 3. [9] *Let $(A, \vee, \wedge, 0, 1, \neg)$ be a Boolean algebra. Then $S(a, b) = (a \vee b, a \wedge b)$ is a solution to the set-theoretical Yang-Baxter equation.*

Proof. Let $(A, \vee, \wedge, 0, 1, \neg)$ be a Boolean algebra. S^{12} and S^{23} are defined in the following forms:

$$S^{12}(a, b, c) = (a \vee b, a \wedge b, c), \qquad S^{23}(a, b, c) = (a, b \vee c, b \wedge c).$$

For all $(a, b, c) \in A^3$, we obtain

$$\begin{array}{rcl}
(S^{12} \circ S^{23} \circ S^{12})(a, b, c) & = & S^{12}(S^{23}(S^{12}(a, b, c))) \\
& = & S^{12}(S^{23}(a \vee b, a \wedge b, c)) \\
& = & S^{12}(a \vee b, (a \wedge b) \vee c, (a \wedge b) \wedge c) \\
& = & ((a \vee b) \vee ((a \wedge b) \vee c), (a \vee b) \wedge ((a \wedge b) \vee c), (a \wedge b) \wedge c) \\
& = & (((a \vee b) \vee (a \wedge b)) \vee c, (a \vee b) \wedge ((a \vee c) \wedge (b \vee c)), (a \wedge b) \wedge c)) \\
& = & ((a \vee (b \vee (a \wedge b))) \vee c, ((a \vee b) \wedge (a \vee c)) \wedge (b \vee c), a \wedge (b \wedge c)) \\
& = & ((a \vee b) \vee c, (a \vee (b \wedge c)) \wedge (b \vee c), a \wedge (b \wedge c)) \\
& = & (a \vee (b \vee c), (a \wedge (b \vee c)) \vee ((b \wedge c) \wedge (b \vee c)), a \wedge (b \wedge c)) \\
& = & (a \vee (b \vee c), (a \wedge (b \vee c)) \vee (b \wedge c), a \wedge (b \wedge c))
\end{array}$$

and

$$\begin{array}{rcl}
(S^{23} \circ S^{12} \circ S^{23})(a, b, c) & = & S^{23}(S^{12}(S^{23}(a, b, c))) \\
& = & S^{23}(S^{12}(a, b \vee c, b \wedge c)) \\
& = & S^{23}(a \vee (b \vee c), a \wedge (b \vee c), b \wedge c) \\
& = & (a \vee (b \vee c), (a \wedge (b \vee c)) \vee (b \wedge c), (a \wedge (b \vee c)) \wedge (b \wedge c)) \\
& = & (a \vee (b \vee c), (a \wedge (b \vee c)) \vee (b \wedge c), a \wedge ((b \vee c) \wedge (b \wedge c))) \\
& = & (a \vee (b \vee c), (a \wedge (b \vee c)) \vee (b \wedge c), a \wedge (b \wedge c)).
\end{array}$$

Then, $S(a, b) = (a \vee b, a \wedge b)$ is a solution to the set-theoretical Yang-Baxter equation. □

Theorem 4. *Let $(A, \longrightarrow, \neg, 1)$ be a W-algebra. If the identities*

$$\neg((\neg a \longrightarrow b) \longrightarrow \neg c) = (a \longrightarrow \neg c) \longrightarrow \neg(b \longrightarrow \neg c) \tag{3}$$

or

$$(a \longrightarrow \neg b) \longrightarrow c = \neg((\neg a \longrightarrow c) \longrightarrow \neg(\neg b \longrightarrow c)) \tag{4}$$

are satisfied for all a, b and c in A, then

$$S(a,b) = ((a \longrightarrow b) \longrightarrow b, \neg(a \longrightarrow \neg(a \longrightarrow b)))$$

and

$$S(a,b) = (\neg a \longrightarrow b, \neg(a \longrightarrow \neg b))$$

are solutions to the set-theoretical Yang-Baxter equation in the W-algebra A.

Proof. Substituting 1 instead of a and b, or $\neg 1$ instead of a and b in identities (3) and (4), respectively, we obtain $c = \neg c \longrightarrow c$ or $\neg c = c \longrightarrow \neg c$ by Lemma 2 (i) and (iii), and (W1). Then A is a Boolean algebra by Lemma 4. From the definitions of join and meet operators and Theorem 3,

$$\begin{aligned} S(a,b) &= ((a \longrightarrow b) \longrightarrow b, \neg(a \longrightarrow \neg(a \longrightarrow b))) \\ &= (a \vee b, a \wedge b) \end{aligned}$$

is a solution to the set-theoretical Yang-Baxter equation in the W-algebra A. Additionally, putting simultaneously $\neg b$ instead of b and b instead of c in identity (4), we get

$$(a \longrightarrow \neg\neg b) \longrightarrow b = \neg((\neg a \longrightarrow b) \longrightarrow \neg(\neg\neg b \longrightarrow b)). \qquad (*_2)$$

Thus, we have

$$\begin{aligned} (a \longrightarrow b) \longrightarrow b &= (a \longrightarrow \neg\neg b) \longrightarrow b && (\textit{Lemma } 2\ (iii)) \\ &= \neg((\neg a \longrightarrow b) \longrightarrow \neg(b \longrightarrow b)) && (*_2) \\ &= \neg((\neg a \longrightarrow b) \longrightarrow \neg 1) && (\mathrm{W5}) \\ &= \neg\neg(\neg a \longrightarrow b) && (\textit{Lemma } 2\ (ii)) \\ &= \neg a \longrightarrow b. && (\textit{Lemma } 2\ (iii)) \end{aligned}$$

By applying Lemma 2 (iii)–(iv) to identity (3), we get

$$\neg(c \longrightarrow \neg(\neg a \longrightarrow b)) = (c \longrightarrow \neg a) \longrightarrow \neg(c \longrightarrow \neg b). \qquad (*_3)$$

Therefore, we obtain

$$\begin{aligned} \neg(a \longrightarrow \neg(a \longrightarrow b)) &= \neg(a \longrightarrow \neg(\neg\neg a \longrightarrow b)) && (\textit{Lemma } 2\ (iii)) \\ &= (a \longrightarrow \neg\neg a) \longrightarrow \neg(a \longrightarrow \neg b) && (*_3) \\ &= (a \longrightarrow a) \longrightarrow \neg(a \longrightarrow \neg b) && (\textit{Lemma } 2\ (iii)) \\ &= 1 \longrightarrow \neg(a \longrightarrow \neg b) && (\mathrm{W5}) \\ &= \neg(a \longrightarrow \neg b). && (\mathrm{W1}) \end{aligned}$$

Hence, $S(a,b) = (\neg a \longrightarrow b, \neg(a \longrightarrow \neg b))$ is a solution to the set-theoretical Yang-Baxter equation in the W-algebra A. □

Remark 1. *Notice that the map $S(a,b) = (a \longrightarrow b, a)$ is a solution to the set-theoretical Yang-Baxter equation in Boolean algebras (Theorem 5.1 in [24]) while it is not a solution in Wajsberg algebras. Indeed, in W-algebras, we have*

$$\begin{aligned}(S^{12} \circ S^{23} \circ S^{12})(a,b,c) &= S^{12}(S^{23}(S^{12}(a,b,c))) \\ &= S^{12}(S^{23}(a \longrightarrow b, a, c)) \\ &= S^{12}(a \longrightarrow b, a \longrightarrow c, a) \\ &= ((a \longrightarrow b) \longrightarrow (a \longrightarrow c), a \longrightarrow b, a)\end{aligned}$$

and

$$\begin{aligned}(S^{23} \circ S^{12} \circ S^{23})(a,b,c) &= S^{23}(S^{12}(S^{12}(a,b,c))) \\ &= S^{23}(S^{12}(a, b \longrightarrow c, b)) \\ &= S^{23}((a \longrightarrow b) \longrightarrow c, a, b) \\ &= (a \longrightarrow (b \longrightarrow c), a \longrightarrow b, a).\end{aligned}$$

Then, we get $(S^{12} \circ S^{23} \circ S^{12})(a,b,c) \neq (S^{23} \circ S^{12} \circ S^{23})(a,b,c)$ for all a, b, c.

However, under the condition that $a \longrightarrow (b \longrightarrow c) = (a \longrightarrow b) \longrightarrow (a \longrightarrow c)$ holds for any a, b, c, we have that $S(a,b) = (a \longrightarrow b, a)$ is a solution to the set-theoretical Yang-Baxter equation in W-algebras.

Lemma 5. *The following equations hold in every W-algebra:*

(i) $(a \longrightarrow \neg b) \longrightarrow (\neg a \longrightarrow b) = \neg a \longrightarrow b$ *which is equivalent to* $\neg(\neg a \longrightarrow b) \longrightarrow \neg(a \longrightarrow \neg b) = \neg a \longrightarrow b$

(ii) $(a \longrightarrow b) \longrightarrow \neg((\neg a \longrightarrow \neg b) \longrightarrow \neg b) = a$

(iii) $(a \longrightarrow \neg b) \longrightarrow \neg((\neg a \longrightarrow b) \longrightarrow \neg c) = (a \longrightarrow \neg c) \longrightarrow \neg((\neg a \longrightarrow c) \longrightarrow \neg b)$

Proof.

(i)

$$\begin{aligned}(a \longrightarrow \neg b) \longrightarrow (\neg a \longrightarrow b) &= 1 \longrightarrow ((a \longrightarrow \neg b) \longrightarrow (\neg a \longrightarrow b)) && \text{(W1)} \\ &= ((\neg a \longrightarrow b) \longrightarrow ((a \longrightarrow \neg b) \longrightarrow \\ &\quad (\neg a \longrightarrow b))) \longrightarrow ((a \longrightarrow \neg b) \longrightarrow \\ &\quad (\neg a \longrightarrow b)) && \text{(W8)} \\ &= (((a \longrightarrow \neg b) \longrightarrow (\neg a \longrightarrow b)) \\ &\quad \longrightarrow (\neg a \longrightarrow b)) \longrightarrow (\neg a \longrightarrow b) && \text{(W3)} \\ &= ((a \longrightarrow \neg b) \vee (\neg a \longrightarrow b)) \\ &\quad \longrightarrow (\neg a \longrightarrow b) && \text{(Proposition 2)} \\ &= 1 \longrightarrow (\neg a \longrightarrow b) && \text{(Proposition 3 (7))} \\ &= \neg a \longrightarrow b && \text{(W1)}\end{aligned}$$

is equivalent to

$$\neg(\neg a \longrightarrow b) \longrightarrow \neg(a \longrightarrow \neg b) = (a \longrightarrow \neg b) \longrightarrow (\neg a \longrightarrow b) = \neg a \longrightarrow b$$

from Lemma 2 (iv).

(*ii*)

$$\begin{aligned}
(a \longrightarrow b) \longrightarrow \neg((\neg a \longrightarrow \neg b) \longrightarrow \neg b) &= (a \longrightarrow b) \longrightarrow \\
&\quad \neg(\neg a \vee \neg b) && (\textit{Proposition } 2) \\
&= (a \longrightarrow b) \longrightarrow (a \wedge b) && (\textit{Proposition } 3\ (1)) \\
&= ((a \longrightarrow b) \longrightarrow a) \wedge \\
&\quad ((a \longrightarrow b) \longrightarrow b) && (\textit{Proposition } 3\ (6)) \\
&= ((a \longrightarrow b) \longrightarrow a) \wedge \\
&\quad ((b \longrightarrow a) \longrightarrow a) && (W3) \\
&= ((a \longrightarrow b) \vee \\
&\quad (b \longrightarrow a)) \longrightarrow a && (\textit{Proposition } 3\ (3)) \\
&= 1 \longrightarrow a && (\textit{Proposition } 3\ (7)) \\
&= a && (W1)
\end{aligned}$$

(*iii*) By Proposition 3 (7), we already have that $(a \longrightarrow \neg b) \vee (\neg b \longrightarrow a) = 1$ and $(a \longrightarrow \neg c) \vee (\neg c \longrightarrow a) = 1$. Thus, from definition of the join operator $\vee$, $a \longrightarrow \neg b = 1$ or $\neg a \longrightarrow b = \neg b \longrightarrow a = 1$ (*Lemma* 2 (*iii*)–(*iv*)) and $a \longrightarrow \neg c = 1$ or $\neg a \longrightarrow c = \neg c \longrightarrow a = 1$ (*Lemma* 2 (*iii*)–(*iv*)).

Case 1. Assume that $a \longrightarrow \neg b = 1$ and $a \longrightarrow \neg c = 1$. Then by using $a \longrightarrow \neg b = 1$, we get

$$(a \longrightarrow \neg b) \longrightarrow \neg((\neg a \longrightarrow b) \longrightarrow \neg c) = \neg((\neg a \longrightarrow b) \longrightarrow \neg c) \qquad (W1)$$

and by using $a \longrightarrow \neg c = 1$ we have

$$(a \longrightarrow \neg c) \longrightarrow \neg((\neg a \longrightarrow c) \longrightarrow \neg b) = \neg((\neg a \longrightarrow c) \longrightarrow \neg b). \qquad (W1)$$

Case 2. Assume that $a \longrightarrow \neg b = 1$ and $\neg a \longrightarrow c = 1$. Then by using $a \longrightarrow \neg b = 1$, we have

$$(a \longrightarrow \neg b) \longrightarrow \neg((\neg a \longrightarrow b) \longrightarrow \neg c) = \neg((\neg a \longrightarrow b) \longrightarrow \neg c) \qquad (W1)$$

and by using $\neg a \longrightarrow c = 1$ we get

$$(a \longrightarrow \neg c) \longrightarrow \neg((\neg a \longrightarrow c) \longrightarrow \neg b) = (a \longrightarrow \neg c) \longrightarrow b. \qquad ((W1) \textit{ and Lemma } 2\ (iii))$$

Case 3. Suppose that $\neg a \longrightarrow b = 1$ and $a \longrightarrow \neg c = 1$. Then by using $\neg a \longrightarrow b = 1$, we get

$$(a \longrightarrow \neg b) \longrightarrow \neg((\neg a \longrightarrow b) \longrightarrow \neg c) = (a \longrightarrow \neg b) \longrightarrow c \qquad ((W1) \textit{ and Lemma } 2\ (iii))$$

and by using $a \longrightarrow \neg c = 1$ we have

$$(a \longrightarrow \neg c) \longrightarrow \neg((\neg a \longrightarrow c) \longrightarrow \neg b) = \neg((\neg a \longrightarrow c) \longrightarrow \neg b). \qquad (W1)$$

Case 4. Suppose that $\neg a \longrightarrow b = 1$ and $\neg a \longrightarrow c = 1$. Then by using $\neg a \longrightarrow b = 1$, we get

$$(a \longrightarrow \neg b) \longrightarrow \neg((\neg a \longrightarrow b) \longrightarrow \neg c) = (a \longrightarrow \neg b) \longrightarrow c \qquad ((W1) \textit{ and Lemma } 2\ (iii))$$

and by using $\neg a \longrightarrow c = 1$ we have

$$(a \longrightarrow \neg c) \longrightarrow \neg((\neg a \longrightarrow c) \longrightarrow \neg b) = (a \longrightarrow \neg c) \longrightarrow b. \qquad ((W1) \textit{ and Lemma } 2\ (iii))$$

Hence, by Case 1, 2, 3 and 4,

$$(a \longrightarrow \neg b) \longrightarrow \neg((\neg a \longrightarrow b) \longrightarrow \neg c) = (a \longrightarrow \neg c) \longrightarrow \neg((\neg a \longrightarrow c) \longrightarrow \neg b)$$

holds in every W-algebra. □

Proposition 4. *The following identity*

$$(a \longrightarrow \neg b) \longrightarrow \neg((\neg a \longrightarrow b) \longrightarrow \neg c) = \neg((\neg a \longrightarrow b) \longrightarrow \neg((a \longrightarrow \neg b) \longrightarrow c))$$

holds for every W-algebra.

Proof. By Proposition 3 (7), we already have that $(a \longrightarrow \neg b) \vee (\neg b \longrightarrow a) = 1$, and from definition of the join operator $\vee$, $a \longrightarrow \neg b = 1$ or $\neg a \longrightarrow b = \neg b \longrightarrow a = 1$ (*Lemma* 2 (*iii*)–(*iv*)).

Assume that $a \longrightarrow \neg b = 1$. Then, we get

$$\begin{aligned}
(a \longrightarrow \neg b) \longrightarrow \neg((\neg a \longrightarrow b) \longrightarrow \neg c) &= 1 \longrightarrow \neg((\neg a \longrightarrow b) \longrightarrow \neg c) && (\textit{Hypothesis}) \\
&= \neg((\neg a \longrightarrow b) \longrightarrow \neg(1 \longrightarrow c)) && (\text{W1}) \\
&= \neg((\neg a \longrightarrow b) \longrightarrow \\
&\quad \neg((a \longrightarrow \neg b) \longrightarrow c)) && (\textit{Hypothesis})
\end{aligned}$$

for all a, b, c in A.

Suppose that $\neg a \longrightarrow b = 1$. Then, we obtain

$$\begin{aligned}
(a \longrightarrow \neg b) \longrightarrow \neg((\neg a \longrightarrow b) \longrightarrow \neg c) &= (a \longrightarrow \neg b) \longrightarrow \neg(1 \longrightarrow \neg c) && (\textit{Hypothesis}) \\
&= \neg(1 \longrightarrow \neg((a \longrightarrow \neg b) \longrightarrow c)) && ((\text{W1}) \textit{ and Lemma } 2\ (\textit{iii})) \\
&= \neg((\neg a \longrightarrow b) \longrightarrow \\
&\quad \neg((a \longrightarrow \neg b) \longrightarrow c)) && (\textit{Hypothesis})
\end{aligned}$$

for all a, b, c in A. □

Corollary 2. *Let* $(A, \longrightarrow, \neg, 1)$ *be a W-algebra. Then* $S(a, b) = (\neg a \longrightarrow b, \neg(a \longrightarrow \neg b))$ *is a solution to the set-theoretical Yang-Baxter equation in the W-algebra A.*

Proof. S^{12} and S^{23} are defined in the following forms:

$$S^{12}(a, b, c) = (\neg a \longrightarrow b, \neg(a \longrightarrow \neg b), c), \quad S^{23}(a, b, c) = (a, \neg b \longrightarrow c, \neg(b \longrightarrow \neg c)).$$

For all $(a,b,c) \in A^3$, we get

$$
\begin{aligned}
(S^{12} \circ S^{23} \circ S^{12})(a,b,c) &= S^{12}(S^{23}(S^{12}(a,b,c))) \\
&= S^{12}(S^{23}(\neg a \longrightarrow b, \neg(a \longrightarrow \neg b), z)) \\
&= S^{12}(\neg a \longrightarrow b, \neg\neg(a \longrightarrow \neg b) \longrightarrow c, \neg(\neg(a \longrightarrow \neg b) \longrightarrow \neg c)) \\
&= (\neg(\neg a \longrightarrow b) \longrightarrow (\neg\neg(a \longrightarrow \neg b) \longrightarrow c), \\
&\quad \neg((\neg a \longrightarrow b) \longrightarrow \neg(\neg\neg(a \longrightarrow \neg b) \longrightarrow c)), \\
&\quad \neg(\neg(a \longrightarrow \neg b) \longrightarrow \neg c)) \\
&= (\neg(\neg a \longrightarrow b) \longrightarrow ((a \longrightarrow \neg b) \longrightarrow c), \\
&\quad \neg((\neg a \longrightarrow b) \longrightarrow \neg((a \longrightarrow \neg b) \longrightarrow c)), \\
&\quad \neg(\neg(a \longrightarrow \neg b) \longrightarrow \neg c)) && (\textit{Lemma } 2\ (iii)) \\
&= ((a \longrightarrow \neg b) \longrightarrow (\neg(\neg a \longrightarrow b) \longrightarrow c), \\
&\quad \neg((\neg a \longrightarrow b) \longrightarrow \neg((a \longrightarrow \neg b) \longrightarrow c)), \\
&\quad \neg(\neg(a \longrightarrow \neg b) \longrightarrow \neg c)) && (\mathrm{W11}) \\
&= ((a \longrightarrow \neg b) \longrightarrow (\neg c \longrightarrow (\neg a \longrightarrow b)), \\
&\quad \neg((\neg a \longrightarrow b) \longrightarrow \neg((a \longrightarrow \neg b) \longrightarrow c)), \\
&\quad \neg(c \longrightarrow (a \longrightarrow \neg b))) && (\textit{Lemma } 2\ (iii)\text{–}(iv)) \\
&= (\neg c \longrightarrow ((a \longrightarrow \neg b) \longrightarrow (\neg a \longrightarrow b)), \\
&\quad \neg((\neg a \longrightarrow b) \longrightarrow \neg((a \longrightarrow \neg b) \longrightarrow c)), \\
&\quad \neg(c \longrightarrow (a \longrightarrow \neg b))) && (\mathrm{W11}) \\
&= (\neg c \longrightarrow (\neg a \longrightarrow b), \\
&\quad \neg((\neg a \longrightarrow b) \longrightarrow \neg((a \longrightarrow \neg b) \longrightarrow c)), \\
&\quad \neg(c \longrightarrow (a \longrightarrow \neg b))) && (\textit{Lemma } 5\ (i)) \\
&= (\neg a \longrightarrow (\neg c \longrightarrow b), \\
&\quad \neg((\neg a \longrightarrow b) \longrightarrow \neg((a \longrightarrow \neg b) \longrightarrow c)), \\
&\quad \neg(c \longrightarrow (a \longrightarrow \neg b))) && (\mathrm{W11}) \\
&= (\neg a \longrightarrow (\neg b \longrightarrow c), \\
&\quad \neg((\neg a \longrightarrow b) \longrightarrow \neg((a \longrightarrow \neg b) \longrightarrow c)), \\
&\quad \neg(c \longrightarrow (a \longrightarrow \neg b))) && (\textit{Lemma } 2\ (iii)\text{–}(iv)) \\
&= (\neg a \longrightarrow (\neg b \longrightarrow c), \\
&\quad (a \longrightarrow \neg b) \longrightarrow \neg((\neg a \longrightarrow b) \longrightarrow \neg c), \\
&\quad \neg(c \longrightarrow (a \longrightarrow \neg b))) && (\textit{Proposition } 4)
\end{aligned}
$$

and we have

$$
\begin{aligned}
(S^{23} \circ S^{12} \circ S^{23})(a,b,c) &= S^{23}(S^{12}(S^{23}(a,b,c))) \\
&= S^{23}(S^{12}(a, \neg b \longrightarrow c, \neg(b \longrightarrow \neg c))) \\
&= S^{23}(\neg a \longrightarrow (\neg b \longrightarrow c), \neg(a \longrightarrow \neg(\neg b \longrightarrow c)), \neg(b \longrightarrow \neg c)) \\
&= (\neg a \longrightarrow (\neg b \longrightarrow c), \\
&\quad \neg\neg(a \longrightarrow \neg(\neg b \longrightarrow c)) \longrightarrow \neg(b \longrightarrow \neg c), \\
&\quad \neg(\neg(a \longrightarrow \neg(\neg b \longrightarrow c)) \longrightarrow \neg\neg(b \longrightarrow \neg c))) \\
&= (\neg a \longrightarrow (\neg b \longrightarrow c), \\
&\quad (b \longrightarrow \neg c) \longrightarrow \neg(a \longrightarrow \neg(\neg b \longrightarrow c)), \\
&\quad \neg(\neg(a \longrightarrow \neg(\neg b \longrightarrow c)) \longrightarrow \\
&\quad (b \longrightarrow \neg c))) && (\textit{Lemma } 2\ (iv)) \\
&= (\neg a \longrightarrow (\neg b \longrightarrow c), \\
&\quad (b \longrightarrow \neg c) \longrightarrow \neg((\neg b \longrightarrow c) \longrightarrow \neg a), \\
&\quad \neg(\neg(a \longrightarrow \neg(\neg b \longrightarrow c)) \longrightarrow \\
&\quad (b \longrightarrow \neg c))) && (\textit{Lemma } 2\ (iii)\text{–}(iv)) \\
&= (\neg a \longrightarrow (\neg b \longrightarrow c), \\
&\quad (b \longrightarrow \neg a) \longrightarrow \neg((\neg b \longrightarrow a) \longrightarrow \neg c), \\
&\quad \neg(\neg(a \longrightarrow \neg(\neg b \longrightarrow c)) \longrightarrow \\
&\quad (b \longrightarrow \neg c))) && (\textit{Lemma } 5\ (iii)) \\
&= (\neg a \longrightarrow (\neg b \longrightarrow c), \\
&\quad (a \longrightarrow \neg b) \longrightarrow \neg((\neg a \longrightarrow b) \longrightarrow \neg c), \\
&\quad \neg(\neg(b \longrightarrow \neg c) \longrightarrow (a \longrightarrow \\
&\quad \neg(\neg b \longrightarrow c)))) && (\textit{Lemma } 2\ (iii)\text{–}(iv)) \\
&= (\neg a \longrightarrow (\neg b \longrightarrow c), \\
&\quad (a \longrightarrow \neg b) \longrightarrow \neg((\neg a \longrightarrow b) \longrightarrow \neg c), \\
&\quad \neg(a \longrightarrow (\neg(b \longrightarrow \neg c) \longrightarrow \\
&\quad \neg(\neg b \longrightarrow c)))) && (W11) \\
&= (\neg a \longrightarrow (\neg b \longrightarrow c), \\
&\quad (a \longrightarrow \neg b) \longrightarrow \neg((\neg a \longrightarrow b) \longrightarrow \neg c), \\
&\quad \neg(a \longrightarrow (b \longrightarrow \neg c))) && (\textit{Lemma } 5\ (i)) \\
&= (\neg a \longrightarrow (\neg b \longrightarrow c), \\
&\quad (a \longrightarrow \neg b) \longrightarrow \neg((\neg a \longrightarrow b) \longrightarrow \neg c), \\
&\quad \neg(a \longrightarrow (c \longrightarrow \neg b))) && (\textit{Lemma } 2\ (iii)\text{–}(iv)) \\
&= (\neg a \longrightarrow (\neg b \longrightarrow c), \\
&\quad (a \longrightarrow \neg b) \longrightarrow \neg((\neg a \longrightarrow b) \longrightarrow \neg c), \\
&\quad \neg(c \longrightarrow (a \longrightarrow \neg b))). && (W11)
\end{aligned}
$$

Then, $S(a,b) = (\neg a \longrightarrow b, \neg(a \longrightarrow \neg b))$ is a solution to the set-theoretical Yang-Baxter equation in the W-algebra A. □

Acknowledgments: The authors thank the academic editor for their valuable comments and suggestions and the anonymous referees for his/her remarks which helped them to improve the presentation of the paper.

Author Contributions: The coauthors, Tahsin Oner and Tugce Katıcan, wrote the paper jointly and contributed equally to this work.

References

1. Perk, J.H.H.; Au, Y.H. Yang-Baxter Equations. In *Encyclopedia of Mathematical Physics*; Françoise, J.-P., Naber, G.L., Tsou, S.T., Eds.; Elseiver: Oxford, UK, 2006; Volume 5, pp. 465–473.
2. Baxter, R.J. *Exactly Solved Models in Statical Mechanics*; Academy Press: London, UK, 1982.
3. Baxter, R.J. Partition Function of the Eight-Vertex Lattice Model. *Ann. Phys.* **1972**, *70*, 193–228.

4. Yang, C.N. Some Exact Results for the Many-Body Problem in One Dimension with Repulsive Delta-Function Interaction. *Phys. Rev. Lett.* **1967**, *19*, 1312–1315.
5. Jimbio, M. *Yang-Baxter Equation in Integrable Systems;* Advanced Series in Mathematical Physics; World Scientific Publishing Co. Inc.: Singapore, 1990; Volume 10.
6. Jimbio, M. Introduction to the Yang-Baxter Equation. *Int. J. Mod. Phys.* **1989**, *4*, 3759.
7. Lu, J.-H.; Yan, M.; Zhu, Y.-C. On the Set-Theoretical Yang-Baxter Equation. *Duke Math. J.* **2000**, *104*, 1–18.
8. Nichita, F.F. Hopf algebras, Quantum Groups and Yang-Baxter Equations, 2014 (Special Issue). Available online: http://www.mdpi.com/journal/axioms/special_issue/hopf_algebras_2014 (accessed on 22 June 2017).
9. Nichita, F.F. On the set-theoretical Yang-Baxter Equation. *Acta Univ. Apulensis Math. Inf.* **2003**, *5*, 97–100.
10. Nichita, F.F. Yang-Baxter Equations, Computational Methods and Applications. *Axioms* **2015**, *4*, 423–435.
11. Oner, T.; Senturk, I.; Oner, G. An Independent Set of Axioms of MV-Algebras and Solutions of the Set-Theoretical Yang-Baxter Equation. *Axioms* **2017**, *6, 17*, doi:10.3390/axioms6030017.
12. Gateva-Ivanova, T. Noetherian Properties of Skew-Polynomial Rings with Binomial Relations. *Trans. Am. Math. Soc.* **1994**, *343*, 203–219.
13. Gateva-Ivanova, T. Skew Polynomial Rings with Binomial Relations. *J. Algebra* **1996**, *185*, 710–753.
14. Gateva-Ivanova, T.; Van den Bergh, M. Semigroups of I-type. *J. Algebra* **1998**, *206*, 97–112.
15. Tate, J.; Van den Bergh, M. Homological Properties of Sklyanin Algebras. *Invent. Math.* **1996**, *124*, 619–647.
16. Etingof, P.; Schedler, T.; Soloviev, A. Set-Theoretical Solutions to the Quantum Yang-Baxter Equation. *Duke Math. J.* **1999**, *100*, 169–209.
17. Etingof, P.; Gelaki, S. A Method of Construction of Finite-Dimensional Triangular Semisimple Hopf Algebras. *Math. Res. Lett.* **1998**, *5*, 551–561.
18. Veselov, A.P. Yang-Baxter Maps and Integral Dynamics. *Phys. Lett. A.* **2003**, *314*, 214–221.
19. Etingof, P. Geometric Crystals and Set-Theoretical Solutions to the Quantum Yang-Baxter Equation. *Comm. Algebra* **2003**, *31*, 1961–1973.
20. Cignoli, R.L.O.; D'ottaviano, I.M.L.; Mundici, D. *Algebraic Foundations of Many-Valued Reasoning;* Kluwer Academic Publishers: Dordrech, The Netherlands, 2000.
21. Radfar, A.; Rezaei, A.; Sadei, A.B. Extensions of BCK-algebras. *Cogent Math.* **2016**, doi:10.1080/23311835.2016.1265297.
22. Pałasiński, M. The variety of all commutative BCK-algebras is generated by its finite members as a quasivariety. *Demonstr. Math.* **2012**, *45*, 495–518.
23. Font, J.M.; Rodriguez, A.J.; Torrens, A. Wajsberg Algebras. *Stochastica* **1984**, *8*, 5–31.
24. Solomon, M.; Nichita, F.F. On Transcendental Numbers: New Results and a Little History. *Mathematics* **2015**, submitted.

3

On Transcendental Numbers: New Results and a Little History

Solomon Marcus [1,2,†] **and Florin F. Nichita** [1,*]

[1] Institute of Mathematics of the Romanian Academy, 21 Calea Grivitei Street, 010702 Bucharest, Romania
[2] Romanian Academy, 125 Calea Victoriei Street, 010071 Bucharest, Romania
[*] Correspondence: Florin.Nichita@imar.ro.
[†] This article is dedicated to the memory of my friend and esteemed colleague, Professor Solomon Marcus, who passed away on 17 March 2016. His enthusiasm and optimism will be sadly missed.

Abstract: Bringing toghether mathematical and philosophical ideas related to transcendental numbers, this paper begins with a survey on transcendence and transcendental numbers, it then presents several properties of the transcendental numbers e and π, and then it gives the proof of a new inequality for transcendental numbers. Also, in relationship with these topics, we study solutions to the Yang-Baxter equation from hyperbolic functions and from logical implication.

Keywords: Euler's identity; Euler's relation; transcendental numbers; transcendental operations/functions; transcendence; Yang-Baxter equation, hyperbolic functions, logical implication

MSC: 01A05; 11D09; 11T23; 33B10; 16T25

---

1. Introduction

One of the most famous formulas in mathematics, Euler's identity,

$$e^{\pi i} + 1 = 0 \tag{1}$$

contains the transcendental numbers e and π, the imaginary number i, the constants 0 and 1, and (transcendental) operations. Beautiful, powerful, and surprising, it has influenced many areas of mathematics and theoretical physics.

Euler's formula states (for $x = \pi$ Euler's formula becomes Euler's identity):

$$e^{ix} = \cos x + i \sin x \quad \forall x \in \mathbb{R}. \tag{2}$$

Another interesting relation appeared in [1]:

$$|e^i - \pi| > e\,. \tag{3}$$

It can also be expressed as: $|e^i - \pi| \approx e + 0.016$, and it has a geometrical interpretation (see Figure 1).

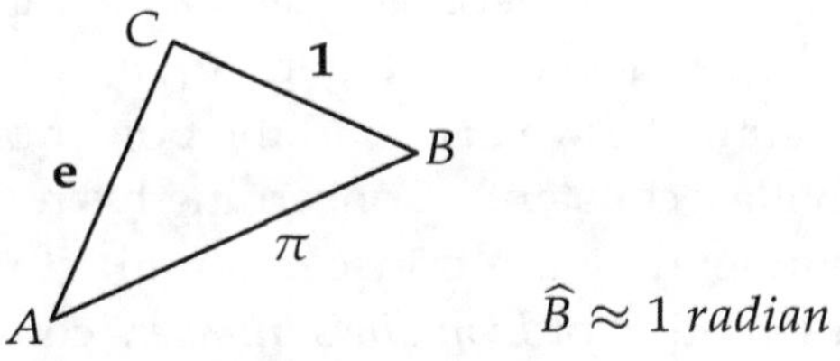

Figure 1. A triangle with edges of length 1, π and e has an angle of approximately 1 radian.

Equations (1) and (3) can be unified, and we will explain how in Section 3.

The next section is a survey on transcendental numbers, transcendental functions, transcendental operations, and transcendence (following the ideas from [2,3]). Section 3 deals with identities and inequalities of transcendental numbers. We prove Equation (3). We also present some open problems. Section 4 contains two important constructions: solutions to the Yang–Baxter equation both from hyperbolic functions and from material implication. The last section provides conclusions.

2. Transcendence and Transcendental Numbers in Mathematics

The term "transcendence" has been introduced to express issues related to divinity. Implying initially the aforementioned attributes, its use has been used widely in movies, novels, jazz albums, corporations, game brands, poems, internet websites, internet blogs, names of all kinds of doctrines, philosophies, etc. The "trans" prefix comes from Latin, implying a passing through a certain area, and it appears in many (common) words: transport, transparent, translation, transmission, transformation, transplant, transgress, transdisciplinarity, etc.

In the 18th century, Leonhard Euler initiated a discussion about *transcendence in mathematics* (see [4]). Leibniz was the first to refer to transcendence in mathematics (see [5]). Euler does not refer directly to transcendental numbers (as it happened, subsequently, in the 19th century), but he refers to *transcendental operations*. The non-transcendental operations are addition, subtraction, multiplication, division, exponentiation, and rooting (applied to integers and to those numbers that are obtained from them by such operations). These operations are considered to be non-transcendent. They are algebraic operations because all numbers generated in this way are roots of some algebraic equations; we refer to those equations expressed by polynomial structures, where all mentioned operations are applied for a finite number of times (focus on finite) and, as soon as this is not the case, we enter in the field of transcendence. Therefore, in mathematics, *the idea of transcendence is essentially related to the idea of infinity*—but not to just any kind of infinity. For Euler, the entrance to mathematical transcendence is through the gate of the differential and integral calculus. Euler considers the exponentiation a to x, the logarithm, and the trigonometric sine/cosine functions as *transcendental* functions and operations because they are associated, in one way or another, with the mathematical idea of integrals and with expressions that include an integral (for instance, the logarithm is associated with the integral of dx/x). For Euler, the function log(x) + 7 is transcendent, but log(7) + x is algebraic.

The passing from non-transcendent to transcendent is gradual. Such an assertion deviates us from mathematical rigor but consolidates our understanding. We have in mathematics the situation of irrational numbers, such as the square root of 2. The word "irrational" would suggest transcendence, but it is immediately completed by the word "algebraic," the negation of "transcendent." We may consider the algebraic irrationals (an algebraic number is a root of a polynomial equation with integer coefficients) as a passing bridge, or intermediary steps, when passing from non-transcendent to transcendent. It is not rigorous, but it is suggestive. We must add that we talk about rational numbers (such as 2/3 or −3/7) and irrational numbers, and we are tempted to associate such irrational numbers as the square root of 2 with something beyond reason. However, the Latin etymology clarifies the fact that "rational" is seen here as a ratio or as a fraction, where, in a number of the form a/b, a and b are integers, and b is not zero.

Joseph Liouville is the first who managed to encounter an example of *transcendental numbers*. This important discovery happened in the year 1844. Liouville introduced a class of eponymously named real numbers. A real number x is a Liouville number if there is an integer b greater than or equal to 2 and an infinite sequence of integers (a_1, a_2, .., a_n, ...) so that x is the sum of the series that has, as a general term, a ratio between a_k and the power of exponent k! of b. (For b = 10 and $a_k = 1$ for any k, x becomes Liouville's constant.) Concerning his numbers (see also Remark 5), Liouville proves that they are not algebraic; they are transcendental. It was the first example of non-algebraic real numbers. Another presentation of *Liouville's numbers* consists in approximation by rational numbers: x is a Liouville number if, for any natural number n, there are integers p and q, where q is greater than or equal to 2, and the absolute value of the difference between x and p/q is strictly between zero and the fraction with 1 as the numerator and the value of q to the power of n as the denominator. It is thus clear that Liouville's numbers have tight approximations by rational numbers. The number e also has tight approximations by rational numbers (see Remark 2). These facts were not expected because they suggest that, *in some respects, transcendental numbers are nearer to rational numbers than algebraic irrationals.*

The manner in which rational numbers can approach transcendent numbers urges us to see the *transcendence process* in the infinite range of approximations of transcendental numbers. This situation suggests us to generally regard *transcendence as a result of an asymptotic process* when all stages are in the terrestrial universe, but the infinite number of stages precludes their being crossed in real time. Such a view increases the likelihood that a sharp border between transcendent and terrestrial is absent.

There are other philosophical enigmas related to mathematical transcendence. One would expect that transcendental numbers, such as π or e, would be farther from rational numbers than irrational algebraic numbers. However, in terms of approximation by rational numbers, it seems that this does not happen. It must also be noted that transcendental numbers are greater than algebraic numbers, but "greater" here is a metaphoric extension of such qualification, from finite to infinite. Mathematicians express this by the following statement: "The algebraic numbers form a countable set, while the transcendental numbers form an uncountable set; it is a set of the power of the continuum."

3. Transcendental Numbers: Identities, Inequalities and Open Problems

The following identities that contain the transcendental numbers e and π are well-known:

$$\int_{-\infty}^{+\infty} e^{-x^2} dx = \sqrt{\pi}\,, \qquad \int_{-\infty}^{+\infty} e^{-ix^2} dx = \sqrt{\frac{\pi}{2}}(1-i).$$

These formulas, unified in [6], can be proved by contour integration.

Other inequalities for e and π (from [1]) are quite new; here, we list three:

$$|\, e^{1-z} + e^{\bar{z}} \,| > \pi \quad \forall z \in \mathbb{C} \tag{4}$$

$$\int_a^b e^{-x^2} dx \;<\; \frac{e^e}{\pi}\Big(\frac{1}{e^{\pi a}} - \frac{1}{e^{\pi b}}\Big) \;\forall a,\, b \in \mathbb{R}$$

$$|e^i - \pi| > e.$$

We now prove the last formula, Inequality (3).

$|e^i - \pi| \;>\; e$ is equivalent to $(\pi + e)(\pi - e) + 1 \;>\; 2\pi \cos 1$. Now, we use the estimations $\pi \;>\; 3.141$, $e \;<\; 2.719$, $(\pi + e)(\pi - e) + 1 \;>\; 3.47292$, and $\cos 1 \;<\; 1 - \frac{1}{2} + \frac{1}{4!} \;=\; \frac{13}{24}$, $2\pi \cos 1 \;<\; 3.142 \times \frac{13}{12} = 3.4038(3)$. (This proof was slightly improved by Dr. Cezar Joita.)

Remark 1. *Let us consider the two variable complex functions $f : \mathbb{C} \times \mathbb{C} \to \mathbb{R}$, $f(z,w) = |e^z + e^w|$. The formulas (1), (3) and (4) can be unified using the function $f(z,w)$:*

$$f(\pi i,\, 0) = 0\,, \qquad f(1-z,\, \bar{z}) > \pi \;\forall z \in \mathbb{C}\,, \qquad f(i,\, \pi i + \ln \pi) > e.$$

Remark 2. *Related to Section 2, one could consider the equation $x^2 - \pi x + (1 + \frac{1}{r})^r = 0\,,\;\; r \in \mathbb{Q}^*$, which has real solutions for r that are sufficiently small, but the "limit" equation (for $r \to \infty$) $x^2 - \pi x + e = 0$ has no real solutions because $\Delta = \pi^2 - 4e < 0$. The "limit" equation can be interpretated as an inequality:*

$$x^2 + e \;>\; \pi x \;\;\forall x \in \mathbb{R} \tag{5}$$

which will be generalized in the next section. The question if $\Delta = \pi^2 - 4e$ is a transcendental number is an open problem.

Figure 2. A circle with diameter π and a square with edges of length e.

Remark 3. *Resembling the problem of squaring the circle, the geometrical interpretation of the formula* $\pi^2 < 4e$ *can be stated as follows: "The perimeter of the circle with diameter* π *is almost equal (and less) to the perimeter of a square with edges of length e" (see Figure 2). In this case, the area of the circle is greater than the area of the square because* $\pi^3 > 4e^2$.

Remark 4. *We now present open problems related to* π.

Let us consider an arbitrary convex closed curve (see, for example, Figure 2), and the center of mass of the body, which corresponds to the domain inside the given curve. The lines going through this center of mass intersect the given curve twice, and they are called diameters. The smallest diameter is represented by d, and the maximum diameter represented by D.

(i) If L is the length of the given curve, then

$$\frac{L}{D} \leq \pi \leq \frac{L}{d} \ .$$

(ii) Moreover, the first inequality becomes equality if and only if the second inequality becomes equality if and only if the given curve is a circle.
(iii) If the area of the domain inside the given curve is A, then $dD > A$.
(iv) The equation $x^2 - \frac{L}{2}x + A = 0$ *is not completely solved; for example, if the given curve is an ellipse, this equation cannot be solved.*

Remark 5. *Solving the equation* $x^2 + \sum_0^\infty 10^{-k!}\, x + \sum_1^\infty 10^{-k!} = 0$*, in which two coefficients are Liouville numbers, by using the quadratic formula is hard; however, one can use the formula* $Ax^2 + (A+C)x + C = (Ax+C)(x+1)$ *in order to solve this equation. The solution* $x = -1$ *can be observed directly.*

4. Transcendental Numbers in Mathematical Physics

In some special issues on Hopf algebras, quantum groups, and Yang–Baxter equations, several papers, as well as a feature paper, have covered many topics related to the Yang–Baxter equation, ranging from mathematical physics to Hopf algebras, from Azumaya Monads to quantum computing, and from mathematical logic to Rota–Baxter equations.

The terminology of this section is consistent with those papers.

V is a complex vector space, and $I_j : V^{\otimes j} \to V^{\otimes j}$ $\forall j \in \{1,2\}$ are identity maps. We consider $J : V^{\otimes 2} \to V^{\otimes 2}$ a linear map that satisfies $J \circ J = -I_2$ and $J^{12} \circ J^{23} = J^{23} \circ J^{12}$, where $J^{12} = J \otimes I_1$, $J^{23} = I_1 \otimes J$.

Thus,

$$R(x) = \cos(x) I_2 + \sin(x) J = \cos x + J \sin x \tag{6}$$

satisfies the colored Yang–Baxter equation:

$$R^{12}(x) \circ R^{23}(x+y) \circ R^{12}(y) = R^{23}(y) \circ R^{12}(x+y) \circ R^{23}(x) \ . \tag{7}$$

Remark 6. *The theory of integrable Hamiltonian systems makes great use of solutions of Equation (7), since coefficients of the power series expansion of such a solution give rise to commuting integrals of motion. For more applications of the mathematical physics of Equation (7), see [7].*

There is a version of Equation (2) for matrices:

$$R(x) = \cos x + J \sin x = e^{x\,J}$$

and it can be used in order to simplify our proof that $R(x)$ is a solution of Equation (7); therefore, that proof reduces to checking the equality $xJ^{12} + (x+y)\,J^{23} + yJ^{12} = yJ^{23} + (x+y)\,J^{12} + xJ^{23}$.

In dimension two, J could have the following matrix form (for $\alpha \in \mathbb{R}$):

$$J = \begin{pmatrix} 0 & 0 & 0 & \frac{1}{\alpha}i \\ 0 & 0 & i & 0 \\ 0 & i & 0 & 0 \\ \alpha i & 0 & 0 & 0 \end{pmatrix}. \tag{8}$$

Thus, we have a generalization of Equation (1):

$$e^{\pi\,J} + I_4 = 0_4 \qquad J,\, I_4, 0_4 \in \mathcal{M}_4(\mathbb{C})\,.$$

In a similar manner, we generalize Inequality (5) for matrices below (we leave this general case without proof):

$$X^2 + eI_2 > \pi X \qquad\qquad \forall\, X \in \mathcal{M}_2(\mathbb{R}^*_+)),\ trace(X) > \pi\,. \tag{9}$$

A generalization of Equation (6), with $J \circ J = pI_2\,,\ J^{12} \circ J^{23} = qJ^{23} \circ J^{12},\ pq = -1\,$, was studied in [8].

Replacing only the condition $J^{12} \circ J^{23} = J^{23} \circ J^{12}$ with $J^{12} \circ J^{23} = -J^{23} \circ J^{12}$, the authors of [9,10] obtained interesting results (a new realization of doubling degeneracy based on an emergent Majorana operator, new solutions for the Yang–Baxter equation, etc).

With these cases, we enter into the world of quaternions and Clifford algebras.

We present two new theorems below. The first one can be interpreted as a version of Euler's formula for hyperbolic functions.

Theorem 1. *Solutions to the Yang–Baxter equation from hyperbolic functions:*
Let $J : V^{\otimes 2} \to V^{\otimes 2}$ be a linear map that satisfies $J \circ J = I_2$ and $J^{12} \circ J^{23} = J^{23} \circ J^{12}$.
Then, $R(x) = \cosh(x) I_2 + \sinh(x) J$ satisfies the colored Yang–Baxter equation (Equation (7)).

Proof. The equality $(\cosh x + \sinh x J^{12})(\cosh x + \sinh x J^{23})(\cosh x + \sinh x J^{12}) = (\cosh x + \sinh x J^{23})(\cosh x + \sinh x J^{12})(\cosh x + \sinh x J^{23})$ follows from properties of hyperbolic functions, $\cosh(x+y) = \cosh x \cosh y + \sinh x \sinh y$ and $\sinh(x+y) = \sinh x \cosh y + \cosh x \sinh y$, and from $J \circ J = I_2$ and $J^{12} \circ J^{23} = J^{23} \circ J^{12}$. □

Remark 7. *In dimension two, the matrix form of the above operator J could be*

$$\begin{pmatrix} 0 & 0 & 0 & 1 \\ 0 & 0 & 1 & 0 \\ 0 & 1 & 0 & 0 \\ 1 & 0 & 0 & 0 \end{pmatrix}. \tag{10}$$

The corresponding matrix for $R(x)$ in this case can be expressed as follows:

$$\frac{1}{2}\begin{pmatrix} e^x + e^{-x} & 0 & 0 & e^x - e^{-x} \\ 0 & e^x + e^{-x} & e^x - e^{-x} & 0 \\ 0 & e^x - e^{-x} & e^x + e^{-x} & 0 \\ e^x - e^{-x} & 0 & 0 & e^x + e^{-x} \end{pmatrix}. \tag{11}$$

Notice that $R(i\pi) = -I_2$. *Therefore, this formula could be understood as a version of Equation (1) for hyperbolic functions.*

Theorem 2. *Solutions to the Yang–Baxter equation from material implication:*
For a Boolean algebra, the map $R(a, b) = (a \to b, a)$ *is a solution for the constant Yang–Baxter equation:*

$$R^{12} \circ R^{23} \circ R^{12} = R^{23} \circ R^{12} \circ R^{23} . \tag{12}$$

Proof. The equality $R^{12} \circ R^{23} \circ R^{12}(a, b, c) = R^{23} \circ R^{12} \circ R^{23}(a, b, c)$ implies $a \to (b \to c) = (a \to b) \to (a \to c)$, but $a \to (b \to c) = \bar{a} \vee (\bar{b} \vee c) = (\bar{a} \vee \bar{b} \vee c) \wedge (\bar{a} \vee a \vee c) = (a \wedge \bar{b}) \vee (\bar{a} \vee c) = (a \to b) \to (a \to c)$. □

Remark 8. *Only the self-distributive law from the definition of racks is used in the above proof.*

Remark 9. *The question if the above theorem holds in more general settings (such as Wajsberg–Algebras, MV–Algebras, and BCK/BCI–Algebras) has led to some interesting investigations (see [11]).*

5. Conclusions

The current paper consists of three main parts: a heuristical–phylosophical introduction, a short essay outlining some questions about π and e, and a section with deep results, where ideas from the previous section may be applied. Our main results are Theorems 1 and 2, the proof by hand of Equation (3), and the unification of Equations (1) and (3). Several open problems are also presented (Remarks 2 and 4).

We used results and concepts from a number of different fields, including algebra, mathematical analysis, mathematical physics, geometry, history of mathematics, numerical analysis, epistemology, and philosophy. The high number of new disciplines behooves one to have a transdisciplinary understanding of the world: a transdisciplinary approach (see [12–15]) attempts to discover what is between disciplines, across different disciplines, and beyond all disciplines. Transcendence is a concept that plays an important role in theology, in science, and in art and can be considered beyond all disciplines. The Yang–Baxter equation appears across different disciplines, and mathematical physics is at the border of two disciplines.

The author of [16] considers two types of scientists: birds, which resemble scientists with a broad vision, who try to unify theories and who obtain results of interest for a large readership, and frogs, which he considers less influential. Solomon Marcus used the terms of Francis Bacon (Novum Organum)— bees and ants—in this matter, and he also refers to spiders, which are scientists that build large networks of results.

We conclude with some historical facts. Euler was born in Basel (on 15 April 1707), and he received his Master of Philosophy from University of Basel. On 17 March 2016, Solomon Marcus passed away.

Acknowledgments: We would like to thank all those who helped us to improve our initial preprint: editors, referees, and coleagues.

Author Contributions: Section 2 was written by Solomon Marcus; Section 4 by Florin F Nichita; and the other parts were written by Florin F. Nichita as a consequence of the authors' meetings.

References

1. Nichita, F.F. On Transcendental Numbers. *Axioms* **2014**, *3*, 64–69.
2. Marcus, S. Transcendence, as a Universal Paradigm. *Int. J. Commun. Res.* **2014**, *4*, 60–72.
3. Marcus, S. Transcendence, as a universal paradigm. *Conv. Lit.* **2014**, 17–28. Available online: http://convorbiri-literare.ro/?p=2131 (accessed on 20 February 2018). (In Romanian)

4. Petrie, B.J. Leonhard Euler's use and understanding of mathematical transcendence. *Hist. Math.* **2012**, *39*, 280–291.
5. Breger, H. Leibniz Einfuhrung des Transzendenten. 300 Jahre "Nova Methodus" von G.W. Leibniz (1684–1984). *Stud. Leibnit.* **1986**, *Sonderheft XIV*, 119–132.
6. Desbrow, D. On Evaluating $\int_{-\infty}^{+\infty} e^{ax(x-2b)} dx$ by Contour Integration Round a Parallelogram. *Am. Math. Month.* **1998**, *105*, 726–731.
7. Links, J. Hopf Algebra Symmetries of an Integrable Hamiltonian for Anionic Pairing. *Axioms* **2012**, *1*, 226–237.
8. Nichita, F.F. *Jordan Algebras, Jordan Coalgebras and Unification Theories*; Editions Universitaires Europeennes; Omniscriptum Publishing Group: Saarbrucken, Germany, 2017, ISBN 978-3-639-62347-5.
9. Yu, L.W.; Ge, M.L. More about the doubling degeneracy operators associated with Majorana fermions and Yang–Baxter equation. *Sci. Rep.* **2015**, *5*, 8102.
10. Ge, M.L.; Yu, L.W.; Xue, K.; Zhao Q. Yang–Baxter equation, Majorana fermions and three body entangling states. *Int. J. Mod. Phys. B* **2014**, *28*, 1450089.
11. Oner, T.; Katican, T. On Solutions to the Set–Theoretical Yang–Baxter Equation in Wajsberg–Algebras. *Axioms* **2018**, *7*, 6.
12. Nicolescu, B. *Manifesto of Transdisciplinarity*; State University of New York (SUNY) Press: New York, NY, USA, 2002.
13. Nicolescu, B. Transdisciplinarity—Past, present and future. In *Moving Worldviews—Reshaping Sciences, Policies and Practices for Endogenous Sustainable Development*; Haverkort, B., Reijntjes, C., Eds.; COMPAS Editions: Amsterdam, The Netherlands, 2006; pp. 142–166.
14. Nicolescu, B. The unexplected way to holiness: Simone Weil (II). *Conv. Lit.* **2017**, *10*, 26–28. (In Romanian)
15. Nichita, F.F. On Models for Transdisciplinarity. *Transdiscipl. J. Eng. Sci.* **2011**, *2011*, 42–46.
16. Dyson, F. Birds and Frogs. *Not. AMS* **2009**, *56*, 212–223.

4

The Hecke Bicategory

Alexander E. Hoffnung

Department of Mathematics, Temple University, 1805 N. Broad Street, Philadelphia, PA 19122, USA;
E-Mail: hoffnung@temple.edu

Abstract: We present an application of the program of groupoidification leading up to a sketch of a categorification of the Hecke algebroid—the category of permutation representations of a finite group. As an immediate consequence, we obtain a categorification of the Hecke algebra. We suggest an explicit connection to new higher isomorphisms arising from incidence geometries, which are solutions of the Zamolodchikov tetrahedron equation. This paper is expository in style and is meant as a companion to *Higher Dimensional Algebra VII: Groupoidification* and an exploration of structures arising in the work in progress, *Higher Dimensional Algebra VIII: The Hecke Bicategory*, which introduces the Hecke bicategory in detail.

Keywords: Hecke algebras; categorification; groupoidification; Yang–Baxter equations; Zamalodchikov tetrahedron equations; spans; enriched bicategories; buildings; incidence geometries

---

1. Introduction

Categorification is, in part, the attempt to shed new light on familiar mathematical notions by replacing a set-theoretic interpretation with a category-theoretic analogue. Loosely speaking, categorification replaces sets, or more generally n-categories, with categories, or more generally $(n+1)$-categories, and functions with functors. By replacing *interesting* equations by isomorphisms, or more generally equivalences, this process often brings to light a new layer of structure previously hidden from view. While categorification is not a systematic process—in other words, finding this new layer of structure may require a certain amount of creativity—the reverse process of *decategorification* should be a systematic way of recovering the original set-theoretic structure or concept. The key idea is that considering a process of categorification requires, as a first step, a Definition of the corresponding decategorification process. We then think of categorification simply as a heuristic tool allowing us to "undo" the process of decategorification.

In *Higher Dimensional Algebra VII: Groupoidification* [1], Baez, Walker and the author introduced a program called *groupoidification* initiated by Baez, Dolan and Trimble, and aimed at categorifying various notions from linear algebra, representation theory and mathematical physics. The very simple idea was that one could replace vector spaces by groupoids, *i.e.*, categories with only isomorphisms, and replace linear operators by spans of groupoids. In fact, what we really did was define a systematic process called *degroupoidification*:

$$\text{groupoids} \mapsto \text{vector spaces}$$

$$\text{spans of groupoids} \mapsto \text{matrices}$$

Thus, groupoidification is a form of categorification. We then suggested some applications of groupoidification to Hall algebras, Hecke algebras, and Feynman diagrams, so that other researchers could begin to categorify *their* favorite notions from representation theory.

In this paper, we give an expository account of a theory of *categorified intertwining operators* or *categorified Hecke operators* for representations of a very basic type: the *permutation representations* of a finite group. Following the description of categorification above, this suggests the study of a 2-dimensional category-theoretic structure and a decategorification functor. We describe this 2-dimensional structure, which we call the *Hecke bicategory*, in Section 4.1. Pairing the Hecke bicategory with the degroupoidification functor we are then able to state a categorification theorem as Claim 8.

The statements of the main results are as follows. For each finite group G, there is an equivalence of categories, or more precisely algebroids, between the category of permutation representations of G—the *Hecke algebroid* of *Hecke operators*—and the degroupoidification of the *Hecke bicategory* of *categorified Hecke operators*, which has G-sets as objects and is enriched over the monoidal bicategory of spans of groupoids. In other words, *the Hecke bicategory categorifies the category of permutation representations*. When G is the simple Lie group over a finite field of q elements attached to a Dynkin diagram, then one can choose a Borel subgroup B, and construct the G-set $X = G/B$, known as the flag complex. The choice of one object X in $\text{Hecke}(G)$ yields a groupoid $\text{Hecke}(G)(X, X)$ and a span called composition. *The groupoid* $\text{Hecke}(G)(X, X)$ *and accompanying span categorify the usual Hecke algebra for the chosen Dynkin diagram and prime power* q.

The term "Hecke algebra" is seen in several areas of mathematics. The Hecke algebras we consider, the *Iwahori–Hecke algebras*, are one-parameter deformations of the group algebras of Coxeter groups. In the theory of modular forms, or more generally, automorphic representations, Hecke algebras are commutative algebras of Hecke operators. These operators can be expressed by means of double cosets in the modular group, or more generally, with respect to certain compact subgroups. Here we use the term "Hecke operator" to highlight the relationship between the intertwining operators between permutation representations of a finite group and the Hecke operators acting on the modular group. We discuss an example of Hecke operators in terms of "flag-flag relations" in the setting of Coxeter groups in Section 6.2. To describe Hecke algebras, one may use relations between varieties of the form G/H for various subgroups H, namely the *discrete* subgroups, or more generally, the *compact* subgroups. So, we think of Hecke algebras as algebras of "Hecke operators" in a slightly generalized sense. We are changing the groups, and making them *finite*, so that instead of varieties G/H, we have certain finite sets G/H. Thus we think of a Hecke algebra as the algebra of intertwining operators from a permutation representation to itself. Generalizing this slightly, we think of intertwining operators between permutation representations in general as *Hecke operators*.

We make use of the techniques of groupoidification along with the machinery of enriched bicategories and some very basic topos theory. Thus, this paper is intended to give an introduction to some concepts which should play a significant role as the subject of categorified representation theory continues to develop. Detailed accounts of the necessary structures of higher category categories of spans are presented in papers by the author [2] and by the author in collaboration with John Baez [3]. We now proceed to give a brief overview and explanation of each section.

We hope that both representation theorists and category theorists might benefit from this exposition in their continued development of suitable higher categorical frameworks for representation theory. While significant attention is given to certain fundamental ideas about Hecke algebras, the reader not well acquainted with these structures should not be hindered greatly as our discussion stems from widely accessible ideas in incidence geometries. We have attempted to include a fair sampling of the higher category theory needed in these constructions, so that the exposition would be largely self-contained for category theorist and representation theorist alike. Of course, the researcher wishing to pursue these ideas further should consult the references within.

1.1. Matrices and Spans

In Section 2, we give a heuristic discussion of a very simple notion of categorification. In particular, we recall the basic notions of *spans*, also known as *correspondences*, and see that *spans of sets categorify matrices of (extended) natural numbers* $\mathbb{N} \cup \{\infty\}$. Decategorification can be defined using just set cardinality with the usual rule of transfinite arithmetic, $\infty + n = \infty$, for any $n \in \mathbb{N}$, and the free vector space construction, and this process is indeed functorial, since composition of spans by pullback corresponds to matrix multiplication.

After discussing the example of linear operators, we pass to the intertwining operators for a finite group G. Then we need to consider not only spans of sets but spans of G-sets. Since the maps in the spans are now taken to be G-equivariant, the corresponding matrices should also be G-equivariant. This prompts us to recall the relationships between G-sets and permutation representations of G. There is a faithful, essentially surjective functor from the category of G-sets to the category of permutation representations of G. However, this functor is not *full*. Spans in a category with pullbacks naturally form a bicategory. Since G-sets and permutation representations of G are closely related, except that there are "not enough" maps of G-sets, the bicategory of spans of G-sets is a first clue in constructing categorified permutation representations. We will return to the role of spans of G-sets in Sections 2, 5 and 6.

1.2. Groupoidification and Enriched Bicategories

To understand groupoidification, we need to recall the construction and basic properties of the *degroupoidification functor* defined in [1]. We discuss this functor and extend it to a functor on the bicategory of spans of groupoids in Section 3.2. Our intention is to define the Hecke bicategory as an enriched structure, but the replacement of functors in the monoidal 2-category of groupoids by spans in the monoidal bicategory of spans of groupoids necessitates a generalization of enriched category theory to enriched bicategory theory. Definitions of enriched bicategories were developed independently by the author and earlier in the unpublished Ph.D. thesis of Carmody [4], which is reproduced in part by Forcey [5]. We present a partial Definition explaining the basic idea in this work and will define the full structure along with a *change of base* theorem in future work. However, the reader well-acquainted with enriched category theory and with some patience may work out the details independently. Change of base for enriched bicategories is analogous to the theorem for enriched categories, and together with the degroupoidification functor, allows us to obtain a categorification theorem for Hecke operators.

The degroupoidification functor takes the monoidal bicategory Span(Grpd) of spans of (tame) groupoids to the monoidal category Vect. The condition that a groupoid be tame ensures that certain sums converge. This condition is defined in [1], but should not be of much concern to the reader at present. We briefly recall the degroupoidification functor here. A groupoid is sent to the free vector space on its set of isomorphism classes of objects. A span of groupoids is sent to a linear operator using the *weak* or *pseudo* pullback of groupoids and the notion of *groupoid cardinality* [6]. That is, we think of a span of groupoids as a categorified or groupoid-valued matrix in much the same way as we think of a span of sets as a set-valued matrix, where set cardinality is replaced by groupoid cardinality.

The notion of enriched bicategories is then used to make the description of our decategorification processes precise. Given a monoidal bicategory $\mathcal{V}$, a $\mathcal{V}$-enriched bicategory consists of a set of objects, hom-objects in $\mathcal{V}$, composition morphisms in $\mathcal{V}$, and further structure and axioms, all of which live in the monoidal bicategory $\mathcal{V}$. The only theorem about enriched bicategories that we will need is a *change of base* theorem. In particular, given a functor $\mathcal{F}\colon \mathcal{V} \to \mathcal{V}'$ and a $\mathcal{V}$-enriched bicategory, then we obtain a $\mathcal{V}'$-enriched bicategory with hom-objects $\mathcal{F}(\hom(x,y))$. If $\mathcal{V}$ is a monoidal category, then a $\mathcal{V}$-enriched bicategory is a $\mathcal{V}$-enriched category in the usual sense [7]. Enriched bicategories and change of base are discussed in more detail in Section 3.3.

1.3. The Hecke Bicategory

The Hecke bicategory Hecke(G) of a finite group G is a categorification of the permutation representations of G. This is the main construction of this paper. Section 4.1 gives the basic structure of this family of enriched bicategories. For each finite group G, Hecke(G) is an enriched bicategory over the monoidal bicategory of spans of groupoids. The structure of this monoidal bicategory is given roughly in Section 3.2. A more complete and general account of monoidal bicategories (and monoidal tricategories) of spans will be given in [2].

The Hecke bicategory is constructed to study the Hecke operators between permutation representations X and Y of a finite group G by keeping track of information about the G-orbits in $X \times Y$ as the hom-groupoid $(X \times Y)//G$. This double-slash notation denotes the *action groupoid*, which we recall in Definition 3. The composition process between such groupoids is closely related to the *pull-tensor-push* construction familiar from geometric representation theory.

Now, applying the change of base theorem of enriched bicategories together with the degroupoidification functor, we obtain a Vect-enriched category from Hecke(G). The resulting Vect-enriched category is equivalent to the Vect-enriched category of permutation representations of G. This is our main theorem and is stated in Claim 8.

1.4. Spans of Groupoids and Cocontinuous Functors

In Section 2.3, the bicategory of spans of G-sets appears in our study of the category of permutation representations. We view spans of G-sets as categorified G-equivariant matrices, but do not specify a decategorification process in this setting, although we discuss the importance of set cardinality. This bicategory of spans Span$(G\text{Set})$ plays two important roles in our attempt to understand categorified representation theory, which we briefly discuss here.

In Section 6.2, we will explain the categorification of the Hecke algebra associated to the A_2 Dynkin diagram as a hom-category in Span$(G\text{Set})$. This makes the categorification explicit in that we see the Hecke algebra relations holding *up to isomorphism* with special spans of G-sets acting as *categorified generators*. The isomorphisms representing the relations come explicitly from incidence geometries—in this example, projective plane geometry—associated to the Dynkin diagram.

To facilitate this point of view, we describe a monoidal functor in Section 5.2 that takes a groupoid to the corresponding presheaf category and linearizes a span of groupoids to a cocontinuous functor by a *pull-tensor-push* process familiar from many geometric constructions in representation theory, but quite generally applicable in the theory of Grothendieck toposes. This is a functor $\mathcal{L}\colon$ Span $\to$ Cocont, where Cocont is the monoidal 2-category of presheaf categories on groupoids, cocontinuous functors, and natural isomorphisms. By change of base, we obtain a bicategory enriched over certain presheaf toposes, the objects of which are interpreted as spans of G-sets in Section 5.3.

We denote the functor taking a groupoid to its category of presheaves $\mathcal{L}$ suggesting that this might be interpreted as *categorified linearization*. In this interpretation one thinks of a groupoid as a *categorified vector space* that is equipped with a chosen basis—its set of isomorphism classes of objects or connected components. From a Grothendieck topos that is equivalent to a category of presheaves on a groupoid, we can always recover the groupoid. Where a basis can always be recovered up to isomorphism from a vector space, a groupoid can be recovered up to equivalence from such a presheaf category. Then we think of such a topos as an *abstract categorified vector space*.

The relationship between categorified representation theory enriched over monoidal bicategories of spans of spaces and enrichment over corresponding 2-categories of sheaves on the spaces and cocontinuous functors between these will be the focus of future work. From our point of view, these monoidal 2-categories are the basic objects of study in categorified linear algebra, and in this sense, *categorified representation theory is enriched over categorified linear algebra*. Part of the motivation for this line of work is to better understand and unify (co)homology theories which arise in geometric representation theory.

1.5. The Categorified Hecke Algebra and Zamolodchikov Equation

As already mentioned, Section 6 discusses the main corollary, Claim 16—a categorification of the Hecke algebra of a Coxeter group—as well as possible future directions in low-dimensional topology and higher-category theory. In Section 6.1, we recall the notion of Hecke algebras associated to Dynkin diagrams and prime powers. We describe how a categorification of the Hecke algebra naturally arises from the Hecke bicategory.

Finally, in Section 6.2, we describe a concrete example of the categorified Hecke algebra in terms of spans of G-sets. We describe solutions to the Zamolodchikov tetrahedron equation, which we hope will lead to constructions of braided monoidal 2-categories as pointed out by Kapranov and Voevodsky [8], and eventually (higher) tangle invariants [9].

Other approaches to categorified Hecke algebras and their representations have been studied by a number of authors, building on Kazhdan–Lusztig theory [10]. One key step was Soergel's introduction of what are nowadays called Soergel bimodules [11,12]. Also important was Khovanov's categorification of the Jones polynomial [13] and work by Bernstein, Frenkel, Khovanov and Stroppel on categorifying Temperley–Lieb algebras, which are quotients of the type A Hecke algebras [14,15]. A diagrammatic interpretation of the Soergel bimodule category was developed by Elias and Khovanov [16], and a geometric approach led Webster and Williamson [17] to deep applications in knot homology theory. This geometric interpretation can be seen as going beyond the simple form of groupoidification we consider here, and requires considering groupoids in the category of schemes.

2. Matrices, Spans and G-Sets

2.1. *Spans as Matrices*

The first tool of representation theory is linear algebra. Vector spaces and linear operators have nice properties, which allow representation theorists to extract a great deal of information about algebraic gadgets ranging from finite groups to Lie groups to Lie algebras and their various relatives and generalizations. We start at the beginning, considering the representation theory of finite groups. Noting the utility of linear algebra in representation theory, this paper is fundamentally based on the idea that the heavy dependence of linear algebra on fields, very often the complex numbers, may at times obscure the combinatorial skeleton of the subject. Then, we hope that by peeling back the soft tissue of the continuum, we may expose and examine the bones, revealing new truths by working directly with the underlying combinatorics. In this section, we consider the notion of spans of sets, a very simple idea, which is at the heart of categorified representation theory.

A **span of sets** from X to Y is a pair of functions with a common domain like so:

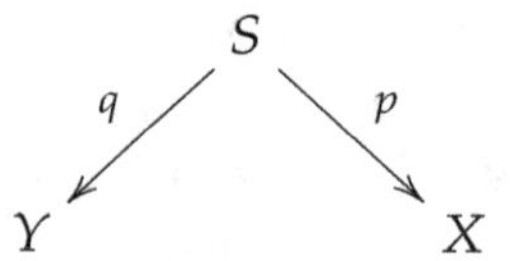

We will often denote a span by its apex, when no confusion is likely to arise.

A span of sets can be viewed as a matrix of sets

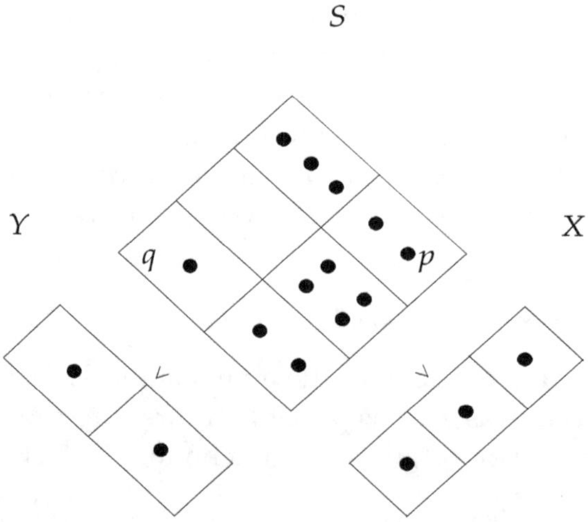

For each pair $(x,y) \in X \times Y$, we have a set $S_{x,y} = p^{-1}(x) \cap q^{-1}(y)$. In other words, there is a "decategorification" process from spans of sets to matrices with values in $\mathbb{N} \cup \{\infty\}$. If all the sets $S_{x,y}$ are finite, we obtain a matrix of natural numbers $|S_{x,y}|$—a very familiar object in linear algebra. In this sense, a span is a "categorification" of a matrix.

In addition to spans giving rise to matrices, composition of spans gives rise to matrix multiplication. Given a pair of composable spans

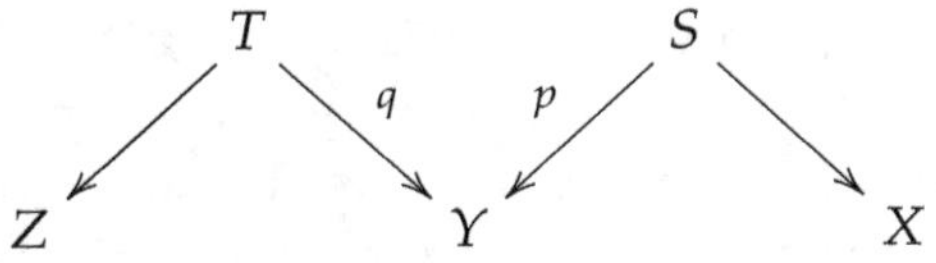

we define the composite to be the **pullback** of the pair of functions $p\colon S \to Y$ and $q\colon T \to Y$, which is a new span

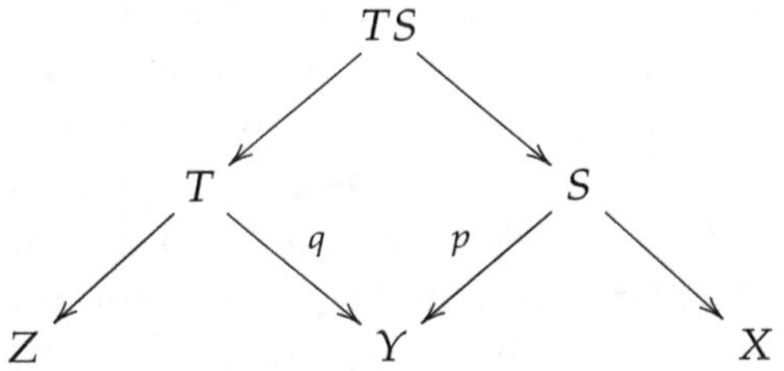

where TS is the subset of $T \times S$

$$\{(t,s) \subseteq T \times S \mid p(s) = q(t)\}$$

with the obvious projections to S and T. It is straightforward to check that this process agrees with matrix multiplication after decategorification.

In the above example, we turn spans of sets into matrices simply by counting the number of elements in each set $S_{x,y}$. Checking that composition of spans and matrix multiplication agree after taking the cardinality is the main step in showing that our decategorification process—from the bicategory of spans of sets to the category of linear operators—is *functorial*.

2.2. Permutation Representations

Again we start with a very simple idea. We want to study the actions of a finite group G on sets—*G-sets*. However, in this article we restrict to those G-sets with a finite number of orbits. These extend to *permutation representations of G*. We fix the field of complex numbers and consider only complex vector spaces throughout this paper.

Definition 1. *A* **permutation representation** *of a finite group G is a finite-dimensional representation of G together with a chosen basis such that the action of G maps basis vectors to basis vectors.*

Definition 2. *An* **intertwining operator** $f\colon V \to W$ *between permutation representations of a finite group G is a linear operator from V to W that is G-equivariant,* i.e., *commutes with the actions of G.*

G-sets can be linearized to obtain permutation representations of G. In fact, this describes a relationship between the objects of the *category* of G-sets and the objects of the *category* of permutation representations of G. Given a finite group G, the category of G-sets has

- G-sets (with finitely many orbits) as objects,
- G-equivariant functions as morphisms,

and the category of permutation representations $\mathrm{PermRep}(G)$ has

- permutation representations of G as objects,
- intertwining operators as morphisms.

One usually wants the morphisms in a category to preserve the structure on the objects. Of course, an intertwining operator does not necessarily preserve the chosen basis of a permutation representation. We can reconcile our choice of intertwining operators as morphisms, by noticing that there is a bijection between objects in PermRep(G) and the category consisting of finite-dimensional representations of G with the *property* that there *exists* a basis preserved by the action of G. Thus, we are justified in working with this Definition of PermRep(G).

A primary goal of this paper is to categorify the q-deformed versions of the group algebras of Coxeter groups known as Hecke algebras. Of course, an algebra is a Vect-enriched category with exactly one object, and the Hecke algebras are isomorphic to certain one-object subcategories of the Vect-enriched category of permutation representations. Thus, we refer to the category PermRep(G) as the *Hecke algebroid*—a many-object generalization of the Hecke algebra. We will construct a bicategory—or more precisely, an *enriched bicategory*—called the *Hecke bicategory* that categorifies the Hecke algebroid for any finite group G.

There is a functor from G-sets to permutation representations of G. A G-set X is linearized to a permutation representation $\widetilde{X}$, which is the free vector space on X. As stated above, the maps between G-sets are G-equivariant functions—that is, functions between G-sets X and Y that respect the actions of G. Such a function $f\colon X \to Y$ gives rise to a G-equivariant linear map (or intertwining operator) $\widetilde{f}\colon \widetilde{X} \to \widetilde{Y}$. However, there are many more intertwining operators from $\widetilde{X}$ to $\widetilde{Y}$ than there are G-equivariant maps from X to Y. In particular, the former is a complex vector space, while the latter is often a finite set. For example, an intertwining operator $\widetilde{f}\colon \widetilde{X} \to \widetilde{Y}$ may take a basis vector $x \in \widetilde{X}$ to any -linear combination of basis vectors in $\widetilde{Y}$, whereas a map of G-sets does not have the freedom of scaling or adding basis elements.

So, in the language of category theory the process of linearizing G-sets to obtain permutation representations is a faithful, essentially surjective functor, which is not at all full.

2.3. *Spans of G-Sets*

In the previous section, we discussed the relationship between G-sets and permutation representations. In Section 2.1, we saw the close relationship between spans of sets and matrices of (extended) natural numbers. There is an analogous relationship between spans of G-sets and G-equivariant matrices of (extended) natural numbers.

A **span of** G**-sets** from a G-set X to a G-set Y is a pair of maps with a common domain

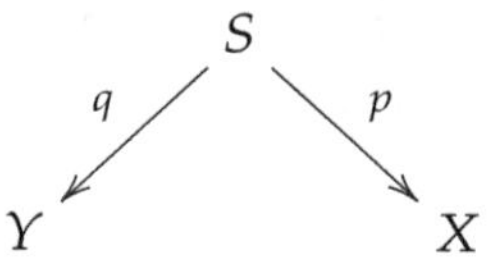

where S is a G-set, and p and q are G-equivariant maps with respective codomains X and Y.

Spans of G-sets are natural structures to consider in categorified representation theory because spans of sets—and, similarly G-sets—naturally form a bicategory.

The development of bicategories by Benabou [18] is an early example of categorification. A (small) category consists of a *set of objects* and a *set of morphisms*. A bicategory is a categorification of this concept, so there is a new layer of structure [19]. In particular, a (small) bicategory $\mathcal{B}$ consists of:

- a set of objects $x, y, z \ldots$,
- for each pair of objects, a set of morphisms,
- for each pair of morphisms, a set of 2-morphisms,

and given any pair of objects x, y, this data forms a hom-category hom(x, y) which has:

- 1-morphisms $x \to y$ of $\mathcal{B}$ as objects,

- 2-morphisms:

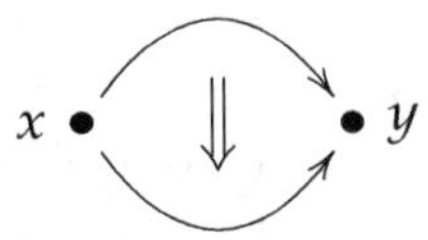

of $\mathcal{B}$ as morphisms, and a *vertical composition*

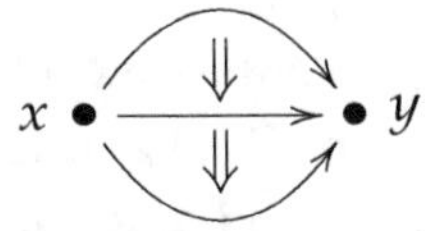

of 2-morphisms of $\mathcal{B}$.

Further, the bicategory $\mathcal{B}$ consists of:

- for each triple of objects x, y, z, a *horizontal composition* functor $\circ_{xyz} \colon \hom(x,y) \times \hom(y,z) \to \hom(x,z)$

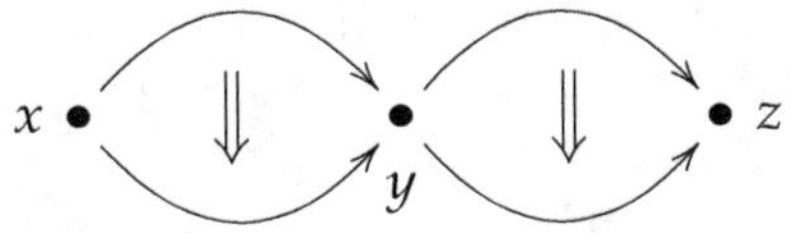

on 1-morphisms and 2-morphisms,
- for each object x, an *identity-assigning* functor $I_x \colon \mathbf{1} \to \hom(x,x)$,
- for objects w, x, y, z, an *associator* natural isomorphism

$$a_{wxyz} \colon \circ_{wyz}(\circ_{wxy} \times 1) \Rightarrow \circ_{wxz}(1 \times \circ_{xyz})$$

- for pairs of objects x, y, *left and right unitor* natural isomorphisms

$$l_{xy} \colon \circ_{xxy}(I_x \times 1) \Rightarrow 1 \text{ and } r_{xy} \colon \circ_{xyy}(1 \times I_y) \Rightarrow 1$$

all of which is required to satisfy *associativity* and *unit* axioms:

$$(1 \circ a_{xyz})a_{w(xy)z}(a_{wxy} \circ 1) = a_{wx(yz)}a_{(wx)yz} \text{ and } (1 \circ r_{yz})a_{xyz} = l_{xy} \circ 1$$

See Leinster's article [19] for a concise working reference on bicategories.

Benabou's Definition followed from several important examples of bicategories, which he presented in [18], and which are very familiar in categorified and geometric representation theory. The first example is the bicategory of spans of sets, which has:

- sets as objects,
- spans of sets as morphisms,
- maps of spans of sets as 2-morphisms.

We defined spans of sets in Section 2.1. A **map of spans of sets** from a span S to a span S' is a function $f \colon S \to S'$ such that the following diagram commutes:

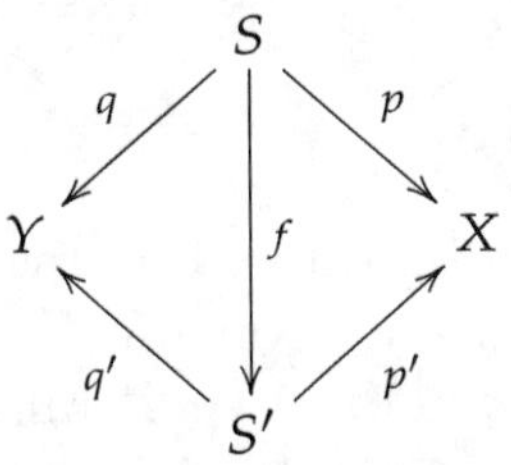

For each finite group G, there is a closely related bicategory Span(GSet), which has:

- G-sets as objects,
- spans of G-sets as morphisms,

- maps of spans of G-sets as 2-morphisms.

The Definitions are the same as in the bicategory of spans of sets, except for the finiteness condition on orbits and that every arrow should be G-equivariant.

While this bicategory is a good candidate for a categorification of the Hecke algebroid PermRep(G), the theory of groupoidification allows for a related, but from our point of view, heuristically nicer categorification. In what follows, we develop the necessary machinery to present this categorification, the Hecke bicategory Hecke(G), in the context of groupoidification. In Sections 5 and 6.2, we return to Span(GSet) and in future work we will make precise the relationship between spans of G-sets and the Hecke bicategory Hecke(G).

3. Groupoidification and Enriched Bicategories

The following sections introduce the necessary machinery to present the Hecke bicategory, a categorification of the Hecke algebroid. Enriched bicategories are developed for use in Section 4 to construct the Hecke bicategory and state the main result in Theorem 4.2, and in Section 5 to make a connection with the bicategory Span(GSet) of Section 2.3.

3.1. Action Groupoids and Groupoid Cardinality

In this section, we draw a connection between G-sets and groupoids via the "action groupoid" construction. We then introduce *groupoid cardinality*, which is the first step in describing the degroupoidification functor in the next section.

For any G-set, there exists a corresponding groupoid, called the *action groupoid* or *weak quotient*:

Definition 3. *Given a group G and a G-set X, the* **action groupoid** $X//G$ *is the category which has:*

- *elements of X as objects,*
- *pairs $(g,x) \in G \times X$ as morphisms $(g,x)\colon x \to x'$, where $g \cdot x = x'$.*

Composition of morphisms is defined by the product in G.

Of course, associativity follows from associativity in G and the construction defines a groupoid since any morphism $(g,x)\colon x \to x'$ has an inverse $(g^{-1},x')\colon x' \to x$.

So every G-set defines a groupoid, and we will see in Section 4.1 that the weak quotient of G-sets plays an important role in understanding categorified permutation representations.

Next, we recall the Definition of groupoid cardinality [6]:

Definition 4. *Given a (small) groupoid $\mathcal{G}$, its* **groupoid cardinality** *is defined as:*

$$|\mathcal{G}| = \sum_{\text{isomorphism classes of objects [x]}} \frac{1}{|\mathrm{Aut}(x)|}$$

If this sum diverges, we say $|\mathcal{G}| = \infty$.

In our examples, in particular, the hom-groupoids of the Hecke bicategory, all groupoids will have finitely many objects and morphisms, since we consider action groupoids of finite groups acting on flag complexes in vector spaces over finite fields. In general, we allow groupoids with infinitely many isomorphism classes of objects, and the cardinality of a groupoid takes values in the non-negative real numbers in case the sum converges. Generalized cardinalities have been studied by a number of authors [20–23].

We can think of groupoid cardinality as a form of categorified division analogous to the quotient of a G-set by its action of G in the case when this action is free. See the paper of Baez and Dolan [6]. In particular, we have the following equation:

$$|X//G| = |X|/|G|$$

whenever G is a finite group acting on a finite set X.

In the next section, we define degroupoidification using the notion of groupoid cardinality.

3.2. *Degroupoidification*

In this section, we recall some of the main ideas of groupoidification. Of course, in practice this means we will discuss the corresponding process of decategorification—the degroupoidification functor.

To define degroupoidification in [1], a functor was constructed from the category of spans of "tame" groupoids to the category of linear operators between vector spaces. In the present setting, we will need to extend degroupoidification to a functor between bicategories.

We extend the functor to a bicategory Span(Grpd) which has:

- (tame) groupoids as objects,
- spans of groupoids as 1-morphisms,
- isomorphism classes of maps of spans of groupoids as 2-morphisms.

We will often drop the adjective tame when confusion is not likely to arise.

Given a pair of parallel spans in the 2-category Grpd of groupoids, functors, and natural isomorphisms:

a **map of spans** is a triple (α, f, β), where $f\colon S \to S'$ is a functor and $\alpha\colon p \Rightarrow p'f$ and $\beta\colon q \Rightarrow q'f$ are natural isomorphisms:

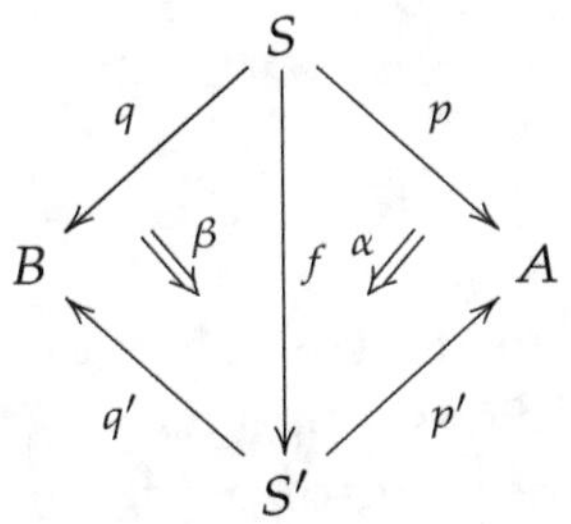

In general, spans of groupoids form a (monoidal) tricategory, which has not only *maps of spans* as 2-morphisms, but also *maps of maps of spans* as 3-morphisms.

Given a parallel pair of maps of spans (α, f, β) and (α', f', β'):

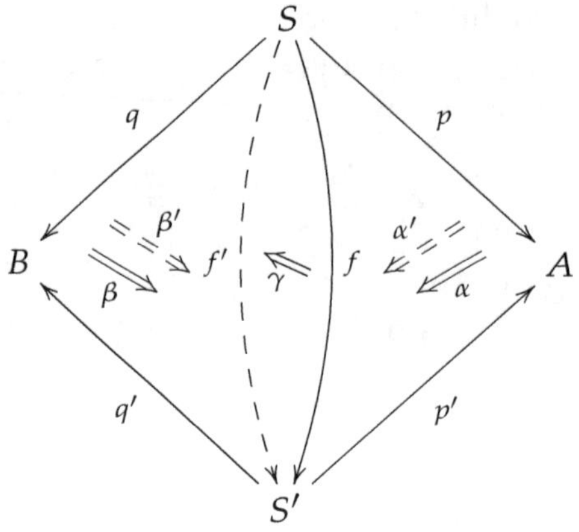

a **map of maps of spans**

$$\gamma\colon (\alpha, f, \beta) \Longrightarrow (\alpha', f', \beta')$$

is a natural isomorphism $\gamma\colon f \Rightarrow f'$ satisfying:

$$(p' \cdot \gamma)\alpha = \alpha' \quad \text{and} \quad (q' \cdot \gamma)\beta = \beta'. \tag{1}$$

For our purposes we restrict this structure to a bicategory. While there may be more sophisticated ways of obtaining such a bicategory, we do so by taking *isomorphism classes of maps of spans* as 2-morphisms [2]. A related span construction is found in [24], where the 2-category of spans corresponds roughly to the local 2-categories in our tricategory of spans.

With appropriate finiteness conditions imposed, spans of groupoids are categorified matrices of non-negative rational numbers in the same way that spans of sets are categorified matrices of natural numbers. A *span of groupoids* is a pair of functors with common domain, and we can picture one of these roughly as follows:

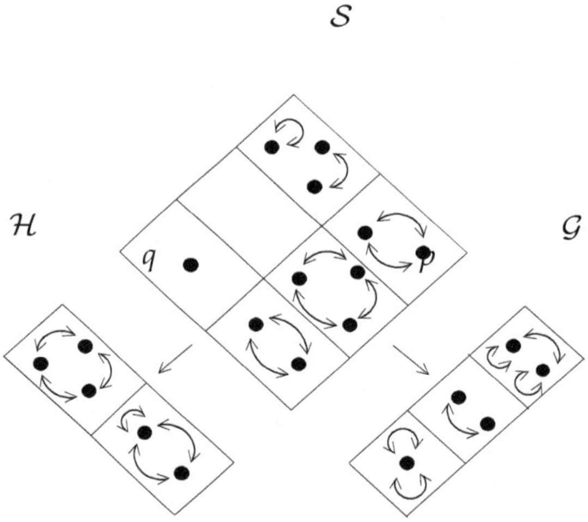

Whereas one uses set cardinality to realize spans of sets as matrices, we can use groupoid cardinality to obtain a matrix from a span of groupoids.

We have seen evidence that spans of groupoids categorify matrices, so we will want to think of a groupoid as a categorified vector space. To make these notions precise, we recall the degroupoidification functor

$$\mathcal{D}\colon \text{Span(Grpd)} \to \text{Vect}$$

Given a groupoid $\mathcal{G}$, we obtain a vector space $\mathcal{D}(\mathcal{G})$, called the **degroupoidification** of $\mathcal{G}$, by taking the free vector space on the set of isomorphism classes of objects of $\mathcal{G}$.

We say a groupoid $\mathcal{V}$ *over* a groupoid $\mathcal{G}$

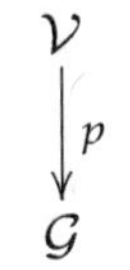

is a **groupoidified vector**. In particular, from the functor p we can produce a vector in $\mathcal{D}(\mathcal{G})$ in the following way.

The **full inverse image** of an object x in $\mathcal{G}$ is the groupoid $p^{-1}(x)$, which has:

- objects v of $\mathcal{V}$, such that $p(v) \cong x$, as objects,
- morphisms $v \to v$ in $\mathcal{V}$ as morphisms.

We note that this construction depends only on the isomorphism class of x. Since the set of isomorphism classes of $\mathcal{G}$ determines a basis of the corresponding vector space, the vector determined by p can be defined as

$$\sum_{\text{isomorphism classes of objects } [x]} |p^{-1}(x)|[x]$$

where $|p^{-1}(x)|$ is the groupoid cardinality of $p^{-1}(x)$. We note that a "groupoidified basis" can be obtained in this way as a set of functors from the terminal groupoid **1** to representative objects of each isomorphism class of $\mathcal{G}$. A **groupoidified basis** of $\mathcal{G}$ is a set of groupoids $\mathcal{V} \to \mathcal{G}$ over $\mathcal{G}$ such that the corresponding vectors give a basis of the vector space $\mathcal{D}(\mathcal{G})$.

Given a span of groupoids

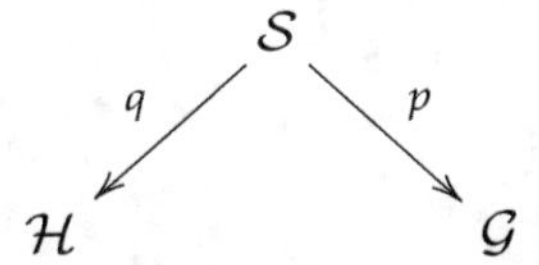

we want to produce a linear map $\mathcal{D}(\mathcal{S})\colon \mathcal{D}(\mathcal{G}) \to \mathcal{D}(\mathcal{H})$. The details are checked in [1]. Here we show only that given a basis vector of $\mathcal{D}(\mathcal{G})$ viewed as a groupoidified basis vector of $\mathcal{G}$, the span $\mathcal{S}$ determines a vector in $\mathcal{D}(\mathcal{H})$. To do this, we need the notion of the weak pullback of groupoids—a categorified version of the pullback of sets.

Given a diagram of groupoids

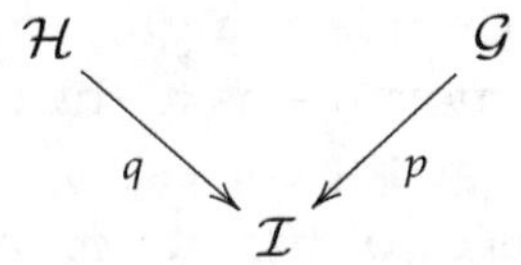

the **weak pullback** of $p\colon \mathcal{G} \to \mathcal{I}$ and $q\colon \mathcal{H} \to \mathcal{I}$ is the diagram

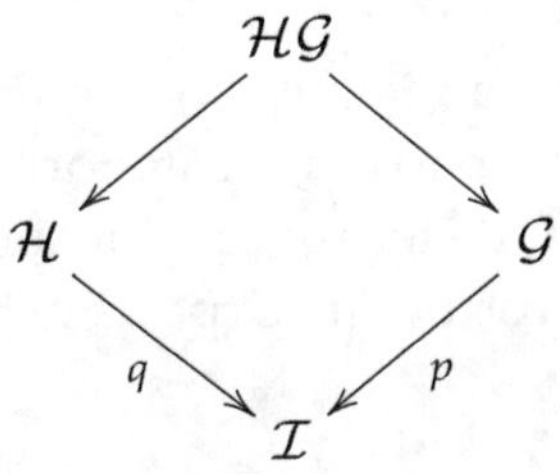

where $\mathcal{HG}$ is a groupoid whose objects are triples (h, g, α) consisting of an object $h \in \mathcal{H}$, an object $g \in \mathcal{G}$, and an isomorphism $\alpha\colon p(g) \to q(h)$ in $\mathcal{I}$. A morphism in $\mathcal{HG}$ from (h, g, α) to (h', g', α') consists of a morphism $f\colon g \to g'$ in $\mathcal{G}$ and a morphism $f'\colon h \to h'$ in $\mathcal{H}$ such that the following square commutes:

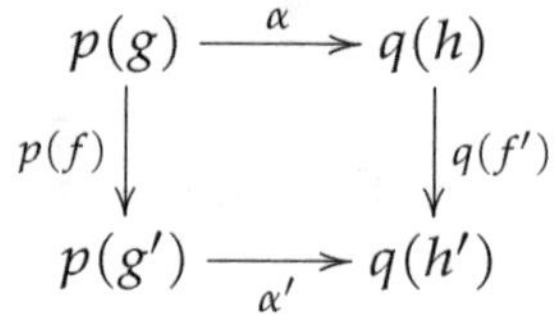

As in the case of the pullback of sets, the maps out of $\mathcal{HG}$ are the obvious projections. Further, this construction satisfies a certain universal property. See [25], for example.

Now, given our span and a chosen *groupoidified basis vector*:

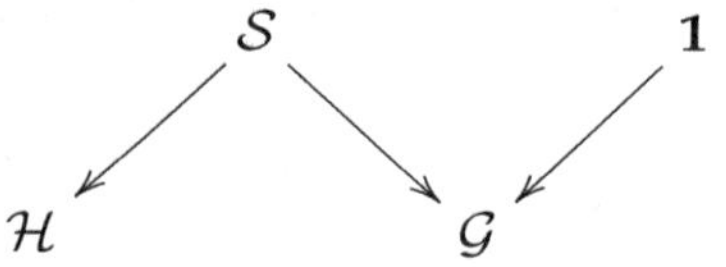

we obtain a groupoid over $\mathcal{H}$ by constructing the weak pullback:

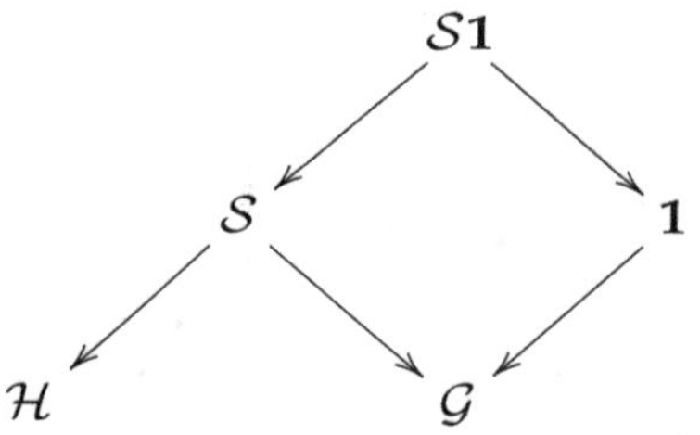

Now, $\mathcal{S}\mathbf{1}$ is a groupoid over $\mathcal{H}$, and we can compute the resulting vector. The tameness condition on groupoids guarantees that the process of passing a groupoidified vector across a span defines a linear operator [1].

One can check that the process described above defines a linear operator from a span of groupoids, and, further, that this process is functorial [1]. This is the degroupoidification functor. Since isomorphic spans are sent to the same linear operator, it is straightforward to extend this to our bicategory of spans of groupoids by adding identity 2-morphisms to the category of vector spaces and sending all 2-morphisms between spans of groupoids to the corresponding identity 2-morphism.

In the next section, we give the basics of the notion of *enriched bicategories*. We will see that constructing an enriched bicategory requires having a monoidal bicategory in hand. The bicategory Span(Grpd) defined above is, in fact, a monoidal bicategory—that is, Span(Grpd) has a tensor product, which is a functor

$$\otimes\colon \mathrm{Span(Grpd)} \times \mathrm{Span(Grpd)} \to \mathrm{Span(Grpd)}$$

along with further structure satisfying some coherence relations. The structure of Span(Grpd), or more generally, monoidal bicategories of spans in 2-categories, is described in detail in [2].

We describe the main components of the tensor product on Span(Grpd). Given a pair of groupoids $\mathcal{G}, \mathcal{H}$, the tensor product $\mathcal{G} \times \mathcal{H}$ is the product in Grpd. Further, for each pair of pairs of groupoids $(\mathcal{G}, \mathcal{H}), (\mathcal{I}, \mathcal{J})$ there is a functor:

$$\otimes\colon \mathrm{Span(Grpd)}(\mathcal{G}, \mathcal{H}) \times \mathrm{Span(Grpd)}(\mathcal{I}, \mathcal{J}) \to \mathrm{Span(Grpd)}(\mathcal{G} \times \mathcal{I}, \mathcal{H} \times \mathcal{J})$$

defined roughly as follows:

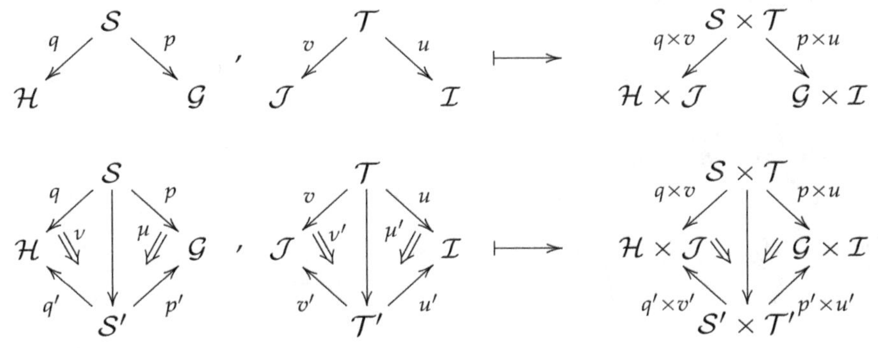

In fact, this tensor product will be not just a functor, but a homomorphism of bicategories. This means that it carries some more structure and satisfies some extra axioms, but we will not give these details here. See the manuscript of Gordon, Power, and Street [26] for the Definition of monoidal bicategory and homomorphism between monoidal bicategories.

3.3. Enriched Bicategories

A monoidal structure, such as the tensor product on Span(Grpd) discussed in the previous section, is the crucial ingredient for defining *enriched bicategories*. In particular, given a monoidal bicategory $\mathcal{V}$ with the tensor product $\otimes$, a $\mathcal{V}$-enriched bicategory has, for each pair of objects x, y, an object $\hom(x, y)$ of $\mathcal{V}$. Composition involves the tensor product in $\mathcal{V}$

$$\circ \colon \hom(x, y) \otimes \hom(y, z) \to \hom(x, z)$$

While writing the final draft of this article, we realized that our Definition of enriched bicategory is almost identical to one previously given by Carmody [4] and recalled in part by Forcey [5].

After giving the basic structure of enriched bicategories, we state a *change of base* theorem, which says which sort of map $f \colon \mathcal{V} \to \mathcal{V}'$ lets us turn a $\mathcal{V}$-enriched bicategory into a $\mathcal{V}'$-enriched bicategory.

Recall that, for each finite group G, we have defined a category of permutation representations PermRep(G). Further, the theory of enriched bicategories allows us to define the Span(Grpd)-enriched bicategory Hecke(G), which we call the Hecke bicategory. In an ordinary bicategory, the composition operation is given by a functor. So while composition in Span(Grpd) is defined as a functor, in the Hecke bicategory, composition is given not by a functor, but rather a general morphism in the enriching bicategory Span(Grpd). Thus, the composition operation in Hecke(G) is given by a span of groupoids. This more general notion of composition forces us to work in the setting of enriched bicategories. An advantage to working in the setting of enriched bicategories is the existence of a change of base theorem, which we will employ as the main tool in proving our categorification theorem via degroupoidification.

Before giving the Definition of an enriched bicategory, we recall the Definition of an enriched category—that is, a category enriched over a monoidal category $\mathcal{V}$ [7]. An **enriched category** (or $\mathcal{V}$-category) consists of:

- a set of objects $x, y, z \ldots$,
- for each pair of objects x, y, an object $\hom(x, y) \in \mathcal{V}$,
- composition and identity-assigning maps that are morphisms in $\mathcal{V}$.

For example, PermRep(G) is a category enriched over the monoidal category of vector spaces.

We now define enriched bicategories:

Definition 5. *Let $\mathcal{V}$ be a monoidal bicategory. An* **enriched bicategory** *(or $\mathcal{V}$-bicategory) $\mathcal{B}$ consists of the following data subject to the following axioms:*

- *a collection of objects* $x, y, z, \ldots,$
- *for every pair of objects* x, y, *a* **hom-object** $\hom(x, y) \in \mathcal{V}$, *which we will often denote* (x, y),
- *a morphism called* **composition**

$$\circ\colon \hom(x, y) \otimes \hom(y, z) \to \hom(x, z)$$

for each triple of objects $x, y, z \in \mathcal{B}$,
- *an* **identity-assigning** *morphism*

$$i_x\colon I \to \hom(x, x)$$

for each object $x \in \mathcal{B}$,
- *an invertible 2-morphism called the* **associator**

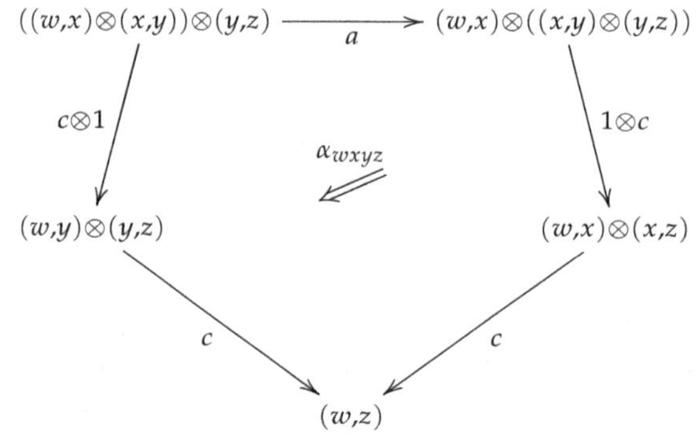

for each quadruple of objects $w, x, y, z \in \mathcal{B}$;
- *and invertible 2-morphisms called the* **right unitor** *and* **left unitor**

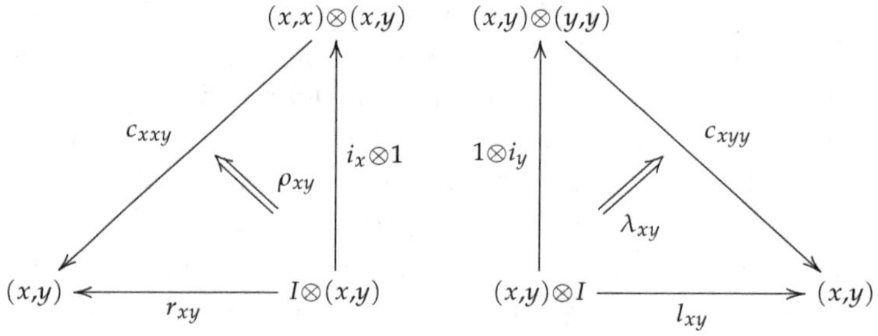

for every pair of objects $x, y \in \mathcal{B}$;
- *and axioms given by closed surface diagrams, the more interesting of the two being the* permutahedron *[4].*

Given a monoidal bicategory $\mathcal{V}$, which has only identity 2-morphisms, every $\mathcal{V}$-bicategory is a $\mathcal{V}$-category in the obvious way, and every $\mathcal{V}$-enriched category can be trivially extended to a $\mathcal{V}$-bicategory. This flexibility will allow us to think of PermRep(G) as either a Vect-enriched category or as a Vect-enriched bicategory.

Now we state a *change of base* construction which allows us to change a $\mathcal{V}$-enriched bicategory to a $\mathcal{V}'$-enriched bicategory.

Claim 6: *Given a lax-monoidal homomorphism of monoidal bicategories* $f\colon \mathcal{V} \to \mathcal{V}'$ *and a* $\mathcal{V}$*-bicategory* $\mathcal{B}_{\mathcal{V}}$, *then there is a* $\mathcal{V}'$*-bicategory*

$$\bar{f}(\mathcal{B}_{\mathcal{V}})$$

A monoidal homomorphism is a map between bicategories preserving the monoidal structure up to isomorphism [26,27]. A *lax*-monoidal homomorphism f is a bit more general: it need not preserve the tensor product up to isomorphism. Instead, it preserves the tensor product only *up to a morphism*:

$$f(x) \otimes' f(y) \to f(x \otimes y)$$

where the symbol $\otimes'$ is the monoidal product in $\mathcal{V}'$.

The data of the enriched bicategory $\bar{f}(\mathcal{B}_{\mathcal{V}})$ is straightforward to write down and the proof of the Claim is an equally straightforward, yet tedious surface diagram chase. Here we just point out the most important idea. The new enriched bicategory $\bar{f}(\mathcal{B}_{\mathcal{V}})$ has the same objects as $\mathcal{B}_{\mathcal{V}}$, and for each pair of objects x, y, the hom-category of $\bar{f}(\mathcal{B}_{\mathcal{V}})$ is

$$\mathrm{hom}_{\bar{f}(\mathcal{B}_{\mathcal{V}})}(x,y) := f(\mathrm{hom}_{\mathcal{B}_{\mathcal{V}}}(x,y))$$

This theorem will allow us to compare the Hecke bicategory, which we define in the next section, to bicategory of spans of finite G-sets Span(GSet).

4. A Categorification Theorem

The following sections are devoted to categorifying the Hecke algebroid. We will show how to obtain the category of permutation representations using the process of *degroupoidification*.

4.1. The Hecke Bicategory

We are now in a position to present the spans of groupoids enriched bicategory Hecke(G)—the *Hecke bicategory*.

Claim 7: *Given a finite group G, there is a* Span(Grpd)*-enriched bicategory* Hecke(G) *which has:*

- *G-sets $X, Y, Z \ldots$ as objects,*
- *for each pair of G-sets X, Y, an object of* Span(Grpd), *the action groupoid:*

$$\mathrm{hom}(X,Y) = (X \times Y)//G$$

- *composition*

$$\circ\colon (X \times Y)//G \times (Y \times Z)//G \to (X \times Z)//G$$

  is the span of groupoids,

$$(X \times Z)//G \xleftarrow{\ \pi_{13}\ } (X \times Y \times Z)//G \xrightarrow{\ \pi_{12} \times \pi_{23}\ } (X \times Y)//G \times (Y \times Z)//G$$

- *for each G-set X, an identity assigning span from the terminal groupoid 1 to $(X \times X)//G$,*
- *invertible 2-morphisms in* Span(Grpd) *assuming the role of the associator and left and right unitors.*

Given this structure one needs to check that the axioms of an enriched bicategory are satisfied; however, we will not prove this here. Combining the degroupoidification functor of Section 3.2, the change of base theorem of Section 3.3, and the enriched bicategory Hecke(G) described above, we can now state the main theorem. This is the content of the next section.

4.2. A Categorification of the Hecke Algebroid

In this section, we describe the relationship between the Hecke algebroid PermRep(G) of permutation representations of a finite group G and the Hecke bicategory Hecke(G). The idea is that for each finite group G, the Hecke bicategory Hecke(G) categorifies PermRep(G).

We recall the functor *degroupoidification*:

$$\mathcal{D}\colon \mathrm{Span(Grpd)} \to \mathrm{Vect}$$

which replaces groupoids with vector spaces and spans of groupoids with linear operators. With this functor in hand, we can apply the change of base theorem to the Span(Grpd)-enriched bicategory Hecke(G). In other words, for each finite group G there is a Vect-enriched bicategory:

$$\bar{\mathcal{D}}\,(\mathrm{Hecke}(G))$$

which has

- permutation representations $X, Y, Z, \ldots$ of G as objects,
- for each pair of permutation representations X, Y, the vector space

$$\hom(X, Y) = \mathcal{D}\,((X \times Y)//G)$$

with G-orbits of $X \times Y$ as basis. Of course, a Vect-enriched bicategory is also a Vect-enriched category. The following is the statement of the main theorem, an equivalence of Vect-enriched categories.

Claim 8: *Given a finite group G,*

$$\bar{\mathcal{D}}\,(\mathrm{Hecke}(G)) \simeq \mathrm{PermRep}(G)$$

as Vect-*enriched categories.*

More explicitly, this says that given two permutation representations $\widetilde{X}$ and $\widetilde{Y}$, the vector space of intertwining operators between them can be constructed as the degroupoidification of the groupoid $(X \times Y)//G$.

An important corollary of Claim 8 is that for certain G-sets—the flag varieties X associated to Dynkin diagrams—the hom-groupoid Hecke$(G)(X, X)$ categorifies the associated Hecke algebra. We will describe these Hecke algebras in Section 6.1 and make the relationship to the Hecke bicategory and some of its applications explicit in Section 6.2.

5. Spans of Groupoids and Cocontinuous Functors

In the following sections, we sketch the beginning of the project of understanding the relationship between the Hecke bicategory and Span(GSet). For this we will need to introduce the monoidal 2-category Cocont of presheaf categories on groupoids and cocontinuous functors.

5.1. *The Monoidal 2-Category* Cocont

In Section 3.2, we considered groupoids as categorified vector spaces. In particular, the isomorphism classes of objects assumed the role of a basis of the corresponding free vector space. A slightly different point of view, which was discussed at length in [1], assigns to a groupoid the vector space of functions on the set of isomorphism classes of that groupoid. Thus, promoting functions to functors, we can think of a categorified vector space as the presheaf category on a groupoid. Given a groupoid $\mathcal{G}$, its category of presheaves $\widehat{\mathcal{G}}$ has:

- presheaves on $\mathcal{G}$ as objects,
- natural transformations as morphisms.

Recall that a (Set-valued) presheaf on a category $\mathcal{C}$ is a contravariant functor $\mathcal{C}^{\mathrm{op}} \to \mathrm{Set}$. The objects of the monoidal bicategory described in this section are categories equivalent to categories of presheaves. We now define such a category, which we call a *nice topos*.

Definition 6. *A* **nice topos** *is a category equivalent to the category of presheaves on a (tame) groupoid.*

By the above Definition, there is a nice topos $\widehat{\mathcal{G}}$ of presheaves corresponding to any groupoid $\mathcal{G}$. However, mapping groupoids to these special presheaf categories suggests that nice toposes should have an intrinsic characterization. To give such a characterization of these toposes we should look to the generalization to toposes in Grothendieck's Galois theory of schemes. The interested reader is pointed to the survey article [28] and references therein, although the present paper may be read independently of this survey. Giving this intrinsic characterization liberates the nice topos from its dependence on a particular groupoid. In particular, this supports the point of view that nice toposes are the objects of a *basis independent* theory of categorified vector spaces.

Following this line of reasoning, the maps between nice toposes are thought of as categorified linear operators. Thus, they should preserve sums, or more accurately, they should preserve a categorified and generalized notion of "sums"—colimits.

Definition 7. *A functor is said to be* **cocontinuous** *if it preserves all (small) colimits.*

This suggests that cocontinuous functors might play the role of categorified linear operators. Indeed, we take such an approach.

In the next section, we will see further support for the analogy: *nice topos is to groupoid* as *abstract vector space is to vector space with chosen basis* and *cocontinuous functor is to span of groupoids* as *linear operator is to matrix*.

The monoidal bicategory Cocont consists of:

- nice toposes $\mathcal{D}, \mathcal{E}, \mathcal{F}, \ldots$ as objects,
- cocontinuous functors as 1-morphisms,
- natural transformations as 2-morphisms.

Objects of Cocont are categories and the morphisms between them are functors. Thus, there is a faithful functor from Cocont to Cat. It follows from the Definition that the product of tame groupoids is again tame [1]. Then the product of a pair of nice toposes is again a nice topos inducing a monoidal structure on Cocont. In particular, the tensor product of nice toposes $\mathcal{E}$ and $\mathcal{F}$ is the Cartesian product $\mathcal{E} \times \mathcal{F}$. Further, the Cartesian product is a cocontinuous functor in each variable and thus the product of cocontinuous functors is again cocontinuous.

In the next section, we will describe the relationship between spans of groupoids and cocontinuous functors between nice toposes. We use some basic notions of topos theory.

5.2. From Spans of Groupoids to Cocontinuous Functors

The change of base construction for enriched bicategories offers a new interpretation of the Hecke bicategory. We have described two closely related monoidal bicategories. The relationship between groupoids and nice toposes is made manifest as a functor between monoidal bicategories. Understanding this functor will be the focus of this section.

It is clear from the Definition that we can obtain a nice topos by assigning a groupoid $\mathcal{G}$ to its corresponding presheaf category $\widehat{\mathcal{G}}$. Continuing our analogy with abstract vector spaces and vector spaces with a chosen basis, we will explain how a span of groupoids gives a cocontinuous functor between the corresponding presheaf categories. In fact, a groupoid can be recovered up to equivalence from its presheaf category, just as a basis can be recovered up to isomorphism from its vector space, but in each case the equivalence or isomorphism is non-canonical.

First, we review some basic ideas from topos theory. A topos is a category which resembles the category of sets. Categories of presheaves are examples of *Grothendieck toposes*. In general, a Grothendieck topos is a category of sheaves on a site. A site is just a category with a notion of a *covering* of objects called a *Grothendieck topology* [29,30]. A familiar example with a particularly simple Grothendieck topology is the category of presheaves on a topological space.

So if a topos is just a special type of category, then how does topos theory differ from category theory? One answer is that while the morphisms between categories are functors, the morphisms between toposes must satisfy extra properties. Such a morphism is called a *geometric morphism* [30]. We define the morphisms between nice toposes, although the Definition is exactly the same in the more general setting of Grothendieck toposes.

Definition 8. *A* **geometric morphism** $e\colon \mathcal{E} \to \mathcal{F}$ *between nice toposes is a pair of functors* $e^*\colon \mathcal{F} \to \mathcal{E}$ *and* $e_*\colon \mathcal{E} \to \mathcal{F}$ *such that* e^* *is left adjoint to* e_* *and* e^* *is left exact,* i.e., *preserves finite limits. A geometric morphism* $e\colon \mathcal{E} \to \mathcal{F}$ *is said to be* **essential** *if there exists a functor* $e_!\colon \mathcal{E} \to \mathcal{F}$ *which is left adjoint to* e^*.

We note a relationship to functors between groupoids, which allows us to define cocontinuous functors from spans. Any functor $f\colon \mathcal{G} \to \mathcal{H}$ defines a geometric morphism between the corresponding presheaf categories:

$$\widehat{f}\colon \widehat{\mathcal{G}} \to \widehat{\mathcal{H}}$$

which consists of the functor:

$$f^*\colon \widehat{\mathcal{H}} \to \widehat{\mathcal{G}}$$

which pulls presheaves back from $\mathcal{H}$ to $\mathcal{G}$, together with the right adjoint of f^*

$$f_*\colon \widehat{\mathcal{G}} \to \widehat{\mathcal{H}}$$

which pushes presheaves forward from $\mathcal{G}$ to $\mathcal{H}$. The particularly important fact is that a geometric morphism induced by a functor between groupoids will always be essential—that is, there exists a *left* adjoint to f^*:

$$f_!\colon \widehat{\mathcal{G}} \to \widehat{\mathcal{H}}$$

Definition 9. *A* **map of geometric morphisms** $\alpha\colon e \Rightarrow f$ *is a natural transformation:*

$$\alpha\colon e^* \Rightarrow f^*$$

Using the fact that a functor between groupoids induces an essential geometric morphism, we see that from a span of groupoids:

$$\begin{array}{ccc} & \mathcal{S} & \\ {}^{q}\swarrow & & \searrow^{p} \\ \mathcal{H} & & \mathcal{G} \end{array}$$

we can define a functor:

$$q_!p^*\colon \widehat{\mathcal{G}} \to \widehat{\mathcal{H}}$$

which is a composite of left adjoint functors, and thus, cocontinuous. Although we do not go into detail at present, using this construction on spans it is not difficult to define a natural transformation between cocontinuous functors from a map of spans. After checking details, the following Claim becomes evident.

Claim 13: *There is a homomorphism of monoidal bicategories*

$$\mathcal{L}\colon \mathrm{Span(Grpd)} \to \mathrm{Cocont}$$

which assigns to each groupoid $\mathcal{G}$ *its category of presheaves* $\widehat{\mathcal{G}}$ *and to each span*

$$\begin{array}{ccc} & \mathcal{S} & \\ {}^{q}\swarrow & & \searrow^{p} \\ \mathcal{H} & & \mathcal{G} \end{array}$$

the cocontinuous functor $q_! p^* \colon \widehat{G} \to \widehat{H}$.

In the setting of spans of sets, the homomorphism $\mathcal{L}$ would be analogous to the functor taking a set X to the free vector space with basis X, and a span

$$Y \xleftarrow{\;q\;} S \xrightarrow{\;p\;} X$$

to a linear operator between the corresponding vector spaces.

Using the map $\mathcal{L}$, we can apply change of base to the Hecke bicategory to obtain a Cocont-enriched bicategory. We will discuss the details and benefits of this new structure in the next section.

5.3. Spans of G-Sets as Nice toposes

In this section we take a closer look at the structure of the Cocont-enriched bicategory $\bar{\mathcal{L}}(\mathrm{Hecke}(G))$. Our goal is to rephrase the construction as a cocompletion of the bicategory $\mathrm{Span}(G\mathrm{Set})$ of spans of G-sets.

Claim 14: *Given a finite group G, there is a* Cocont*-enriched bicategory $\bar{\mathcal{L}}(\mathrm{Hecke}(G))$ which has:*

- *G-sets $X, Y, Z \dots$ as objects,*
- *for each pair of G-sets X, Y, an object of* Cocont:

$$\hom(X, Y) = (X \widehat{\times Y)//}G$$

- *composition*

$$\circ \colon (X \widehat{\times Y)//}G \times (Y \widehat{\times Z)//}G \to (X \widehat{\times Z)//}G$$

is the cocontinuous functor $(\pi_{13})_!(\pi_{12} \times \pi_{23})^*$,
- *for each G-set X, an identity assigning cocontinuous functor from the topos $\widehat{1} \simeq$* Set *to* $(X \widehat{\times X)//}G$,
- *invertible 2-morphisms in* Cocont *assuming the role of the associator and left and right unitors.*

The proof of this Claim is immediate from the proofs of Claim 6 and Claim 7.

It turns out that the hom-categories of the Cocont-enriched bicategory $\bar{\mathcal{L}}(\mathrm{Hecke}(G))$ have a very simple description as spans of G-sets and maps of spans. Given G-sets X and Y, the product $X \times Y$ is again a G-set in an obvious way, so we can construct the action groupoid $(X \times Y)//G$. The category of presheaves on this groupoid will be a nice topos.

Lemma 1. *Given a pair of G-sets X, Y, the category whose objects are spans of G-sets from X to Y and whose morphisms are maps of these spans of G-sets is equivalent to the nice topos $(X \widehat{\times Y)//}G$.*

The construction which proves this lemma is sometimes called the *Grothendieck construction*. The construction says that given a pair of G-sets X and Y, presheaves on $(X \times Y)//G$ are spans from X to Y and natural transformations are maps of spans. We sketch the proof of this lemma now.

Proof. (Sketch) Given a span of G-sets:

$$Y \xleftarrow{\;q\;} S \xrightarrow{\;p\;} X$$

there is a presheaf on $(X \times Y)//G$, which we can think of approximately as a categorified matrix of natural numbers, *i.e.*, a matrix of sets. Each object (x, y) determines an entry in the matrix, and the entries are the sets $S_{x,y} = p^{-1}(x) \cap q^{-1}(y)$ defined in Section 2.1. For each morphism

$(g,(x,y))\colon (x,y) \to (x',y')$, we define a function from $S_{x',y'}$ to $S_{x,y}$ by the action of g^{-1} on the G-set $S_{x',y'}$. Thus, we obtain a presheaf from the span S.

Now from a map of spans of G-sets:

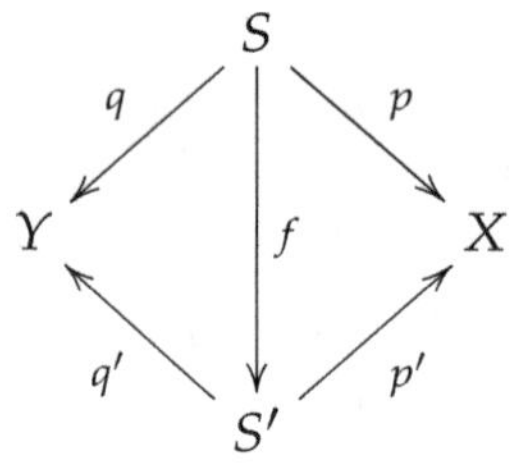

we construct a natural transformation between the presheaves corresponding to S and S'. For each object (x,y) of $(X \times Y)//G$, the component of the natural transformation takes an element $s \in S_{x,y}$ to $f(s) \in S'_{x,y}$. Since f is G-equivariant, the naturality squares commute.

It is not difficult to check that this process defines an equivalence of categories. We only need to build a functor in the other direction such that the respective composites are naturally isomorphic to the identity. We construct such a functor here.

Consider a presheaf P on the groupoid $(X \times Y)//G$. We construct a span with apex the disjoint union of the sets $P(x,y)$. Given a morphism $(g,(x,y))\colon (x,y) \to (g \cdot x, g \cdot y) \in (X \times Y)//G$, there is an induced action on the set $P(x,y)$ coming from $P(g,(x,y))\colon P(x,y) \to P(g \cdot x, g \cdot y)$. The projections onto X and Y are G-equivariant maps yielding the span

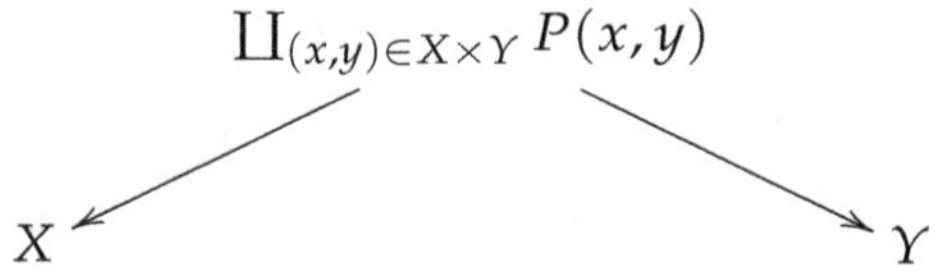

Next, consider a natural transformation from a presheaf P to a presheaf Q. We define a map between the corresponding spans denoted by their apex sets S_P and S_Q. Each element $p \in S_P$ is assigned a pair (x,y) by the projections, and to this pair, the natural transformation assigns a function $f\colon P(x,y) \to Q(x,y)$. Since $p \in P(x,y) \subseteq S_P$ and $f(p) \in Q(x,y) \subseteq S_Q$, we have defined a map of spans. □

We have seen that the hom-categories of the Cocont-enriched bicategory $\bar{\mathcal{L}}(\text{Hecke}(G))$ are actually categories whose objects are spans of G-sets and whose morphisms are maps between such spans. In particular, this bicategory is the local cocompletion of Span$(G\text{Set})$. In describing categorified Hecke algebras we will present a groupoid and a span of groupoids. It will be useful when we consider a specific example to consider the presheaf category in place of the groupoid and to think of each presheaf as a span of G-sets.

6. The Categorified Hecke Algebra and Zamolodchikov Equation

The main Claim of this paper is the existence of a categorification of the Hecke algebroid by the Hecke bicategory Hecke(G). This is, in fact, a statement about categorified Hecke algebras. In attempting to make connections between the categorified Hecke algebra and knot theory, the bicategory Span$(G\text{Set})$ allows a hands-on approach to the categorified generators and the isomorphisms of spans arising from the defining equations of the Hecke algebra. We show that in certain cases these are Yang–Baxter operators that satisfy the Zamolodchikov tetrahedron equation. The hope is that these Yang–Baxter equations will lead to interesting braided monoidal 2-categories [8].

6.1. The Hecke Algebra

There are several well-known equivalent descriptions of the Hecke algebra $\mathcal{H}(\Gamma, q)$ obtained from a Dynkin diagram Γ and a prime power q. One description of the Hecke algebra is as a q-deformation of the group algebra of the Coxeter group of Γ. A standard example of a Coxeter group associated to a Dynkin diagram is the symmetric group on n letters S_n, which is the Coxeter group of the A_{n-1} Dynkin diagram. We will return to this Definition in Section 6.2 and see that it lends itself to combinatorial applications of the Hecke algebra. This combinatorial aspect comes from the close link between the Coxeter group and its associated Coxeter complex, a finite simplicial complex that plays an essential role in the theory of buildings [31].

Hecke algebras have an alternative Definition as algebras of intertwining operators between certain coinduced representations [32]. Given a Dynkin diagram Γ and prime power q, there is an associated simple algebraic group $G = G(\Gamma, q)$. Choosing a Borel subgroup $B \subset G$, *i.e.*, a maximal solvable subgroup, we can construct the corresponding flag complex $X = G/B$, a transitive G-set.

Now, for a finite group G and a representation V of a subgroup $H \subset G$, the *induced representation* of G from H is defined as the V-valued functions on G, which commute with the action of H:

$$\mathrm{Ind}_H^G(V) = \{f \colon G \to V \mid h \cdot f(x) = f(hx), \text{ for } h \in H\}$$

The action of $g \in G$ is defined on a function $f \colon G \to V$ as $(g \cdot f)(x) = f(xg)$. A standard fact about finite groups says that the representation induced from the trivial representation of any subgroup is isomorphic to the permutation representation on the cosets of that subgroup. Thus, from the trivial representation of a Borel subgroup B, we obtain the permutation representation $\widetilde{X}$ on the cosets of B, *i.e.*, the flag complex X. Then the Hecke algebra is defined as the algebra of intertwining operators from $\widetilde{X}$ to itself:

$$\mathcal{H}(\Gamma, q) := \mathrm{PermRep}(G)(\widetilde{X}, \widetilde{X})$$

where $G = G(\Gamma, q)$.

Given this Definition of the Hecke algebra, we have an immediate corollary to Claim 8:

Claim 16: *Given a Dynkin diagram Γ and prime power q, denote $G = G(\Gamma, q)$. Then the* hom-*category* $\mathrm{Hecke}(G)(X, X)$ *of the Hecke bicategory categorifies the Hecke algebra $\mathcal{H}(\Gamma, q)$.*

6.2. Categorified Generators and the Zamolodchikov Equation

Now that we have seen a categorification of Hecke algebras abstractly as a corollary, we look at a concrete example. The categorified Hecke algebra is particularly easy to understand as living inside the bicategory $\mathrm{Span}(G\mathrm{Set})$. While we found it useful earlier to view Hecke algebras as algebras of intertwining operators, viewing the Hecke algebra by its presentation as a q-deformation of a Coxeter group [33] is helpful in examples.

Any Dynkin diagram gives rise to a simple Lie group, and the Weyl group of this simple Lie group is a Coxeter group. Let Γ be a Dynkin diagram. We write $d \in \Gamma$ to mean that d is a dot in this diagram. Associated to each unordered pair of dots $d, d' \in \Gamma$ is a number $m_{dd'} \in \{2, 3, 4, 6\}$. In the usual Dynkin diagram conventions:

- $m_{dd'} = 2$ is drawn as no edge at all,
- $m_{dd'} = 3$ is drawn as a single edge,
- $m_{dd'} = 4$ is drawn as a double edge,
- $m_{dd'} = 6$ is drawn as a triple edge.

For any prime power q, our Dynkin diagram Γ yields a Hecke algebra. The *Hecke algebra* $\mathcal{H}(\Gamma, q)$ corresponding to this data is the associative algebra with one generator σ_d for each $d \in \Gamma$, and relations:

$$\sigma_d^2 = (q-1)\sigma_d + q$$

for all $d \in \Gamma$, and

$$\sigma_d \sigma_{d'} \sigma_d \cdots = \sigma_{d'} \sigma_d \sigma_{d'} \cdots$$

for all $d, d' \in \Gamma$, where each side has $m_{dd'}$ factors.

When $q = 1$, this Hecke algebra is simply the group algebra of the Coxeter group associated to Γ—that is, the group with one generator s_d for each dot $d \in \Gamma$, and relations

$$s_d^2 = 1, \qquad (s_d s_{d'})^{m_{dd'}} = 1$$

So, the Hecke algebra can be thought of as a q-deformation of this Coxeter group.

We recall the flag complex $X = G/B$ from Section 6.1 is a finite set equipped with a transitive action of the finite group G. Starting from just this G-set X, we can see an explicit picture of the categorified Hecke algebra of spans of G-sets from X to X.

The key is that for each dot $d \in \Gamma$ there is a special span of G-sets that corresponds to the generator $\sigma_d \in \mathcal{H}(\Gamma, q)$. To illustrate these ideas, let us consider the simplest nontrivial example, the Dynkin diagram A_2:

The Hecke algebra associated to A_2 has two generators, which we call P and L, for reasons soon to be revealed:

$$P = \sigma_1, \qquad L = \sigma_2$$

To make the connection to the description of the Hecke algebra as an algebra of intertwining operators explicit, we can choose a basis of $\widetilde{X}$ corresponding to the elements of the Coxeter group S_3 and write these generators as the following matrices, or intertwining operators.

$$P = \begin{pmatrix} 0 & q & 0 & 0 & 0 & 0 \\ 1 & q-1 & 0 & 0 & 0 & 0 \\ 0 & 0 & 0 & q & 0 & 0 \\ 0 & 0 & 1 & q-1 & 0 & 0 \\ 0 & 0 & 0 & 0 & q-1 & 1 \\ 0 & 0 & 0 & 0 & q & 0 \end{pmatrix}$$

$$L = \begin{pmatrix} 0 & 0 & 0 & 0 & 0 & q \\ 0 & 0 & q & 0 & 0 & 0 \\ 0 & 1 & q-1 & 0 & 0 & 0 \\ 0 & 0 & 0 & q-1 & 1 & 0 \\ 0 & 0 & 0 & q & 0 & 0 \\ 1 & 0 & 0 & 0 & 0 & q-1 \end{pmatrix}$$

The relations are

$$P^2 = (q-1)P + q, \qquad L^2 = (q-1)L + q, \qquad PLP = LPL$$

It follows that this Hecke algebra is a quotient of the group algebra of the 3-strand braid group, which has two generators P and L, which we can draw as braids in 3-dimensional space:

$P =$ 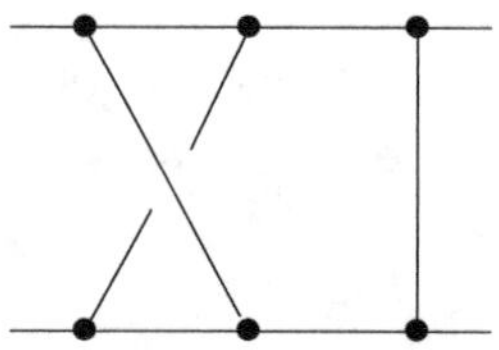 $L =$ 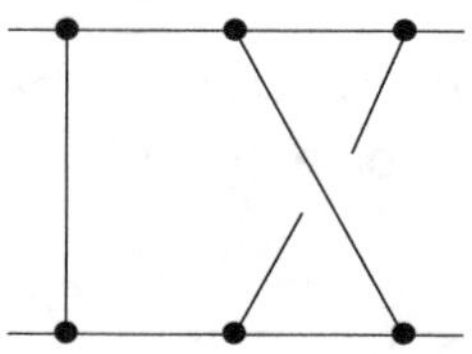

and one relation $PLP = LPL$:

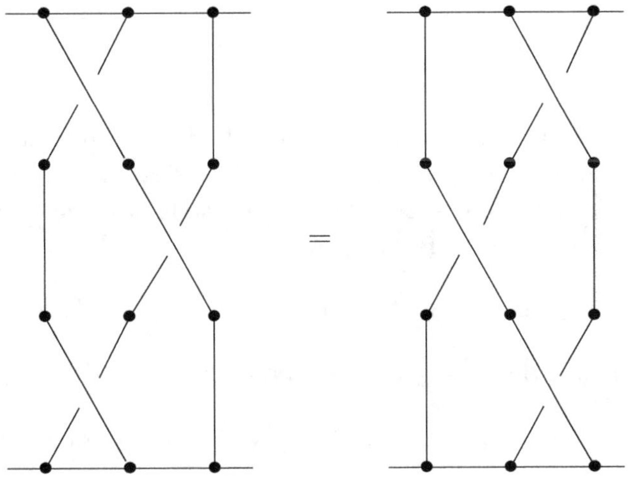

called the *Yang–Baxter equation* or *third Reidemeister move*. This is why Jones could use traces on the A_n Hecke algebras to construct invariants of knots [34]. In light of the success of Khovanov homology—a categorification of the Jones polynomial—this connection to knot theory makes it especially interesting to categorify Hecke algebras.

So, let us see what the categorified Hecke algebra looks like, and where the Yang–Baxter equation comes from. The algebraic group corresponding to the A_2 Dynkin diagram and the prime power q is $G = \mathrm{SL}(3, \mathbb{F}_q)$, and we can choose the Borel subgroup B to consist of upper triangular matrices in $\mathrm{SL}(3, \mathbb{F}_q)$. Recall that a complete flag in the vector space $\mathbb{F}_q^3$ is a pair of subspaces

$$0 \subset V_1 \subset V_2 \subset \mathbb{F}_q^3$$

The subspace V_1 must have dimension one, while V_2 must have dimension two. Since G acts transitively on the set of complete flags and B is the subgroup stabilizing a chosen flag, the flag variety $X = G/B$ in this example is just the set of complete flags in $\mathbb{F}_q^3$—hence its name.

We can think of $V_1 \subset \mathbb{F}_q^3$ as a point in the projective plane $\mathbb{F}_q\mathrm{P}^2$, and $V_2 \subset \mathbb{F}_q^3$ as a line in this projective plane. From this viewpoint, a complete flag is a chosen point lying on a chosen line in $\mathbb{F}_q\mathrm{P}^2$. This viewpoint is natural in the theory of "buildings", where each Dynkin diagram corresponds to a type of geometry [31]. Each dot in the Dynkin diagram then stands for a "type of geometric figure", while each edge stands for an "incidence relation". The A_2 Dynkin diagram corresponds to projective plane geometry. The dots in this diagram stand for the figures "point" and "line":

$$\text{point} \bullet \!\!-\!\!\!-\!\!\!-\!\!\bullet\ \text{line}$$

The edge in this diagram stands for the incidence relation "the point p lies on the line ℓ".

We can think of P and L as special elements of the A_2 Hecke algebra, as already described. But when we categorify the Hecke algebra, P and L correspond to irreducible spans of G-sets—that is, spans that are not coproducts of two non-trivial spans of G-sets. Let us describe these spans and explain how the Hecke algebra relations arise in this categorified setting.

The objects P and L can be defined by giving irreducible spans of G-sets:

In general, any span of G-sets

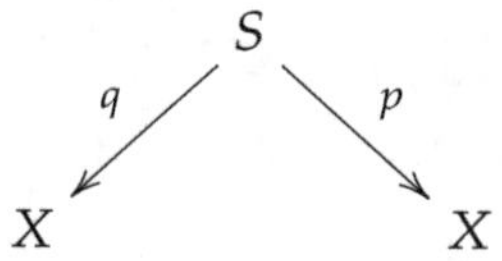

such that $q \times p \colon S \to X \times X$ is injective can be thought of as G-invariant binary relation between elements of X. Irreducible G-invariant spans are always injective in this sense. So, such spans can also be thought of as G-invariant relations between flags. In these terms, we define P to be the relation that says two flags have the same line, but different points:

$$P = \{((p,\ell),(p',\ell)) \in X \times X \mid p \neq p'\}$$

Similarly, we think of L as a relation saying two flags have different lines, but the same point:

$$L = \{((p,\ell),(p,\ell')) \in X \times X \mid \ell \neq \ell'\}$$

Given this, we will check:

$$P^2 \cong (q-1) \times P + q \times 1, \qquad L^2 \cong (q-1) \times L + q \times 1, \qquad PLP \cong LPL$$

Here both sides refer to spans of G-sets. Addition of spans is defined using the coproduct and 1 denotes the identity span from X to X. We use "q" to stand for a fixed q-element set, and similarly for "$q-1$". We compose spans of G-sets using the ordinary pullback.

To check the existence of the first two isomorphisms above, we just need to count. In $\mathbb{F}_q\mathrm{P}^2$, there are $q+1$ points on any line. So, given a flag we can change the point in q different ways. To change it again, we have a choice: we can either send it back to the original point, or change it to one of the $q-1$ other points. So, $P^2 \cong (q-1) \times P + q \times 1$. Since there are also $q+1$ lines through any point, similar reasoning shows that $L^2 \cong (q-1) \times L + q \times 1$.

The Yang–Baxter isomorphism

$$PLP \cong LPL$$

is more interesting. We construct it as follows. First consider the left-hand side, PLP. Start with a complete flag (p_1, ℓ_1):

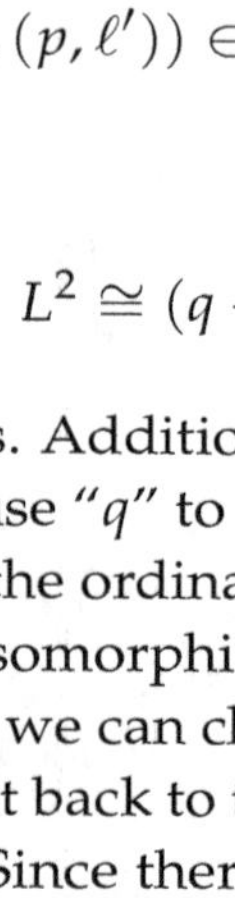

Then, change the point to obtain a flag (p_2, ℓ_1). Next, change the line to obtain a flag (p_2, ℓ_2). Finally, change the point once more, which gives us the flag (p_3, ℓ_2):

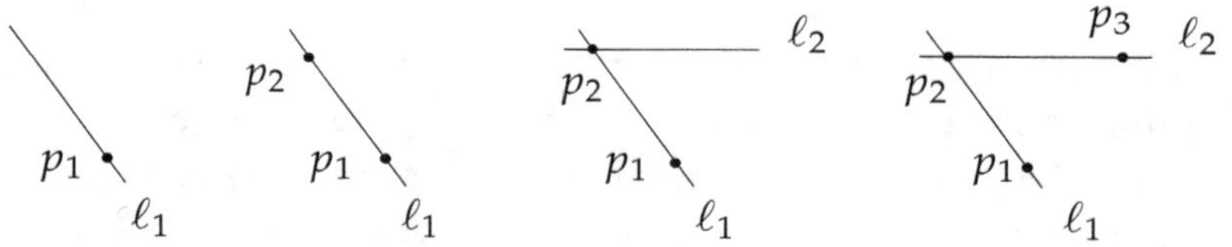

The figure on the far right is a typical element of PLP:

$$((p_1,\ell_1),(p_2,\ell_1),(p_2,\ell_2),(p_3,\ell_2)) \text{ such that } p_1 \neq p_2, p_2 \neq p_3, \ell_1 \neq \ell_2$$

On the other hand, consider LPL. Start with the same flag as before, but now change the line, obtaining (p_1, ℓ_2'). Next change the point, obtaining the flag (p_2', ℓ_2'). Finally, change the line once more, obtaining the flag (p_2', ℓ_3'):

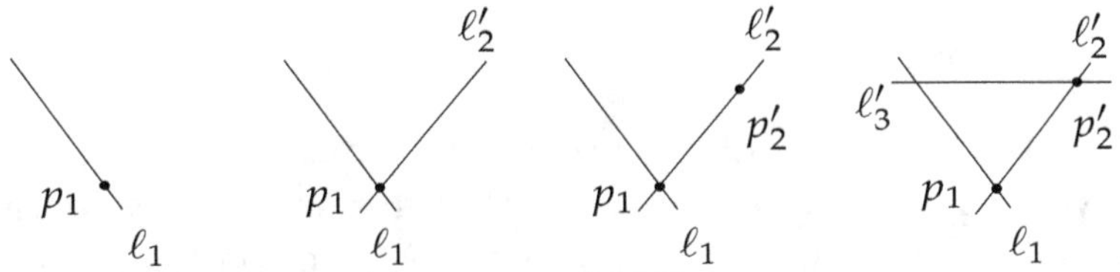

The figure on the far right is a typical element of LPL.

Now, the axioms of projective plane geometry say that any two distinct points lie on a unique line, and any two distinct lines intersect in a unique point. So, any figure of the sort shown on the left below determines a unique figure of the sort shown on the right, and vice versa:

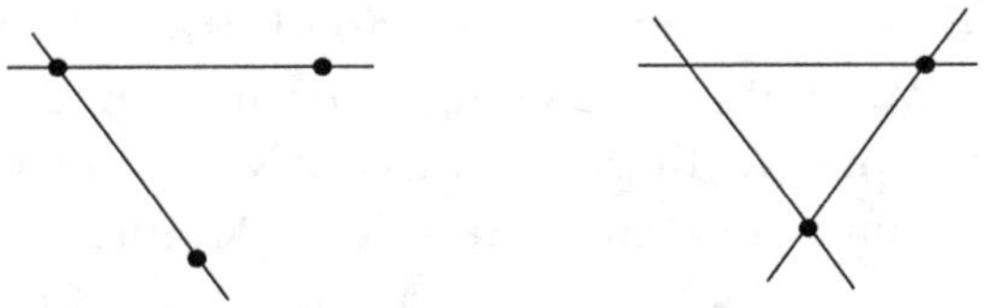

Comparing this with the pictures above, we see this bijection induces an isomorphism of spans $PLP \cong LPL$. So, we have derived the Yang–Baxter isomorphism from the axioms of projective plane geometry!

The above discussion helps illuminate the occurrence of the Yang–Baxter *equation* in the generators and relations description of the Hecke algebra. We have seen that the categorified setting allows us to view these equations as *isomorphisms* of spans of G-sets. As such, these *Yang–Baxter operators* satisfy an equation of their own—the *Zamolodchikov tetrahedron equation* [8]. However, this equation appears in the categorified A_n Hecke algebra, only for $n \geq 3$. We can assign braids on four strands to the generators of the A_3 Hecke algebra:

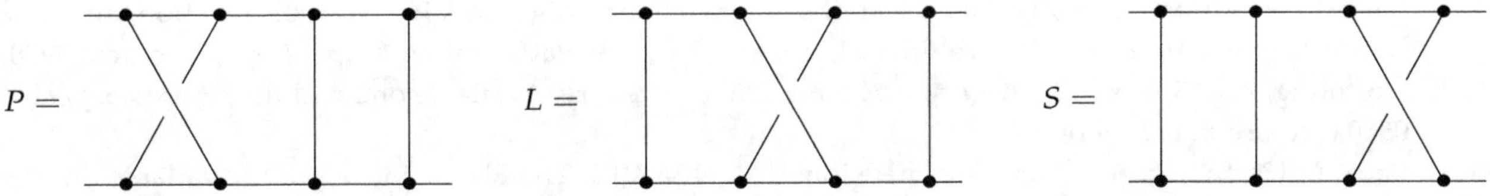

where composition of spans, or multiplication in the Hecke algebra, corresponds to stacking of braid diagrams. Then we can express the Zamolodchikov equation—as an equation in the categorified Hecke algebra—in the form of a commutative diagram of braids [35,36]:

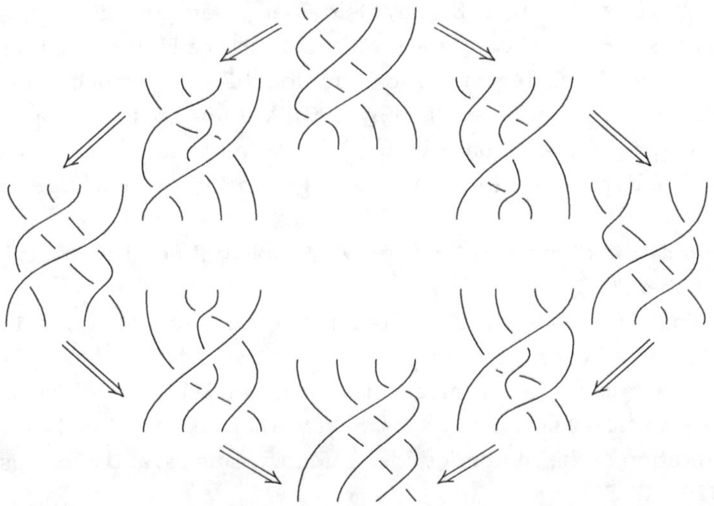

This is just the beginning of a wonderful story involving Dynkin diagrams of more general types, incidence geometries, logic, braided monoidal 2-categories [37,38], knot invariants, topological quantum field theories, geometric representation theory, and more!

Acknowledgments

A tremendous debt is owed to John Baez, James Dolan and Todd Trimble for initiating an interesting and creative long-term endeavour, as well as for generously sharing their time and ideas. Also to the contributors to the n-Category Café, especially Bruce Bartlett, Denis-Charles Cisinski, Tom Leinster, Mike Shulman, and Simon Willerton for taking an interest and sharing and clarifying so many ideas. Thanks to David Ben-Zvi, Anthony Licata, Urs Schreiber for introducing the author to the larger world of geometric function theories. Thanks to Nick Gurski for patiently explaining the basics of tricategories; Mikhail Khovanov and Aaron Lauda for generously hosting visits to Columbia University while many of these ideas were developed; Catharina Stroppel for helpful comments regarding the historical development of Hecke algebras and categorification along with enlightening conversations and generosity while hosting a visit to the Max Planck Institute in Bonn; Jim Stasheff for comments and encouragement on a very early draft; Julia Bergner and Christopher Walker for patiently listening and contributing to many useful conversations; Rick Blute, Alistair Savage, Pieter Hofstra, Phil Scott and others for patiently listening and commenting week after week over which a number of ideas were worked out in the Ottawa Logic and Foundations of Computing Group seminar; and to Dorette Pronk for her invitation to Dalhousie University, where some central ideas regarding the span construction in higher categories were clarified. Finally, the author thanks the University of California, Riverside where a first version of this paper was completed and where the author was supported as a graduate student in large part by the National Science Foundation under Grant No. 0653646 and the Centre de Recherche en Mathématiques for their support of the author as a postdoctoral fellow at the University of Ottawa while a final draft was completed.

References

1. Baez, J.; Hoffnung, A.E.; Walker, C. Higher Dimensional Algebra VII: Groupoidification. In *Theory and Applications of Categories*; 2010; Volume 24, pp. 489–553.; Available online: http://arxiv.org/abs/0908.4305.
2. Hoffnung, A.E. Spans in 2-categories: A monoidal tricategory. Available online: http://arxiv.org/abs/1112.0560 (accessed on 22 September 2012).
3. Baez, J.; Hoffnung, A.E. Higher dimensional Algebra VIII: The Hecke bicategory. Available online: http://math.ucr.edu/home/baez/hecke.pdf (accessed on 19 September 2012).
4. Carmody, S.M. Cobordism Categories. Ph.D. Thesis, University of Cambridge, Cambridge, MA, USA, 1995.
5. Forcey, S. Quotients of the multiplihedron as categorified associahedra. Available online: http://arxiv.org/PS_cache/arxiv/pdf/0803/0803.2694v4.pdf (accessed on 19 September 2012).
6. Baez, J.; Dolan, J. From Finite Sets to Feynman Diagrams. In *Mathematics Unlimited—2001 and Beyond*; Engquist, B., Schmid, W., Eds.; Springer: Berlin/Heidelberg, Germany, 2001; pp. 29–50.
7. Kelly, G.M. *Basic Concepts of Enriched Category Theory*; Cambridge University Press: Cambridge, UK, 1982.
8. Kapranov, M.; Voevodsky, V. 2-Categories and Zamolodchikov Tetrahedra Equations. In *Proceedings of Symposia in Pure Mathematics*, Providence, RI, USA, 1994; Volume 56, Part 2, pp. 177–260.
9. Baez, J.; Langford, L. Higher-Dimensional Algebra IV: 2-tangles. *Adv. Math.* **2003**, *180* , 705–764.
10. Kazhdan, D.; Lusztig, G. Representations of Coxeter groups and Hecke algebras, *Invent. Math.* **1979**, *53*, 165–184.
11. Rouquier, R. Categorification of the braid groups. Available online: http://arxiv.org/abs/math/0409593 (accessed on 19 September 2012).
12. Soergel, W. The combinatorics of Harish-Chandra bimodules, *J. Reine Angew. Math.* **1992**, *429*, 49–74.
13. Khovanov, M. A categorification of the Jones polynomial. *Duke Math. J.* **2000**, *101*, 359–426.
14. Bernstein, J.; Frenkel, I.; Khovanov, M. A categorification of the Temperley-Lieb algebra and Schur quotients of $U(sl_2)$ via projective and Zuckerman functors. *Selecta Math.* **1999**, *5*, 199–241.
15. Stroppel, C. Categorification of the Temperley-Lieb category, tangles, and cobordisms via projective functors. *Duke Math. J.* **2005**, *126*, 547–596.
16. Elias, B.; Khovanov, M. Diagrammatics for Soergel Categories, 2009. Available online: http://arxiv.org/abs/0902.4700 (accessed on 19 September 2012).

17. Webster, B.; Williamson, G. A geometric model for Hochschild homology of Soergel bimodules. *Geom. Topol.* **2008**, *12*, 1243–1263.
18. Bénabou, J. Introduction to Bicategories. In *Reports of the Midwest Category Seminar*; Springer-Verlag: Berlin/Heidelberg, Germany, 1967; pp. 1–77.
19. Leinster, T. Basic bicategories. Available online: http://arxiv.org/abs/math/9810017 (accessed on 19 September'2012).
20. Fiore, M.; Leinster, T. Objects of categories as complex numbers. *Adv. Math.* **2005**, *190*, 264–277.
21. Kim, M.A. Lefschetz trace formula for equivariant cohomology. *Ann. Sci. École Norm. Sup.* **1995**, *28*, 669–688.
22. Leinster, T. The Euler characteristic of a category. *Doc. Math.* **2008**, *13*, 21–49.
23. Weinstein, A. The Volume of a Differentiable Stack. In *Proceedings of the Poisson*, Lausanne, Switzerland, July 2008; Available online: http://arxiv.org/abs/0809.2130 (accessed on 19 September 2012).
24. Street, R. Fibrations and Yoneda's lemma in a 2-category. In *Category Seminar*; Springer: Berlin/Heidelberg, Germany, 1974; Volume 420, pp. 104–133.
25. Weber, M. Strict 2-toposes. Available online: http://arxiv.org/abs/math/0606393 (accessed on 19 September 2012).
26. Gordon, R.; Power, A.J.; Street, R. Coherence for tricategories. *Mem. Am. Math. Soc.* **1995**, *117*, pp.1–81.
27. Gurski, N. An algebraic theory of tricategories. Ph.D. Thesis, University of Chicago, Chicago, IL, USA, June 2006.
28. Moerdijk, I. Toposes and Groupoids. In *Categorical Algebra and Its Applications*; Springer: Berlin/Heidelberg, Germany, 1988; Volume 1348, pp. 280–298.
29. Johnstone, P.T. *Sketches of an Elephant: A Topos Theory Compendium*; Oxford University Press: Oxford, UK, 2002.
30. Mac Lane, S.; Moerdijk, I. *Sheaves in Geometry and Logic: A First Introduction to Topos Theory*; Springer: Berlin/Heidelberg, Germany, 1992.
31. Brown, K. *Buildings*; Springer: Berlin/Heidelberg, Germany, 1989.
32. Bump, D. *Lie Groups*; Springer: New York, NY, USA, 2004.
33. Humphreys, J. *Reflection Groups and Coxeter Groups*; Cambridge University Press: Cambridge, UK, 1992.
34. Jones, V. Hecke algebra representations of braid groups and link polynomials. *Ann. Math.* **1987**, *126*, 335–388.
35. Carter, J.S.; Saito, M. *Knotted Surfaces and Their Diagrams*; American Mathematical Society: Providence, RI, USA, 1998.
36. Baez, J.; Crans, A. Higher-dimensional algebra VI: Lie 2-algebras. *Theory Appl. Categ.* **2004**, *12*, 492–528.
37. Baez, J.; Neuchl, M. Higher-dimensional algebra I: Braided monoidal categories. *Adv. Math.* **1996**, *121*, 196–244.
38. McCrudden, P. Balanced coalgebroids. *Theory Appl. Categ.* **2000**, *7*, 71–147.

5

Valued Graphs and the Representation Theory of Lie Algebras

Joel Lemay

Department of Mathematics and Statistics, University of Ottawa, Ottawa, K1N 6N5, Canada;
E-Mail: jlema072@uottawa.ca

Abstract: Quivers (directed graphs), species (a generalization of quivers) and their representations play a key role in many areas of mathematics including combinatorics, geometry, and algebra. Their importance is especially apparent in their applications to the representation theory of associative algebras, Lie algebras, and quantum groups. In this paper, we discuss the most important results in the representation theory of species, such as Dlab and Ringel's extension of Gabriel's theorem, which classifies all species of finite and tame representation type. We also explain the link between species and K-species (where K is a field). Namely, we show that the category of K-species can be viewed as a subcategory of the category of species. Furthermore, we prove two results about the structure of the tensor ring of a species containing no oriented cycles. Specifically, we prove that two such species have isomorphic tensor rings if and only if they are isomorphic as "crushed" species, and we show that if K is a perfect field, then the tensor algebra of a K-species tensored with the algebraic closure of K is isomorphic to, or Morita equivalent to, the path algebra of a quiver.

Keywords: quiver; species; lie algebra; representation theory; root system; valued graph; modulated quiver; tensor algebra; path algebra; Ringel–Hall algebra

---

1. Introduction

I would like to thank Alistair Savage for introducing me to this topic and for his invaluable guidance and encouragement. Furthermore, I would like to thank Erhard Neher and Vlastimil Dlab for their helpful comments and advice.

Species and their representations were first introduced in 1973 by Gabriel in [1]. Let K be a field. Let A be a finite-dimensional, associative, unital, basic K-algebra and let $\operatorname{rad} A$ denote its Jacobson radical. Then $A/\operatorname{rad} A \cong \prod_{i\in\mathcal{I}} K_i$, where $\mathcal{I}$ is a finite set and K_i is a finite-dimensional K-division algebra for each $i \in \mathcal{I}$. Moreover, $\operatorname{rad} A/(\operatorname{rad} A)^2 \cong \bigoplus_{i,j\in\mathcal{I}} {}_jM_i$, where ${}_jM_i$ is a finite-dimensional (K_j, K_i)-bimodule for each $i, j \in \mathcal{I}$. We then associate to A a valued graph Δ_A with vertex set $\mathcal{I}$ and valued arrows $i \xrightarrow{(d_{ij},d_{ji})} j$ for each ${}_jM_i \neq 0$, where $d_{ij} = \dim_{K_j}({}_jM_i)$. The valued graph Δ_A, the division algebras K_i ($i \in \mathcal{I}$) and the bimodules ${}_jM_i$ ($i, j \in \mathcal{I}$) constitute a *species* and contain a great deal of information about the representation theory of A (in some cases, *all* the information). When working over an algebraically closed field, a species is simply a quiver (directed graph) in the sense that all $K_i \cong K$ and all ${}_jM_i \cong K^n$ so only Δ_A is significant. In this case, Gabriel was able to classify all quivers of finite representation type (that is, quivers with only finitely many non-isomorphic indecomposable representations); they are precisely those whose underlying graph is a (disjoint union of) Dynkin diagram(s) of type A, D or E. Moreover, he discovered that the isomorphism classes of indecomposable representations of these quivers are in bijection with the positive roots of the Kac–Moody Lie algebra associated to the corresponding diagram. Gabriel's theorem is the starting point of a series of remarkable results such as the construction of Kac–Moody Lie algebras and quantum groups via Ringel–Hall algebras, the geometry of quiver varieties, and Lusztig's categorification of quantum groups via perverse sheaves. Lusztig, for example, was able to give a geometric interpretation of the positive part of quantized enveloping algebras using quiver varieties (see [2]).

While quivers are useful tools in representation theory, they have their limitations. In particular, their application to the representation theory of associative unital algebras, in general, only holds when working over an algebraically closed field. Moreover, the Lie theory that is studied by quiver theoretic methods is naturally that of symmetric Kac–Moody Lie algebras. However, many of the fundamental examples of Lie algebras of interest to mathematicians and physicists are symmetrizable Kac–Moody Lie algebras which are not symmetric. Species allow us to relax these limitations.

In his paper [1], Gabriel outlined how one could classify all species of finite representation type over non-algebraically closed fields. However, it was Dlab and Ringel in 1976 (see [3]) who were ultimately able to generalize Gabriel's theorem and show that a species is of finite representation type if and only if its underlying valued graph is a Dynkin diagram of finite type. They also showed that, just as for quivers, there is a bijection between the isomorphism classes of the indecomposable representations and the positive roots of the corresponding Kac–Moody Lie algebra.

Species and quivers also lead naturally to the construction of the Ringel–Hall algebra. One can show that Ringel–Hall algebras are self-dual Hopf algebras (see, for example, [4,5]). Hopf algebras provide solutions to the Yang–Baxter equation (see [6]), which has a number of applications in statistical mechanics, differential equations, knot theory, and other disciplines. Moreover, the generic composition algebra of a Ringel–Hall algebra is isomorphic to the positive part of the quantized enveloping algebra of the corresponding Kac–Moody Lie algebra. Thus, species and quivers provide constructions of important examples of quantum groups.

Despite having been introduced at the same time, the representation theory of quivers is much more well-known and well-developed than that of species. In fact, the very definition of species varies from text to text; some use the more "general" definition of a species (e.g., [3]) while others use the alternate definition of a K-species (e.g., [7]). Yet the relationship between these two definitions is rarely discussed. Moreover, while there are many well-known results in the representation theory of quivers, such as Gabriel's theorem or Kac's theorem, it is rarely mentioned whether or not these results generalize for species. Indeed, there does not appear to be any single comprehensive reference for species in the literature. The main goal of this paper is to compare the current literature and collect all the major, often hard to find, results in the representation theory of species into one text.

This paper is divided into seven sections. In the first, we give all the preliminary material on quivers and valued quivers that will be needed for the subsequent sections. In particular, we address the fact that two definitions of valued quivers exist in the literature.

In Section 2, we define both species and K-species and discuss how the definitions are related. Namely, we define the categories of species and K-species and show that the category of K-species can be thought of as a subcategory of the category of species. That is, via an appropriate functor, all K-species are species. There are, however, species that are not K-species for any field K.

The third section deals with the tensor ring (resp. algebra) $T(\mathcal{Q})$ associated to a species (resp. K-species) $\mathcal{Q}$. This is a generalization of the path algebra of a quiver. If K is a perfect field, then for any finite-dimensional associative unital K-algebra A, the category of A-modules is equivalent to the category of $T(\mathcal{Q})/I$-modules for some K-species $\mathcal{Q}$ and some ideal I. Also, it will be shown in Section 6 that the category of representations of $\mathcal{Q}$ is equivalent to the category of $T(\mathcal{Q})$-modules. These results show why species are such important tools in representation theory; modulo an ideal, they allow us to understand the representation theory of finite-dimensional associative unital algebras.

In Section 4, we follow the work of [7] to show that, when working over a finite field, one can simply deal with quivers (with automorphism) rather than species. That is, we show that if $\mathcal{Q}$ is an $\mathbb{F}_q$-species, then the tensor algebra of $\mathcal{Q}$ is isomorphic to the fixed point algebra of the path algebra of a quiver under the Frobenius morphism.

In the fifth section, we further discuss the link between a species and its tensor ring. In particular, we prove two results that do not seem to appear in the literature.

Theorem 4 *Let $\mathcal{Q}$ and $\mathcal{Q}'$ be two species with no oriented cycles. Then $T(\mathcal{Q}) \cong T(\mathcal{Q}')$ if and only if $\mathcal{Q}^C \cong \mathcal{Q}'^C$ (where $\mathcal{Q}^C$ and $\mathcal{Q}'^C$ denote the* crushed *species of $\mathcal{Q}$ and $\mathcal{Q}'$).*

Theorem 5 and Corollary 2 *Let K be a perfect field and $\mathcal{Q}$ a K-species containing no oriented cycles. Then $\overline{K} \otimes_K T(\mathcal{Q})$ is isomorphic to, or Morita equivalent to, the path algebra of a quiver (where $\overline{K}$ denotes the algebraic closure of K).*

Section 6 deals with representations of species. We discuss many of the most important results in the representation theory of quivers, such as the theorems of Gabriel and Kac, and their generalizations for species.

The seventh and final section deals with the Ringel–Hall algebra of a species. It is well-known that the generic composition algebra of a quiver is isomorphic to the positive part of the quantized enveloping algebra of the associated Kac–Moody Lie algebra. Also, Sevenhant and Van Den Bergh have shown that the Ringel–Hall algebra itself is isomorphic to the positive part of the quantized enveloping algebra of a generalized Kac–Moody Lie algebra (see [8]). We show that these results hold for species as well. While this is not a new result, it does not appear to be explained in detail in the literature.

We assume throughout that all algebras (other than Lie algebras) are associative and unital.

2. Valued Quivers

In this section, we present the preliminary material on quivers and valued quivers that will be used throughout this paper. In particular, we begin with the definition of a quiver and then discuss valued quivers. There are two definitions of valued quivers that can be found in the literature; we present both and give a precise relationship between the two in terms of a functor between categories (see Lemma 1). We also discuss the idea of "folding", which allows one to obtain a valued quiver from a quiver with automorphism.

Definition 1 (Quiver). *A* quiver *Q is a directed graph. That is, $Q = (Q_0, Q_1, t, h)$, where Q_0 and Q_1 are sets and t and h are set maps $Q_1 \to Q_0$. The elements of Q_0 are called* vertices *and the elements of Q_1 are called* arrows. *For every $\rho \in Q_1$, we call $t(\rho)$ the* tail *of ρ and $h(\rho)$ the* head *of ρ. By an abuse of notation, we often simply write $Q = (Q_0, Q_1)$ leaving the maps t and h implied. The sets Q_0 and Q_1 may well be infinite; however we will deal exclusively with quivers having only finitely many vertices and arrows. We will also restrict ourselves to quivers whose underlying undirected graphs are connected.*

A quiver morphism *$\varphi : Q \to Q'$ consists of two set maps, $\varphi_0 : Q_0 \to Q_0'$ and $\varphi_1 : Q_1 \to Q_1'$, such that $\varphi_0(t(\rho)) = t(\varphi_1(\rho))$ and $\varphi_0(h(\rho)) = h(\varphi_1(\rho))$ for each $\rho \in Q_1$.*

For $\rho \in Q_1$, we will often use the notation $\rho : i \to j$ to mean $t(\rho) = i$ and $h(\rho) = j$.

Definition 2 (Absolute valued quiver). *An* absolute valued quiver *is a quiver $\Gamma = (\Gamma_0, \Gamma_1)$ along with a positive integer d_i for each $i \in \Gamma_0$ and a positive integer m_ρ for each $\rho \in \Gamma_1$ such that m_ρ is a common multiple of $d_{t(\rho)}$ and $d_{h(\rho)}$ for each $\rho \in \Gamma_1$. We call $(d_i, m_\rho)_{i \in \Gamma_0, \rho \in \Gamma_1}$ an* (absolute) valuation *of Γ. By a slight abuse of notation, we often refer to Γ as an absolute valued quiver, leaving the valuation implied.*

An absolute valued quiver morphism *is a quiver morphism $\varphi : \Gamma \to \Gamma'$ respecting the valuations. That is, $d'_{\varphi_0(i)} = d_i$ for each $i \in \Gamma_0$ and $m'_{\varphi_1(\rho)} = m_\rho$ for each $\rho \in \Gamma_1$.*

Let $\mathfrak{Q}_{\text{abs}}$ denote the category of absolute valued quivers.

A (non-valued) quiver can be viewed as an absolute valued quiver with trivial values (*i.e.*, all $d_i = m_\rho = 1$). Thus, valued quivers are a generalization of quivers.

Given a quiver Q and an automorphism σ of Q, we can construct an absolute valued quiver Γ with valuation $(d_i, m_\rho)_{i\in\Gamma_0,\rho\in\Gamma_1}$ by "folding" Q as follows:

- $\Gamma_0 = \{\text{vertex orbits of } \sigma\}$,
- $\Gamma_1 = \{\text{arrow orbits of } \sigma\}$,
- for each $i \in \Gamma_0$, d_i is the number of vertices in the orbit i,
- for each $\rho \in \Gamma_1$, m_ρ is the number of arrows in the orbit ρ.

Given $\rho \in \Gamma_1$, let $m = m_\rho$ and $d = d_{t(\rho)}$. The orbit ρ consists of m arrows in Q_0, say $\{\rho_i = \sigma^{i-1}(\rho_1)\}_{i=1}^{m}$. Because σ is a quiver morphism, we have that each $t(\rho_i) = t(\sigma^{i-1}(\rho_1))$ is in the orbit $t(\rho)$ and that $t(\rho_i) = t(\sigma^{i-1}(\rho_1)) = \sigma^{i-1}(t(\rho_1))$. The value d is the least positive integer such that $\sigma^d(t(\rho_1)) = t(\rho_1)$ and since $\sigma^m(t(\rho_1)) = t(\rho_1)$ (because $\sigma^m(\rho_1) = \rho_1$), then $d \mid m$. By the same argument, $d_{h(\rho)} \mid m$. Thus, this construction does in fact yield an absolute valued quiver.

Conversely, given an absolute valued quiver Γ with valuation $(d_i, m_\rho)_{i\in\Gamma_0,\rho\in\Gamma_1}$, it is possible to construct a quiver with automorphism (Q, σ) that folds into Γ in the following way. Let $\overline{x}_y$ be the unique representative of $(x \bmod y)$ in the set $\{1, 2, \ldots, y\}$ for x, y positive integers. Then define:

- $Q_0 = \{v_i(j) \mid i \in \Gamma_0,\ 1 \le j \le d_i\}$,
- $Q_1 = \{a_\rho(k) \mid \rho \in \Gamma_1,\ 1 \le k \le m_\rho\}$,
- $t\left(a_\rho(k)\right) = v_{t(\rho)}\left(\overline{k}_{d_{t(\rho)}}\right)$ and $h\left(a_\rho(k)\right) = v_{h(\rho)}\left(\overline{k}_{d_{h(\rho)}}\right)$,
- $\sigma\left(v_i(j)\right) = v_i\left(\overline{(j+1)}_{d_i}\right)$,
- $\sigma\left(a_\rho(k)\right) = a_\rho\left(\overline{(k+1)}_{m_\rho}\right)$.

It is clear that Q is a quiver. It is easily verified that σ is indeed an automorphism of Q. Given the construction, we see that (Q, σ) folds into Γ. However, we do not have a one-to-one correspondence between absolute valued quivers and quivers with automorphism since, in general, several non-isomorphic quivers with automorphism can fold into the same absolute valued quiver, as the following example demonstrates.

Example 1. *([7, Example 3.4]) Consider the following two quivers.*

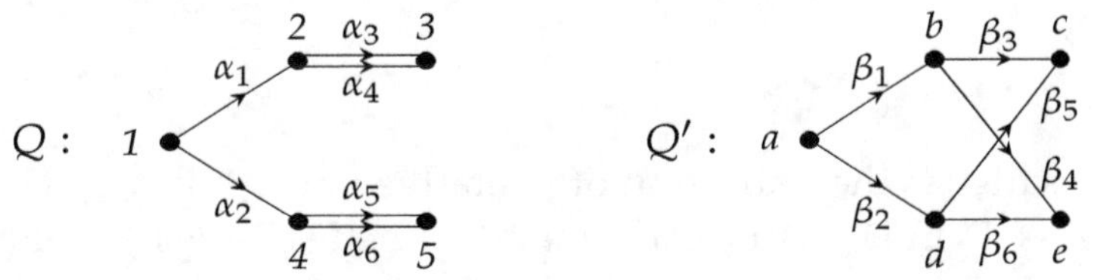

Define $\sigma \in \mathrm{Aut}(Q)$ *and* $\sigma' \in \mathrm{Aut}(Q')$ *by*

$$\sigma : \begin{pmatrix} 1 & 2 & 3 & 4 & 5 \\ 1 & 4 & 5 & 2 & 3 \end{pmatrix}, \begin{pmatrix} \alpha_1 & \alpha_2 & \alpha_3 & \alpha_4 & \alpha_5 & \alpha_6 \\ \alpha_2 & \alpha_1 & \alpha_5 & \alpha_6 & \alpha_3 & \alpha_4 \end{pmatrix}$$

$$\sigma' : \begin{pmatrix} a & b & c & d & e \\ a & d & e & b & c \end{pmatrix}, \begin{pmatrix} \beta_1 & \beta_2 & \beta_3 & \beta_4 & \beta_5 & \beta_6 \\ \beta_2 & \beta_1 & \beta_6 & \beta_5 & \beta_4 & \beta_3 \end{pmatrix}.$$

Then, both (Q, σ) *and* (Q', σ') *fold into*

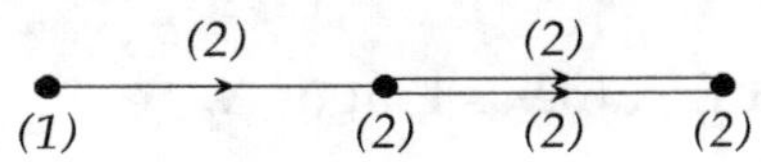

yet Q *and* Q' *are not isomorphic as quivers.*

Definition 3 (Relative valued quiver). *A* relative valued quiver *is a quiver* $\Delta = (\Delta_0, \Delta_1)$ *along with positive integers* d^ρ_{ij}, d^ρ_{ji} *for each arrow* $\rho : i \to j$ *in* Δ_1 *such that there exist positive integers* f_i, $i \in \Delta_0$, *satisfying*

$$d^\rho_{ij} f_j = d^\rho_{ji} f_i$$

for all arrows $\rho : i \to j$ *in* Δ_1. *We call* $(d^\rho_{ij}, d^\rho_{ji})_{(\rho:i\to j)\in\Delta_1}$ *a* (relative) valuation *of* Δ. *By a slight abuse of notation, we often refer to* Δ *as a relative valued quiver, leaving the valuation implied. We will use the notation:*

$$\underset{i}{\bullet} \xrightarrow[\rho]{(d^\rho_{ij},\, d^\rho_{ji})} \underset{j}{\bullet}$$

In the case that $(d^\rho_{ij}, d^\rho_{ji}) = (1, 1)$, *we simply omit it.*

A relative valued quiver morphism *is a quiver morphism* $\varphi : \Delta \to \Delta'$ *satisfying:*

$$(d')^{\varphi_1(\rho)}_{\varphi_0(i)\varphi_0(j)} = d^\rho_{ij} \quad \text{and} \quad (d')^{\varphi_1(\rho)}_{\varphi_0(j)\varphi_0(i)} = d^\rho_{ji}$$

for all arrows $\rho : i \to j$ *in* Δ_1.

Let $\mathfrak{Q}_{\text{rel}}$ *denote the category of relative valued quivers.*

Note that the definition of a relative valued quiver closely resembles the definition of a symmetrizable Cartan matrix. We will explore the link between the two in Section 6, which deals with representations.

As with absolute valued quivers, one can view (non-valued) quivers as relative valued quivers with trivial values (*i.e.*, all $(d^\rho_{ij}, d^\rho_{ji}) = (1, 1)$). Thus, relative valued quivers are also a generalization of quivers.

It is natural to ask, then, how the two categories $\mathfrak{Q}_{\text{abs}}$ and $\mathfrak{Q}_{\text{rel}}$ are related. Given $\Gamma \in \mathfrak{Q}_{\text{abs}}$ with valuation $(d_i, m_\rho)_{i\in\Gamma_0, \rho\in\Gamma_1}$, define $\mathbf{F}(\Gamma) \in \mathfrak{Q}_{\text{rel}}$ with valuation $(d^\rho_{ij}, d^\rho_{ji})_{(\rho:i\to j)\in\mathbf{F}(\Gamma)_1}$ as follows:

- the underlying quiver of $\mathbf{F}(\Gamma)$ is equal to that of Γ,
- the values $(d^\rho_{ij}, d^\rho_{ji})$ are given by:

$$d^\rho_{ij} = \frac{m_\rho}{d_j} \quad \text{and} \quad d^\rho_{ji} = \frac{m_\rho}{d_i}$$

  for all arrows $\rho : i \to j$ in $\mathbf{F}(\Gamma)_1$.

It is clear that $\mathbf{F}(\Gamma)$ satisfies the definition of a relative valued quiver (simply set all the $f_i = d_i$). Given a morphism $\varphi : \Gamma \to \Gamma'$ in $\mathfrak{Q}_{\text{abs}}$, one can simply define $\mathbf{F}(\varphi) : \mathbf{F}(\Gamma) \to \mathbf{F}(\Gamma')$ to be the morphism given by φ, since Γ and Γ' have the same underlying quivers as $\mathbf{F}(\Gamma)$ and $\mathbf{F}(\Gamma')$, respectively. By construction of $\mathbf{F}(\Gamma)$ and $\mathbf{F}(\Gamma')$, it is clear then that $\mathbf{F}(\varphi)$ is a morphism in $\mathfrak{Q}_{\text{rel}}$. Thus, $\mathbf{F}$ is a functor from $\mathfrak{Q}_{\text{abs}}$ to $\mathfrak{Q}_{\text{rel}}$.

Lemma 1. *The functor* $\mathbf{F} : \mathfrak{Q}_{\text{abs}} \to \mathfrak{Q}_{\text{rel}}$ *is faithful and surjective.*

Proof. Suppose $\mathbf{F}(\varphi) = \mathbf{F}(\psi)$ for two morphisms $\varphi, \psi : \Gamma \to \Gamma'$ in $\mathfrak{Q}_{\text{abs}}$. By definition, $\mathbf{F}(\varphi) = \varphi$ on the underlying quivers of Γ and Γ'. Likewise for $\mathbf{F}(\psi)$ and ψ. Thus, $\varphi = \psi$ and $\mathbf{F}$ is faithful.

Suppose Δ is a relative valued quiver. By definition, there exist positive integers f_i, $i \in \Delta_0$, such that $d^\rho_{ij} f_j = d^\rho_{ji} f_i$ for each arrow $\rho : i \to j$ in Δ_1. Fix a particular choice of these f_i. Define $\Gamma \in \mathfrak{Q}_{\text{abs}}$ as follows:

- the underlying quiver of Γ is the same as that of Δ,
- set $d_i = f_i$ for each $i \in \Gamma_0 = \Delta_0$,
- set $m_\rho = d^\rho_{ij} f_j = d^\rho_{ji} f_i$ for each arrow $\rho : i \to j$ in $\Gamma_1 = \Delta_1$.

Then, Γ is an absolute valued quiver and $\mathbf{F}(\Gamma) = \Delta$. Thus, $\mathbf{F}$ is surjective.

□

Note that $\mathbf{F}$ is not full, and thus not an equivalence of categories, as the following example illustrates.

Example 2. *Consider the following two non-isomorphic absolute valued quivers.*

Γ: 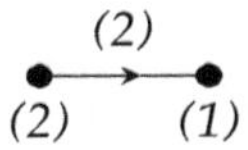

Γ': 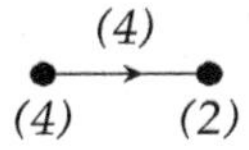

Both Γ and Γ' are mapped to:

$\mathbf{F}(\Gamma) = \mathbf{F}(\Gamma')$: 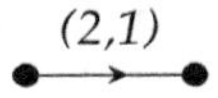

One sees that $\mathrm{Hom}_{\mathfrak{Q}_{\mathrm{abs}}}(\Gamma, \Gamma')$ is empty whereas $\mathrm{Hom}_{\mathfrak{Q}_{\mathrm{rel}}}(\mathbf{F}(\Gamma), \mathbf{F}(\Gamma'))$ is not (it contains the identity). Thus,

$$\mathbf{F} : \mathrm{Hom}_{\mathfrak{Q}_{\mathrm{abs}}}(\Gamma, \Gamma') \to \mathrm{Hom}_{\mathfrak{Q}_{\mathrm{rel}}}(\mathbf{F}(\Gamma), \mathbf{F}(\Gamma'))$$

is not surjective, and hence $\mathbf{F}$ is not full.

It is tempting to think that one could remedy this by restricting $\mathbf{F}$ to the full subcategory of $\mathfrak{Q}_{\mathrm{abs}}$ consisting of objects Γ with valuations $(d_i, m_\rho)_{i \in \Gamma_0, \rho \in \Gamma_1}$ such that the greatest common divisor of all d_i is 1. While one can show that $\mathbf{F}$ restricted to this subcategory is injective on objects, it would still not be full, as the next example illustrates.

Example 3. *Consider the following two absolute valued quivers.*

Γ: 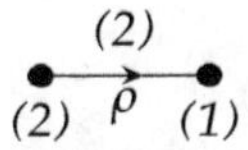

Γ': 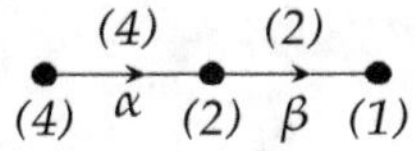

The values of the vertices of Γ have greatest common divisor 1. The same is true of Γ'. By applying $\mathbf{F}$ we get:

$\mathbf{F}(\Gamma)$: 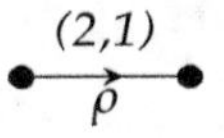

$\mathbf{F}(\Gamma')$: 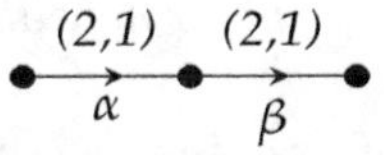

One sees that $\mathrm{Hom}_{\mathfrak{Q}_{\mathrm{abs}}}(\Gamma, \Gamma')$ contains only one morphism (induced by $\rho \mapsto \beta$), while on the other hand $\mathrm{Hom}_{\mathfrak{Q}_{\mathrm{rel}}}(\mathbf{F}(\Gamma), \mathbf{F}(\Gamma'))$ contains two morphisms (induced by $\rho \mapsto \alpha$ and $\rho \mapsto \beta$). Thus,

$$\mathbf{F} : \mathrm{Hom}_{\mathfrak{Q}_{\mathrm{abs}}}(\Gamma, \Gamma') \to \mathrm{Hom}_{\mathfrak{Q}_{\mathrm{rel}}}(\mathbf{F}(\Gamma), \mathbf{F}(\Gamma'))$$

is not surjective, and hence $\mathbf{F}$ is not full, even when restricted to the subcategory of objects with vertex values having greatest common divisor 1.

Note that there is no similar functor $\mathfrak{Q}_{\mathrm{rel}} \to \mathfrak{Q}_{\mathrm{abs}}$. Following the proof of Lemma 1, one sees that finding a preimage under $\mathbf{F}$ of a relative valued quiver Δ is equivalent to making a choice of f_i (from Definition 3). One can show that there is a unique such choice satisfying $\gcd(f_i)_{i \in \Delta_0} = 1$ (so long as Δ is connected). Thus, there is a natural and well-defined way to map objects of $\mathfrak{Q}_{\mathrm{rel}}$ to objects of $\mathfrak{Q}_{\mathrm{abs}}$ by mapping a relative valued quiver Δ to the unique absolute valued quiver Γ with valuation $(d_i, m_\rho)_{i \in \Gamma_0, \rho \in \Gamma_1}$ satisfying $\mathbf{F}(\Gamma) = \Delta$ and $\gcd(d_i)_{i \in \Gamma_0} = 1$. However, there is no such natural mapping on the morphisms of $\mathfrak{Q}_{\mathrm{rel}}$. For instance, under this natural mapping on objects, in Example 3, the relative valued quivers $\mathbf{F}(\Gamma)$ and $\mathbf{F}(\Gamma')$ are mapped to Γ and Γ', respectively. However, there is no natural way to map the morphism $\mathbf{F}(\Gamma) \to \mathbf{F}(\Gamma')$ induced by $\rho \mapsto \alpha$ to a morphism $\Gamma \to \Gamma'$ since there is no morphism $\Gamma \to \Gamma'$ such that $\rho \mapsto \alpha$. Thus, there does not appear to be a functor similar to $\mathbf{F}$ from $\mathfrak{Q}_{\text{rel}}$ to $\mathfrak{Q}_{\text{abs}}$.

3. Species and K-Species

The reason for introducing two different definitions of valued quivers in the previous section is that there are two different definitions of species in the literature: one for each of the two versions of valued quivers. In this section, we introduce both definitions of species and discuss how they are related (see Proposition 1 as well as Examples 5, 6 and 7).

First, we begin with the more general definition of species (see for example [1,3]). Recall that if R and S are rings and M is an (R,S)-bimodule, then $\text{Hom}_R(M,R)$ is an (S,R)-bimodule via $(s\cdot\varphi\cdot r)(m) = \varphi(m\cdot s)r$ and $\text{Hom}_S(M,S)$ is an (S,R)-bimodule via $(s\cdot\psi\cdot r)(m) = s\psi(r\cdot m)$ for all $r\in R$, $s\in S$, $m\in M$, $\varphi\in\text{Hom}_R(M,R)$ and $\psi\in\text{Hom}_S(M,S)$.

Definition 4 (Species). *Let Δ be a relative valued quiver with valuation $(d^{\rho}_{ij}, d^{\rho}_{ji})_{(\rho:i\to j)\in\Delta_1}$. A* modulation *$\mathbb{M}$ of Δ consists of a division ring K_i for each $i\in\Delta_0$, and a $(K_{h(\rho)},K_{t(\rho)})$-bimodule M_ρ for each $\rho\in\Delta_1$ such that the following two conditions hold:*

a. $\text{Hom}_{K_{t(\rho)}}(M_\rho, K_{t(\rho)}) \cong \text{Hom}_{K_{h(\rho)}}(M_\rho, K_{h(\rho)})$ *as $(K_{t(\rho)},K_{h(\rho)})$-bimodules, and*
b. $\dim_{K_{t(\rho)}}(M_\rho) = d^{\rho}_{h(\rho)t(\rho)}$ *and* $\dim_{K_{h(\rho)}}(M_\rho) = d^{\rho}_{t(\rho)h(\rho)}$.

A species *(also called a* modulated quiver*) $\mathcal{Q}$ is a pair $(\Delta,\mathbb{M})$, where Δ is a relative valued quiver and $\mathbb{M}$ is a modulation of Δ.*

A species morphism *$\mathcal{Q}\to\mathcal{Q}'$ consists of a relative valued quiver morphism $\varphi:\Delta\to\Delta'$, a division ring morphism $\psi_i : K_i \to K'_{\varphi_0(i)}$ for each $i\in\Delta_0$, and a compatible Abelian group homomorphism $\psi_\rho : M_\rho \to M'_{\varphi_1(\rho)}$ for each $\rho\in\Delta_1$. That is, for every $\rho\in\Delta_1$ we have $\psi_\rho(a\cdot m) = \psi_{h(\rho)}(a)\cdot\psi_\rho(m)$ and $\psi_\rho(m\cdot b) = \psi_\rho(m)\cdot\psi_{t(\rho)}(b)$ for all $a\in K_{h(\rho)}$, $b\in K_{t(\rho)}$ and $m\in M_\rho$.*

Let $\mathfrak{M}$ denote the category of species.

Remark 1. *Notice that we allow parallel arrows in our definition of valued quivers and thus in our definition of species. However, many texts only allow for single arrows in their definition of species. We will see in Sections 5 and 6 that we can always assume, without loss of generality, that we have no parallel arrows. Thus our definition of species is consistent with the other definitions in the literature.*

Another definition of species also appears in the literature (see for example [7]). This definition depends on a central field K, and so to distinguish between the two definitions, we will call these objects K-species.

Definition 5 (K-species). *Let Γ be an absolute valued quiver with valuation $(d_i, m_\rho)_{i\in\Gamma_0,\rho\in\Gamma_1}$. A* K-modulation *$\mathbb{M}$ of Γ consists of a K-division algebra K_i for each $i\in\Gamma_0$, and a $(K_{h(\rho)},K_{t(\rho)})$-bimodule M_ρ for each $\rho\in\Gamma_1$, such that the following two conditions hold:*

a. K acts centrally on M_ρ (i.e., $k\cdot m = m\cdot k \quad \forall\, k\in K,\ m\in M_\rho$), and
b. $\dim_K(K_i) = d_i$ *and* $\dim_K(M_\rho) = m_\rho$.

A K-species *(also called a* K-modulated quiver*) $\mathcal{Q}$ is a pair $(\Gamma,\mathbb{M})$, where Γ is an absolute valued quiver and $\mathbb{M}$ is a K-modulation of Γ.*

A K-species morphism *$\mathcal{Q}\to\mathcal{Q}'$ consists of an absolute valued quiver morphism $\varphi:\Gamma\to\Gamma'$, a K-division algebra morphism $\psi_i : K_i\to K'_{\varphi_0(i)}$ for each $i\in\Gamma_0$, and a compatible K-linear map $\psi_\rho : M_\rho\to M'_{\varphi_1(\rho)}$ for each $\rho\in\Gamma_1$. That is, for every $\rho\in\Gamma_1$ we have $\psi_\rho(a\cdot m) = \psi_{h(\rho)}(a)\cdot\psi_\rho(m)$ and $\psi_\rho(m\cdot b) = \psi_\rho(m)\cdot\psi_{t(\rho)}(b)$ for all $a\in K_{h(\rho)}$, $b\in K_{t(\rho)}$ and $m\in M_\rho$.*

Let $\mathfrak{M}_K$ denote the category of K-species.

Note that, given a base field K, not every absolute valued quiver has a K-modulation. For example, it is well-known that the only division algebras over $\mathbb{R}$ are $\mathbb{R}$, $\mathbb{C}$ and $\mathbb{H}$, which have dimension 1, 2 and 4, respectively. Thus, any absolute valued quiver containing a vertex with value 3 (or any value not equal to 1, 2 or 4) has no $\mathbb{R}$-modulation. However, given an absolute valued quiver, we can always find a base field K for which there exists a K-modulation. For example, $\mathbb{Q}$ admits field extensions (thus division algebras) of arbitrary dimension, thus $\mathbb{Q}$-modulations always exist.

It is also worth noting that given a valued quiver, relative or absolute, there may exist several non-isomorphic species or K-species (depending on the field K).

Example 4. *Consider the following absolute valued quiver Γ and its image under* $\mathbf{F}$.

One can construct the following two $\mathbb{Q}$-species of Γ.

Then $\mathcal{Q} \not\cong \mathcal{Q}'$ as $\mathbb{Q}$-species, since $\mathbb{Q}(\sqrt{2}) \not\cong \mathbb{Q}(\sqrt{3})$ as algebras. Also, one can show that $\mathcal{Q}$ and $\mathcal{Q}'$ are species of $\mathbf{F}(\Gamma)$ (indeed, this will follow from Proposition 1). But again, $\mathcal{Q} \not\cong \mathcal{Q}'$ as species since $\mathbb{Q}(\sqrt{2}) \not\cong \mathbb{Q}(\sqrt{3})$ as rings.

It is natural to ask how species and K-species are related, *i.e.*, how the categories $\mathfrak{M}$ and $\mathfrak{M}_K$ are related. To answer this question, we first need the following lemma.

Lemma 2. *Let F and G be finite-dimensional (nonzero) division algebras over a field K and let M be a finite-dimensional (F,G)-bimodule on which K acts centrally. Then $\mathrm{Hom}_F(M,F) \cong \mathrm{Hom}_G(M,G)$ as (G,F)-bimodules.*

Proof. A proof can be found in [9, Lemma 3.7], albeit using slightly different terminology. For convenience, we present a brief sketch of the proof.

Let $\tau : F \to K$ be a nonzero K-linear map such that $\tau(ab) = \tau(ba)$ for all $a, b \in F$. Such a map is known to exist; one can take the reduced trace map $F \to Z(F)$, where $Z(F)$ is the centre of F (see [10, Chapter IX, Section 2, Proposition 6]) and compose it with any nonzero map $Z(F) \to K$. Then $T : \mathrm{Hom}_F(M,F) \to \mathrm{Hom}_K(M,K)$ defined by $\varphi \mapsto \tau \circ \varphi$ is a (G,F)-bimodule isomorphism. By an analogous argument, $\mathrm{Hom}_G(M,G) \cong \mathrm{Hom}_K(M,K)$ completing the proof. □

Given Lemma 2, we see that if $\mathcal{Q}$ is a K-species with absolute valued quiver Γ, then $\mathcal{Q}$ is a species with underlying relative valued quiver $\mathbf{F}(\Gamma)$. Also, a K-species morphism $\mathcal{Q} \to \mathcal{Q}'$ is a species morphism when viewing $\mathcal{Q}$ and $\mathcal{Q}'$ as species (because an algebra morphism is a ring morphism and a linear map is a group homomorphism). Thus, we may define a forgetful functor $\mathbf{U}_K : \mathfrak{M}_K \to \mathfrak{M}$, which forgets the underlying field K and views absolute valued quivers as relative valued quivers via the functor $\mathbf{F}$. This yields the following result.

Proposition 1. *The functor $\mathbf{U}_K : \mathfrak{M}_K \to \mathfrak{M}$ is faithful and injective on objects. Hence, we may view $\mathfrak{M}_K$ as a subcategory of $\mathfrak{M}$.*

Proof. Faithfulness is clear, since $\mathbf{F}$ is faithful and $\mathbf{U}_K$ then simply forgets the underlying field K.

To see that $\mathbf{U}_K$ is injective on objects, suppose $\mathcal{Q}$ and $\mathcal{Q}'$ are K-species with modulations $(K_i, M_\rho)_{i \in \Gamma_0, \rho \in \Gamma_1}$ and $(K_i', M_\rho')_{i \in \Gamma_0', \rho \in \Gamma_1'}$, respectively, such that $\mathbf{U}_K(\mathcal{Q}) = \mathbf{U}_K(\mathcal{Q}')$. Then, the underlying (non-valued) quivers of $\mathcal{Q}$ and $\mathcal{Q}'$ are equal. Moreover, $K_i = K_i'$ for all $i \in \Gamma_0 = \Gamma_1'$ and $M_\rho = M_\rho'$ for all $\rho \in \Gamma_1 = \Gamma_1'$. So, $\mathcal{Q} = \mathcal{Q}'$ and thus $\mathbf{U}_K$ is injective. □

Note that $\mathbf{U}_K$ is not full (and hence we cannot view $\mathfrak{M}_K$ as a full subcategory of $\mathfrak{M}$) nor is it essentially surjective. In fact, there are objects in $\mathfrak{M}$ which are not of the form $\mathbf{U}_K(\mathcal{Q})$ for $\mathcal{Q} \in \mathfrak{M}_K$ for any field K. The following examples illustrate these points.

Example 5. *Consider $\mathbb{C}$ as a $\mathbb{C}$-species, that is $\mathbb{C}$ is a $\mathbb{C}$-modulation of the trivially valued quiver consisting of one vertex and no arrows. Then, the only morphism in $\mathrm{Hom}_{\mathbb{C}}(\mathbb{C},\mathbb{C})$ is the identity, since any such morphism must send 1 to 1 and be $\mathbb{C}$-linear. However, $\mathrm{Hom}_{\mathbf{U}_{\mathbb{C}}(\mathbb{C})}(\mathbf{U}_{\mathbb{C}}(\mathbb{C}),\mathbf{U}_{\mathbb{C}}(\mathbb{C}))$ contains more than just the identity. Indeed, let $\varphi : \mathbb{C} \to \mathbb{C}$ given by $z \mapsto \bar{z}$. Then, φ is a ring morphism and thus defines a species morphism. Hence,*

$$\mathbf{U}_{\mathbb{C}} : \mathrm{Hom}_{\mathbb{C}}(\mathbb{C},\mathbb{C}) \to \mathrm{Hom}_{\mathbf{U}_{\mathbb{C}}(\mathbb{C})}(\mathbf{U}_{\mathbb{C}}(\mathbb{C}),\mathbf{U}_{\mathbb{C}}(\mathbb{C}))$$

is not surjective and so $\mathbf{U}_{\mathbb{C}}$ is not full.

Example 6. *There exist division rings which are not finite-dimensional over their centres; such division rings are called* centrally infinite. *Hilbert was the first to construct such a ring (see for example [11, Proposition 14.2]). Suppose R is a centrally infinite ring. Then, for any field K contained in R such that R is a K-algebra, $K \subseteq Z(R)$ and so R is not finite-dimensional K-algebra. Thus, any species containing R as part of its modulation is not isomorphic to any object in the image of $\mathbf{U}_K$ for any field K.*

One might think that we could eliminate this problem by restricting ourselves to modulations containing only centrally finite rings. In other words, one might believe that if $\mathcal{Q}$ is a species whose modulation contains only centrally finite rings, then we can find a field K and a K-species $\mathcal{Q}'$ such that $\mathcal{Q} \cong U_K(\mathcal{Q}')$. However, this is not the case as we see in the following example.

Example 7. *Let p be a prime. Consider:*

$\mathcal{Q}$: 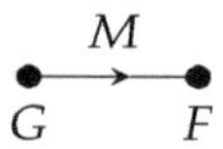

where $F = G = \overline{\mathbb{F}}_p$ (F and G are then centrally finite since they are fields) and $M = \overline{\mathbb{F}}_p$ is an (F,G)-bimodule with actions:

$$f \cdot m \cdot g = fmg^p$$

for all $f \in F$, $g \in G$ and $m \in M$. We claim that $\mathcal{Q}$ is a species. The dimension criterion is clear, as $\dim_F M = \dim_G M = 1$. Thus, it remains to show that

$$\mathrm{Hom}_F(M,F) \cong \mathrm{Hom}_G(M,G)$$

as (G,F)-bimodules. Recall that in $\overline{\mathbb{F}}_p$, p-th roots exist and are unique. Indeed, for any $a \in \overline{\mathbb{F}}_p$, the p-th roots of a are the roots of the polynomial $x^p - a$. Because $\overline{\mathbb{F}}_p$ is algebraically closed, this polynomial has a root, say α. Because $\mathrm{char}\,\overline{\mathbb{F}}_p = p$ we have

$$(x-\alpha)^p = x^p - \alpha^p = x^p - a.$$

Hence, α is the unique p-th root of a. Therefore, we have a well-defined map:

$$\Phi : \mathrm{Hom}_F(M,F) \to \mathrm{Hom}_G(M,G)$$
$$\varphi \mapsto \rho \circ \varphi$$

where ρ is the p-th root map. It is straightforward to show that Φ is a (G,F)-bimodule isomorphism.

Therefore, $\mathcal{Q}$ is a species. Yet the field $\overline{\mathbb{F}}_p$ does not act centrally on M. Indeed, take an element $a \notin \mathbb{F}_p$, then

$$a \cdot 1 = a \neq a^p = 1 \cdot a$$

In fact, the only subfield that does act centrally on M is $\mathbb{F}_p$ since $a^p = a$ if and only if $a \in \mathbb{F}_p$. But, F and G are infinite-dimensional over $\mathbb{F}_p$. Thus, there is no field K for which $\mathcal{Q}$ is isomorphic to an object of the form $\mathbf{U}_K(\mathcal{Q}')$ with $\mathcal{Q}' \in \mathfrak{M}_K$.

4. The Path and Tensor Algebras

In this section we will define the path and tensor algebras associated to quivers and species, respectively. These algebras play an important role in the representation theory of finite-dimensional algebras (see Theorems 1 and 2, and Corollary 1). In subsequent sections, we will give a more in-depth study of these algebras (Sections 4 and 5) and we will show that modules of path and tensor algebras are equivalent to representations of quivers and species, respectively (Section 6).

Recall that a *path* of length n in a quiver Q is a sequence of n arrows in Q_1, $\rho_n\rho_{n-1}\cdots\rho_1$, such that $h(\rho_i) = t(\rho_{i+1})$ for all $i = 1, 2, \ldots, n-1$. For every vertex, we have a trivial path of length 0 (beginning and ending at that vertex).

Definition 6 (Path algebra). *The* path algebra, *KQ, of a quiver Q is the K-algebra with basis the set of all the paths in Q and multiplication given by:*

$$(\beta_n\beta_{n-1}\cdots\beta_1)(\alpha_m\alpha_{m-1}\cdots\alpha_1) = \begin{cases} \beta_n\beta_{n-1}\cdots\beta_1\alpha_m\alpha_{m-1}\cdots\alpha_1, & \text{if } t(\beta_1) = h(\alpha_m), \\ 0, & \text{otherwise.} \end{cases}$$

Remark 2. *According to the convention used, a path $i_1 \xrightarrow{\rho_1} i_2 \xrightarrow{\rho_2} \cdots \xrightarrow{\rho_{n-1}} i_n \xrightarrow{\rho_n} i_{n+1}$ is written from "right to left" $\rho_n\rho_{n-1}\cdots\rho_1$. However, some texts write paths from "left to right" $\rho_1\rho_2\cdots\rho_n$. Using the "left to right" convention yields a path algebra that is opposite to the one defined here.*

Note that KQ is associative and unital (its identity is $\sum_{i\in Q_0}\varepsilon_i$, where ε_i is the path of length zero at i). Also, KQ is finite-dimensional precisely when Q contains no oriented cycles.

Definition 7 (Admissible ideal). *Let Q be a quiver and let $P^n(Q) = \operatorname{span}_K\{$all paths in Q of length $\geq n\}$. An* admissible ideal *I of the path algebra KQ is a two-sided ideal of KQ satisfying*

$$P^n(Q) \subseteq I \subseteq P^2(Q), \quad \text{for some positive integer } n.$$

If Q has no oriented cycles, then any ideal $I \subseteq P^2(Q)$ of KQ is an admissible ideal, since $P^n(Q) = 0$ for sufficiently large n.

There is a strong relationship between path algebras and finite-dimensional algebras, touched upon by Brauer [12], Jans [13] and Yoshii [14], but fully explored by Gabriel [1]. Let A be a finite-dimensional K-algebra. We recall a few definitions. An element $\varepsilon \in A$ is called an *idempotent* if $\varepsilon^2 = \varepsilon$. Two idempotents ε_1 and ε_2 are called *orthogonal* if $\varepsilon_1\varepsilon_2 = \varepsilon_2\varepsilon_1 = 0$. An idempotent ε is called *primitive* if it cannot be written as a sum $\varepsilon = \varepsilon_1 + \varepsilon_2$, where ε_1 and ε_2 are orthogonal idempotents. A set of idempotents $\{\varepsilon_1, \ldots, \varepsilon_n\}$ is called *complete* if $\sum_{i=1}^n \varepsilon_i = 1$. If $\{\varepsilon_1, \ldots, \varepsilon_n\}$ is a complete set of primitive (pairwise) orthogonal idempotents of A, then $A = A\varepsilon_1 \oplus \cdots \oplus A\varepsilon_n$ is a decomposition of A (as a left A-module) into indecomposable modules; this decomposition is unique up to isomorphism and permutation of the terms. We say that A is *basic* if $A\varepsilon_i \not\cong A\varepsilon_j$ as (left) A-modules for all $i \neq j$ (or, alternatively, the decomposition of A into indecomposable modules admits no repeated factors). Finally, A is called *hereditary* if every A-submodule of a projective A-module is again projective.

Theorem 1. *Let K be an algebraically closed field and let A be a finite-dimensional K-algebra.*

a. If A is basic and hereditary, then $A \cong KQ$ (as K-algebras) for some quiver Q.
b. If A is basic, then $A \cong KQ/I$ (as K-algebras) for some quiver Q and some admissible ideal I of KQ.

For a proof of Theorem 1 see [15, Sections II and VII] or [16, Propositions 4.1.7 and 4.2.4] (though it also follows from [17, Proposition 10.2]. The above result is powerful, but it does not necessarily hold over fields which are not algebraically closed. If we want to work with algebras over non-algebraically closed fields, we need to generalize the notion of a path algebra. We look, then, at the analogue of the path algebra for a K-species.

Let $\mathcal{Q}$ be a species of a relative valued quiver Δ with modulation $(K_i, M_\rho)_{i\in\Delta_0,\rho\in\Delta_1}$. Let $D = \prod_{i\in\Delta_0} K_i$ and let $M = \bigoplus_{\rho\in\Delta_1} M_\rho$. Then D is a ring and M naturally becomes a (D, D)-bimodule. If $\mathcal{Q}$ is a K-species, then D is a K-algebra.

Definition 8 (Tensor ring/algebra). *The* tensor ring, *$T(\mathcal{Q})$, of a species $\mathcal{Q}$ is defined by*

$$T(\mathcal{Q}) = \bigoplus_{n=0}^{\infty} T^n(M)$$

where

$$T^0(M) = D \text{ and } T^n(M) = T^{n-1}(M) \otimes_D M \text{ for } n \geq 1$$

Multiplication is determined by the composition

$$T^m(M) \times T^n(M) \twoheadrightarrow T^m(M) \otimes_D T^n(M) \xrightarrow{\cong} T^{m+n}(M)$$

If $\mathcal{Q}$ is a K-species, then $T(\mathcal{Q})$ is a K-algebra. In this case we call $T(\mathcal{Q})$ the tensor algebra *of $\mathcal{Q}$.*

Admissible ideals for tensor rings/algebras are defined in the same way as admissible ideals for path algebras by setting $P^n(\mathcal{Q}) = \bigoplus_{m=n}^{\infty} T^m(M)$.

Suppose that Γ is an absolute valued quiver with trivial valuation (all d_i and m_ρ are equal to 1) and $\mathcal{Q}$ is a K-species of Γ. Then, for each $i \in \Gamma_0$, $\dim_K K_i = 1$, which implies that $K_i \cong K$ (as K-algebras). Likewise, $\dim_K M_\rho = 1$ implies that $M_\rho \cong K$ (as (K,K)-bimodules). Therefore, it follows that $T(\mathcal{Q}) \cong KQ$ where $Q = (\Gamma_0, \Gamma_1)$. Thus, when viewing non-valued quivers as absolute valued quivers with trivial valuation, the tensor algebra of the K-species becomes simply the path algebra (over K) of the quiver. Therefore, the tensor algebra is indeed a generalization of the path algebra. Additionally, the tensor algebra allows us to generalize Theorem 1.

Recall that a field K is called *perfect* if either $\mathrm{char}(K) = 0$ or, if $\mathrm{char}(K) = p > 0$, then $K^p = \{a^p \mid a \in K\} = K$.

Theorem 2. *Let K be a perfect field and let A be a finite-dimensional K-algebra.*

a. If A is basic and hereditary, then $A \cong T(\mathcal{Q})$ (as K-algebras) for some K-species $\mathcal{Q}$.
b. If A is basic, then $A \cong T(\mathcal{Q})/I$ (as K-algebras) for some K-species $\mathcal{Q}$ and some admissible ideal I of $T(\mathcal{Q})$.

For a proof of Theorem 2, see [17, Proposition 10.2] or [16, Corollary 4.1.11 and Proposition 4.2.5] or [18, Section 8.5]. Note that Theorem 2 does not necessarily hold over non-perfect fields. To see why, we first introduce a useful tool in the study of path and tensor algebras.

Definition 9 (Jacobson radical). *The* Jacobson radical *of a ring R is the intersection of all maximal left ideals of R. We denote the Jacobson radical of R by* rad R .

Remark 3. *The intersection of all maximal left ideals coincides with the intersection of all maximal right ideals (see, for example, [11, Corollary 4.5]), so the Jacobson radical could alternatively be defined in terms of right ideals.*

Lemma 3. *Let $\mathcal{Q}$ be a species.*

a. *If $\mathcal{Q}$ contains no oriented cycles, then* $\operatorname{rad} T(\mathcal{Q}) = \bigoplus_{n=1}^{\infty} T^n(M)$.
b. *Let I be an admissible ideal of $T(\mathcal{Q})$. Then,* $\operatorname{rad}(T(\mathcal{Q})/I) = \left(\bigoplus_{n=1}^{\infty} T^n(M)\right)/I$.

Proof. It is well known that if R is a ring and J is a two-sided nilpotent ideal of R such that R/J is semisimple, then $\operatorname{rad} R = J$ (see for example [11, Lemma 4.11 and Proposition 4.6] together with the fact that the radical of a semisimple ring is 0). Let $J = \bigoplus_{n=1}^{\infty} T^n(M)$. If $\mathcal{Q}$ contains no oriented cycles, then $T^n(M) = 0$ for some positive integer n. Thus, $J^n = 0$ and J is then nilpotent. Then $T(\mathcal{Q})/J \cong T^0(M) = D$, which is semisimple. Therefore, $\operatorname{rad} T(\mathcal{Q}) = J$, proving Part 1.

If I is an admissible ideal of $T(\mathcal{Q})$, let $J = \left(\bigoplus_{n=1}^{\infty} T^n(M)/I\right)$. By definition, $P^n(\mathcal{Q}) \subseteq I$ for some n and so $J^n = 0$. Thus, Part 2 follows by a similar argument. □

Remark 4. *Part 1 of Lemma 3 is false if $\mathcal{Q}$ contains oriented cycles. One does not need to look beyond quivers to see why. For example, following [15, Section II, Chapter 1], we can consider the path algebra of the Jordan quiver over an infinite field K. That is, we consider KQ, where:*

Q: 

Then it is clear that $KQ \cong K[t]$, the polynomial ring in one variable. For each $\alpha \in K$, let I_α be the ideal generated by $t + \alpha$. Each I_α is a maximal ideal and $\bigcap_{\alpha \in K} I_\alpha = 0$ since K is infinite. Thus $\operatorname{rad} KQ = 0$ whereas $\bigoplus_{n=1}^{\infty} T^n(Q) \cong (t)$ (the ideal generated by the lone arrow of Q).

With the concept of the Jacobson radical and Lemma 3, we are ready to see why Theorem 2 fails over non-perfect fields. Recall that a K-algebra epimorphism $\varphi : A \twoheadrightarrow B$ is said to *split* if there exists a K-algebra morphism $\mu : B \to A$ such that $\varphi \circ \mu = \mathrm{id}_B$. We see that if $A = T(\mathcal{Q})/I$ for a K-species $\mathcal{Q}$ and admissible ideal I, then the canonical projection $A \twoheadrightarrow A/\operatorname{rad} A$ splits (since $A \cong D \oplus \operatorname{rad} A$). Thus, to construct an example where Theorem 2 fails, it suffices to find an algebra where this canonical projection does not split. This is possible over a non-perfect field.

Example 8. *[16, Remark (ii) following Corollary 4.1.11] Let K_0 be a field of characteristic $p \neq 0$ and let $K = K_0(t)$, which is not a perfect field. Let $A = K[x,y]/(x^p, y^p - x - t)$. A quick calculation shows that $\operatorname{rad} A = (x)$ and thus $A/\operatorname{rad} A \cong K[y]/(y^p - t)$. One can easily verify that the projection $A \twoheadrightarrow A/\operatorname{rad} A$ does not split. Hence, A is not isomorphic to the quotient of the tensor algebra of a species by some admissible ideal.*

Theorems 1 and 2 require our algebras to be basic. There is a slightly weaker property that holds in the case of non-basic algebras, that of *Morita equivalence*.

Definition 10 (Morita equivalence)**.** *Two rings R and S are said to be* Morita equivalent *if their categories of (left) modules, R-Mod and S-Mod, are equivalent.*

Corollary 1. *Let K be a field and let A be a finite-dimensional K-algebra.*

a. *If K is algebraically closed and A is hereditary, then A is Morita equivalent to KQ for some quiver Q.*
b. *If K is algebraically closed, then A is Morita equivalent to KQ/I for some quiver Q and some admissible ideal I of KQ.*
c. *If K is perfect and A is hereditary, then A is Morita equivalent to $T(\mathcal{Q})$ for some K-species $\mathcal{Q}$.*

d. *If K is perfect, then A is Morita equivalent to $T(\mathcal{Q})/I$ for some K-species $\mathcal{Q}$ and some admissible ideal I of $T(\mathcal{Q})$.*

Proof. Every algebra is Morita equivalent to a basic algebra (see [16, Section 2.2]) and Morita equivalence preserves the property of being hereditary (indeed, an equivalence of categories preserves projective modules). Thus, the result follows as a consequence of Theorems 1 and 2. □

5. The Frobenius Morphism

When working over the finite field of q elements, $\mathbb{F}_q$, it is possible to avoid dealing with species altogether and deal only with quivers with automorphism. This is achieved by using the Frobenius morphism (described below).

Definition 11 (Frobenius morphism). *Let $K = \overline{\mathbb{F}}_q$ (the algebraic closure of $\mathbb{F}_q$). Given a quiver with automorphism (Q, σ), the* Frobenius morphism $F = F_{Q,\sigma,q}$ *is defined as*

$$\begin{aligned} F : KQ &\to KQ \\ \sum_i \lambda_i p_i &\mapsto \sum_i \lambda_i^q \sigma(p_i) \end{aligned}$$

for all $\lambda_i \in K$ and paths p_i in Q. The F-fixed point algebra is

$$(KQ)^F = \{x \in KQ \mid F(x) = x\}$$

Note that while KQ is an algebra over K, the fixed point algebra $(KQ)^F$ is an algebra over $\mathbb{F}_q$. Indeed, suppose $0 \neq x \in (KQ)^F$, then $F(\lambda x) = \lambda^q F(x) = \lambda^q x$. Thus, $\lambda x \in (KQ)^F$ if and only if $\lambda^q = \lambda$, which occurs if and only if $\lambda \in \mathbb{F}_q$.

Suppose Γ is the absolute valued quiver obtained by folding (Q, σ). For each $i \in \Gamma_0$ and each $\rho \in \Gamma_1$ define

$$A_i = \bigoplus_{a \in i} K\varepsilon_a \text{ and } A_\rho = \bigoplus_{\tau \in \rho} K\tau$$

where ε_a is the trivial path at vertex a. Then as an $\mathbb{F}_q$-algebra, $A_i^F \cong \mathbb{F}_{q^{d_i}}$. Indeed, fix some $a \in i$, then,

$$A_i^F = \left\{ x = \sum_{j=0}^{d_i - 1} \lambda_j \varepsilon_{\sigma^j(a)} \;\middle|\; \lambda_j \in K \text{ and } F(x) = x \right\}$$

Applying F to an arbitrary $x = \sum_{j=0}^{d_i-1} \lambda_j \varepsilon_{\sigma^j(a)} \in A_i^F$, we obtain:

$$F(x) = \sum_{j=0}^{d_i-1} \lambda_j^q \sigma(\varepsilon_{\sigma^j(a)}) = \sum_{j=0}^{d_i-1} \lambda_j^q \varepsilon_{\sigma^{j+1}(a)}$$

The equality $F(x) = x$ yields $\lambda_j^q = \lambda_{j+1}$ for $j = 0, 1, \ldots, d_i - 2$ and $\lambda_{d_i-1}^q = \lambda_0$. By successive substitution, we get $\lambda_0^{q^{d_i}} = \lambda_0$, which occurs if and only if $\lambda_0 \in \mathbb{F}_{q^{d_i}}$, and $\lambda_j = \lambda_0^{q^j}$. Thus, A_i^F can be rewritten as:

$$\begin{aligned} A_i^F &= \left\{ \sum_{j=0}^{d_i-1} \lambda_0^{q^j} \varepsilon_{\sigma^j(a)} \;\middle|\; \lambda_0 \in \mathbb{F}_{q^{d_i}} \right\} \\ &\cong \mathbb{F}_{q^{d_i}} \qquad \text{(as fields)} \end{aligned}$$

It is easy to see that A_ρ is an $(A_{h(\rho)}, A_{t(\rho)})$-bimodule via multiplication, thus A_ρ^F is an $(A_{h(\rho)}^F, A_{t(\rho)}^F)$-bimodule (on which $\mathbb{F}_q$ acts centrally). Since $A_i^F \cong \mathbb{F}_{q^{d_i}}$ for each $i \in \Gamma_0$, A_ρ^F is then an $(\mathbb{F}_{q^{d_{h(\rho)}}}, \mathbb{F}_{q^{d_{t(\rho)}}})$-bimodule. Over fields, we make no distinction between left and right modules because of commutativity. Thus, A_ρ^F is an $\mathbb{F}_{q^{d_{h(\rho)}}}$-module and an $\mathbb{F}_{q^{d_{t(\rho)}}}$-module, and hence A_ρ^F is a module of the composite field of $\mathbb{F}_{q^{d_{h(\rho)}}}$ and $\mathbb{F}_{q^{d_{t(\rho)}}}$, which in this case is simply the bigger of the two fields (recall that the composite of two fields is the smallest field containing both fields). Over fields, all modules are free and thus A_ρ^F is a free module of the composite field (this fact will be useful later on). Also, $\dim_{\mathbb{F}_q} A_i^F = d_i$ and $\dim_{\mathbb{F}_q} A_\rho^F = m_\rho$ (the dimensions are the number of vertices/arrows in the corresponding orbits). Therefore, $\mathbb{M} = (A_i^F, A_\rho^F)_{i \in \Gamma_0, \rho \in \Gamma_1}$ defines an $\mathbb{F}_q$-modulation of Γ. We will denote the $\mathbb{F}_q$-species $(\Gamma, \mathbb{M})$ by $\mathcal{Q}_{Q,\sigma,q}$. This leads to the following result.

Theorem 3. [7, Theorem 3.25] *Let (Q, σ) be a quiver with automorphism. Then $(KQ)^F \cong T(\mathcal{Q}_{Q,\sigma,q})$ as $\mathbb{F}_q$-algebras.*

In light of Theorem 3, the natural question to ask is: given an arbitrary $\mathbb{F}_q$-species, is its tensor algebra isomorphic to the fixed point algebra of a quiver with automorphism? And if so, to which one?

Suppose $\mathcal{Q}$ is an $\mathbb{F}_q$-species with underlying absolute valued quiver Γ and $\mathbb{F}_q$-modulation $(K_i, M_\rho)_{i \in \Gamma_0, \rho \in \Gamma_1}$. Each K_i is, by definition, a division algebra containing q^{d_i} elements. According to the well-known Wedderburn's little theorem, all finite division algebras are fields. Thus, $K_i \cong \mathbb{F}_{q^{d_i}}$. Similar to the above discussion, M_ρ is then a free module of the composite field of $\mathbb{F}_{q^{d_{h(\rho)}}}$ and $\mathbb{F}_{q^{d_{t(\rho)}}}$. Therefore, by unfolding Γ (as in Section 1, say) to a quiver with automorphism (Q, σ), we get $\mathcal{Q} \cong \mathcal{Q}_{Q,\sigma,q}$ as $\mathbb{F}_q$-species. This leads to the following result.

Proposition 2. [7, Proposition 3.37] *For any $\mathbb{F}_q$-species $\mathcal{Q}$, there exists a quiver with automorphism (Q, σ) such that $T(\mathcal{Q}) \cong (KQ)^F$ as $\mathbb{F}_q$-algebras.*

Note that, given an $\mathbb{F}_q$-species $\mathcal{Q} = (\Gamma, \mathbb{M})$ and a quiver with automorphism (Q, σ) such that $T(\mathcal{Q}) \cong (KQ)^F$, we cannot conclude that (Q, σ) folds into Γ as the following example illustrates.

Example 9. *Consider the following quiver.*

Q: 

There are two possible automorphisms of Q: $\sigma = \mathrm{id}_Q$ and σ', the automorphism defined by interchanging the two arrows of Q. By folding Q with respect to σ and σ', we obtain the following two absolute valued quivers.

Γ: 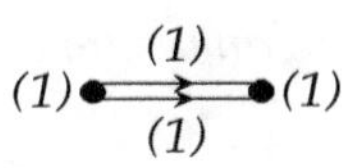

Γ': 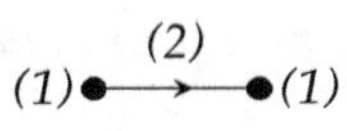

It is clear that Γ and Γ' are not isomorphic. Now, construct $\mathbb{F}_q$-species of Γ and Γ' with the following $\mathbb{F}_q$-modulations.

$\mathcal{Q}$: 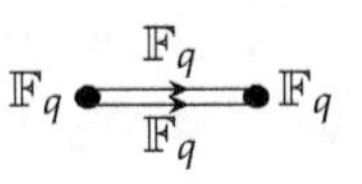

$\mathcal{Q}'$: 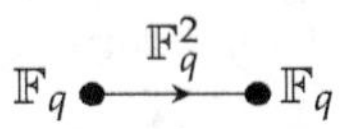

Then we have that $T(\mathcal{Q}) \cong T(\mathcal{Q}')$. Thus, $(KQ)^{F_{Q,\sigma',q}} \cong T(\mathcal{Q}')$, yet (Q, σ) does not fold into Γ'.

The above example raises an interesting question. Notice that the two $\mathbb{F}_q$-species $\mathcal{Q}$ and $\mathcal{Q}'$ are not isomorphic, but their tensor algebras $T(\mathcal{Q})$ and $T(\mathcal{Q}')$ are isomorphic. This phenomenon is not restricted to finite fields either; if we replaced $\mathbb{F}_q$ with some arbitrary field K, we still get $\mathcal{Q} \not\cong \mathcal{Q}'$ as

K-species, but $T(\mathcal{Q}) \cong T(\mathcal{Q}')$ as K-algebras. So we may ask: under what conditions are the tensor algebras (or rings) of two K-species (or species) isomorphic? We answer this question in the following section.

It is worth noting that over infinite fields, there are no (known) methods to extend the results of this section. It is tempting to think that, given an infinite field K and a quiver with automorphism (Q, σ) that folds into an absolute valued quiver Γ, Theorem 3 might be extended by saying that the fixed point algebra $(KQ)^\sigma$ (σ extends to an automorphism of KQ) is isomorphic to the tensor algebra of a K-species of Γ. This is, however, not the case as the following example illustrates.

Example 10. *Take $K = \mathbb{R}$ to be our (infinite) base field. Consider the following quiver.*

$Q:$ 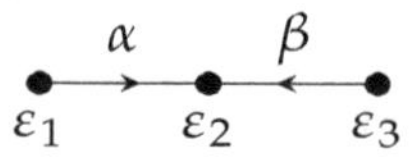

Let σ be the automorphism of Q given by:

$$\sigma : \begin{pmatrix} \varepsilon_1 & \varepsilon_2 & \varepsilon_3 & \alpha & \beta \\ \varepsilon_3 & \varepsilon_2 & \varepsilon_1 & \beta & \alpha \end{pmatrix}$$

Then (Q, σ) folds into the following absolute valued quiver.

$\Gamma:$ 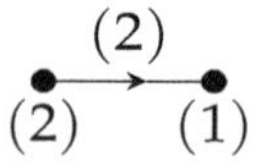

So, we would like for $(KQ)^\sigma$ to be isomorphic to the tensor algebra of a K-species of Γ. However, this does not happen.

An element $x = a_1\varepsilon_1 + a_2\varepsilon_2 + a_3\varepsilon_3 + a_4\alpha + a_5\beta \in KQ$ is fixed by σ if and only if $\sigma(x) = x$, that is, if and only if

$$a_3\varepsilon_1 + a_2\varepsilon_2 + a_1\varepsilon_1 + a_5\alpha + a_4\beta = a_1\varepsilon_1 + a_2\varepsilon_2 + a_3\varepsilon_3 + a_4\alpha + a_5\beta$$

which occurs if and only if $a_1 = a_3$ and $a_4 = a_5$. Hence, $(KQ)^\sigma$ has basis $\{\varepsilon_1 + \varepsilon_3, \varepsilon_2, \alpha + \beta\}$ and we see that it is isomorphic to the path algebra (over K) of the following quiver.

$Q':$ 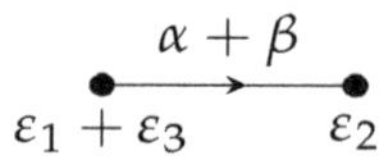

The algebra KQ' is certainly not isomorphic to the tensor algebra of any K-species of Γ. Indeed, over K, KQ' has dimension 3 whereas the tensor algebra of any K-species of Γ has dimension 5.

6. A Closer Look at Tensor Rings

In this section, we find necessary and sufficient conditions for two tensor rings/algebras to be isomorphic. We show that the isomorphism of tensor rings/algebras corresponds to an equivalence on the level of species (see Theorem 4). Furthermore, we show that if $\mathcal{Q}$ is a K-species, where K is a perfect field, then $\overline{K} \otimes_K T(\mathcal{Q})$ is either isomorphic to, or Morita equivalent to, the path algebra of a quiver (see Theorem 5 and Corollary 2). This serves as a partial generalization to [19, Lemma 21] in which Hubery proved a similar result when K is a finite field. We begin by introducing the notion of "crushing".

Definition 12 (Crushed absolute valued quiver)**.** *Let Γ be an absolute valued quiver. Define a new absolute valued quiver, which we will denote Γ^C, as follows:*

- $\Gamma_0^C = \Gamma_0$,

- *# arrows* $i \to j = \begin{cases} 1, & \text{if } \exists\, \rho : i \to j \in \Gamma_1, \\ 0, & \text{otherwise,} \end{cases}$
- $d_i^C = d_i$ *for all* $i \in \Gamma_0^C = \Gamma_0$,
- $m_\rho^C = \sum_{(\alpha : t(\rho) \to h(\rho)) \in \Gamma_1} m_\alpha$ *for all* $\rho \in \Gamma_1^C$.

Intuitively, one "crushes" all parallel arrows of Γ *into a single arrow and sums up the values.*

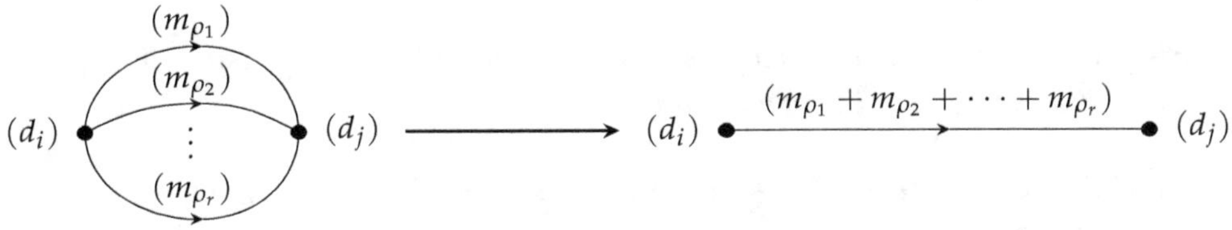

The absolute valued quiver Γ^C *will be called the* crushed (absolute valued) quiver *of* Γ.

Note that Γ^C does indeed satisfy the definition of an absolute valued quiver. Take any $\rho : i \to j \in \Gamma_1^C$. Then $d_i^C = d_i \mid m_\alpha$ for all $\alpha : i \to j$ in Γ_1. Thus, $d_i^C \mid \left(\sum_{(\alpha:i\to j) \in \Gamma_1} m_\alpha\right) = m_\rho^C$. The same is true for d_j^C. Therefore, Γ^C is an absolute valued quiver.

The notion of crushing can be extended to relative valued quivers via the functor $\mathbf{F}$. Recall that if Γ is an absolute valued quiver with valuation $(d_i, m_\rho)_{i \in \Gamma_0, \rho \in \Gamma_1}$, then $\mathbf{F}(\Gamma)$ is a relative valued quiver with valuation $(d_{ij}^\rho, d_{ji}^\rho)_{(\rho:i\to j)\in \mathbf{F}(\Gamma)_1}$ given by $d_{ij}^\rho = m_\rho / d_j$ and $d_{ji}^\rho = m_\rho / d_i$. So, the valuation of $\mathbf{F}(\Gamma^C)$ is given by

$$(d^C)_{ij}^\rho = \left(\sum_{(\alpha:i\to j) \in \Gamma_1} m_\alpha \right) / d_j = \sum_{(\alpha:i\to j) \in \Gamma_1} d_{ij}^\alpha$$

and likewise

$$(d^C)_{ji}^\rho = \left(\sum_{(\alpha:i\to j) \in \Gamma_1} m_\alpha \right) / d_i = \sum_{(\alpha:i\to j) \in \Gamma_1} d_{ji}^\alpha$$

for each $\rho : i \to j$ in $\mathbf{F}(\Gamma^C)_1$. We take this to be the definition of the crushed (relative valued) quiver of a relative valued quiver.

Definition 13 (Crushed relative valued quiver)**.** *Let* Δ *be a relative valued quiver. Define a new relative valued quiver, which we will denote* Δ^C*, as follows:*

- $\Delta_0^C = \Delta_0$,
- *# arrows* $i \to j = \begin{cases} 1, & \text{if } \exists\, \rho : i \to j \in \Delta_1, \\ 0, & \text{otherwise,} \end{cases}$
- $(d^C)_{ij}^\rho = \sum_{(\alpha:i\to j) \in \Delta_1} d_{ij}^\alpha$ *and* $(d^C)_{ji}^\rho = \sum_{(\alpha:i\to j) \in \Delta_1} d_{ji}^\alpha$ *for all* $\rho : i \to j$ *in* Δ_1^C.

Again, the intuition is to "crush" all parallel arrows in Δ *into a single arrow and sum the values.*

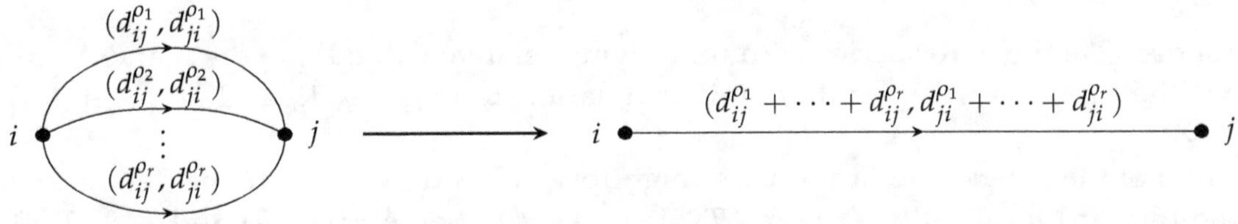

The relative valued quiver Δ^C *will be called the* crushed (relative valued) quiver *of* Δ.

Definition 14 (Crushed species)**.** *Let* $\mathcal{Q}$ *be a species with underlying relative valued quiver* Δ*. Define a new species, which we will denote* $\mathcal{Q}^C$*, as follows:*

- *the underlying valued quiver of* $\mathcal{Q}^C$ *is* Δ^C,
- $K_i^C = K_i$ *for all* $i \in \Delta_0^C = \Delta_0$,
- $M_\rho^C = \bigoplus_{(\alpha:i\to j)\,\in\,\Delta_1} M_\alpha$ *for all* $\rho : i \to j$ *in* Δ_1^C.

The intuition here is similar to that of the previous definitions; one "crushes" all bimodules along parallel arrows into a single bimodule by taking their direct sum. The species $\mathcal{Q}^C$ *will be called the* crushed species *of* $\mathcal{Q}$.

Remark 5. *A* crushed K-species *is defined in exactly the same way, only using the crushed quiver of an absolute valued quiver instead of a relative valued quiver.*

Note that in the above definition $\mathcal{Q}^C$ is indeed a species of Δ^C. Clearly, M_ρ^C is a (K_j^C, K_i^C)-bimodule for all $\rho : i \to j$ in Δ_1^C. Moreover,

$$\begin{aligned}
\operatorname{Hom}_{K_j^C}(M_\rho^C, K_j^C) &= \operatorname{Hom}_{K_j}\left(\bigoplus_{(\alpha:i\to j)\in\Delta_1} M_\alpha, K_j\right) \cong \bigoplus_{(\alpha:i\to j)\in\Delta_1} \operatorname{Hom}_{K_j}(M_\alpha, K_j)\\
&\cong \bigoplus_{(\alpha:i\to j)\in\Delta_1} \operatorname{Hom}_{K_i}(M_\alpha, K_i) \cong \operatorname{Hom}_{K_i}\left(\bigoplus_{(\alpha:i\to j)\in\Delta_1} M_\alpha, K_i\right)\\
&= \operatorname{Hom}_{K_i}(M_\rho^C, K_i^C),
\end{aligned}$$

where all isomorphisms are (K_j^C, K_i^C)-bimodule isomorphisms (in the second isomorphism we use the fact that $\mathcal{Q}$ is a species). Thus the duality condition for species holds. As for the dimension condition:

$$\dim_{K_j^C}(M_\rho^C) = \dim_{K_j}\left(\bigoplus_{(\alpha:i\to j)\in\Delta_1} M_\alpha\right) = \sum_{(\alpha:i\to j)\in\Delta_1} \dim_{K_j}(M_\alpha) = \sum_{(\alpha:i\to j)\in\Delta_1} d_{ij}^\alpha = (d^C)_{ij}^\rho$$

where in the third equality we use the fact that $\mathcal{Q}$ is a species. Likewise, $\dim_{K_i^C}(M_\rho^C) = (d^C)_{ji}^\rho$. Thus, $\mathcal{Q}^C$ is a species of Δ^C.

Note also that if $\mathcal{Q}$ is a K-species of an absolute valued quiver Γ, then $\mathcal{Q}^C$ is a K-species of Γ^C. Indeed, it is clear that M_ρ^C is a (K_j^C, K_i^C)-bimodule on which K acts centrally for all $\rho : i \to j$ in Γ_1^C (since each summand satisfies this condition). Moreover, $\dim_K(K_i^C) = \dim_K(K_i) = d_i = d_i^C$ for all $i \in \Gamma_0^C$, thus the dimension criterion for the vertices is satisfied. Also, $\dim_K(M_\rho^C) = m_\rho^C$ by a computation similar to the above.

With the concept of crushed species/quivers, we obtain the following result, which gives a necessary and sufficient condition for two tensor rings/algebras to be isomorphic.

Theorem 4. *Let* $\mathcal{Q}$ *and* $\mathcal{Q}'$ *be two species containing no oriented cycles. Then* $T(\mathcal{Q}) \cong T(\mathcal{Q}')$ *as rings if and only if* $\mathcal{Q}^C \cong \mathcal{Q}'^C$ *as species. Moreover, if* $\mathcal{Q}$ *and* $\mathcal{Q}'$ *are* K*-species, then* $T(\mathcal{Q}) \cong T(\mathcal{Q})$ *as* K*-algebras if and only if* $\mathcal{Q}^C \cong \mathcal{Q}'^C$ *as* K*-species.*

Proof. Suppose $\mathcal{Q}$ and $\mathcal{Q}'$ are species with underlying relative valued quivers Δ and Δ', respectively. Throughout the proof we will use the familiar notation $D := \Pi_{i\in\Delta_0} K_i$ and $M := \bigoplus_{\rho\in\Delta_1} M_\rho$ (add primes for $\mathcal{Q}'$).

The proof of the reverse implication is straightforward and so we leave the details to the reader.

For the forward implication, assume $T(\mathcal{Q}) \cong T(\mathcal{Q}')$. Let $A = T(\mathcal{Q})$ and $B = T(\mathcal{Q}')$ and let $\varphi : A \to B$ be a ring isomorphism. Then, there exists an induced ring isomorphism $\overline{\varphi} : A/\operatorname{rad} A \to B/\operatorname{rad} B$. By Lemma 3, $A/\operatorname{rad} A \cong D$ and $B/\operatorname{rad} B \cong D'$. Thus, we have an isomorphism $\widetilde{\varphi}_D : D \to D'$. It is not difficult to show that $\{1_{K_i}\}_{i\in\Delta_0}$ is the only complete set of primitive orthogonal idempotents in D and that, likewise, $\{1_{K_i'}\}_{i\in\Delta_0'}$ is the only complete set of primitive orthogonal idempotents of D'.

Since any isomorphism must bijectively map a complete set of primitive orthogonal idempotents to a complete set of primitive orthogonal idempotents, we may identify Δ_0 and Δ_0' and assume, without loss of generality, that $\widetilde{\varphi}_D(1_{K_i}) = 1_{K_i'}$ for each $i \in \Delta_0 = \Delta_0'$. Since $1_{K_i} \cdot D = K_i$ and $1_{K_i'} \cdot D' = K_i'$, we have that $\widetilde{\varphi}_D|_{K_i}$ is a ring isomorphism $K_i \to K_i'$ for each $i \in \Delta_0 = \Delta_0'$.

Now, $\varphi|_{\mathrm{rad}\,A}$ is a ring isomorphism from rad A to rad B. Thus, as before, we have an induced ring isomorphism (and hence an Abelian group isomorphism) $\overline{\varphi|_{\mathrm{rad}\,A}} : \mathrm{rad}\,A/(\mathrm{rad}\,A)^2 \to \mathrm{rad}\,B/(\mathrm{rad}\,B)^2$. Since $\mathrm{rad}\,A/(\mathrm{rad}\,A)^2 \cong M$ and $\mathrm{rad}\,B/(\mathrm{rad}\,B)^2 \cong M'$, we have an isomorphism $\widetilde{\varphi}_M : M \to M'$. For any $i, j \in \Delta_0$, $1_{K_j} \cdot M \cdot 1_{K_i} = \bigoplus_{(\rho:i\to j)\in\Delta_0} M_\rho =: {}_jM_i$ and $1_{K_j'} \cdot M' \cdot 1_{K_i'} = \bigoplus_{(\rho:i\to j)\in\Delta_0'} M_\rho' =: {}_jM_i'$. Therefore, $\widetilde{\varphi}_M|_{{}_jM_i}$ is an Abelian group isomorphism ${}_jM_i \to {}_jM_i'$.

Hence, $\{\widetilde{\varphi}_D|_{K_i}, \widetilde{\varphi}_M|_{{}_jM_i}\}_{i\in\Delta_0^C=\Delta_0,(i\to j)\in\Delta_1^C}$ defines an isomorphism of species from $\mathcal{Q}^C$ to $\mathcal{Q}'^C$.

In the case of K-species, one simply has to replace the terms "ring" with "K-algebra", "ring morphism" with "K-algebra morphism" and "Abelian group homomorphism" with "K-linear map" and the proof is the same. □

If $\mathcal{Q}$ (and $\mathcal{Q}'$) contain oriented cycles, the arguments in the proof of Theorem 4 fail since, in general, it is not true that rad $T(\mathcal{Q}) = \bigoplus_{n=1}^{\infty} T^n(M)$. However, it seems likely that one could modify the proof to avoid using the radical. Hence, we offer the following conjecture.

Conjecture 6.6. *Theorem 4 holds even if $\mathcal{Q}$ and $\mathcal{Q}'$ contain oriented cycles.*

Remark 6. *Theorem 4 serves as a first step in justifying Remark 1 (i.e., that we can always assume, without loss of generality, that we have no parallel arrows in our valued quivers) since a species with parallel arrows can always be crushed to one with only single arrows and its tensor algebra remains the same.*

Theorem 4 shows that there does not exist an equivalence on the level of valued quivers (relative or absolute) such that

$$T(\mathcal{Q}) \cong T(\mathcal{Q}') \iff \Delta \text{ is equivalent to } \Delta'$$

since there are species (respectively K-species), with identical underlying valued quivers, that are not isomorphic as crushed species (respectively K-species) and hence have non-isomorphic tensor rings (respectively algebras) (see Example 4).

In the case of K-species, one may wonder what happens when we tensor $T(\mathcal{Q})$ with the algebraic closure of K. Indeed, maybe we can find an equivalence on the level of absolute valued quivers such that

$$\overline{K} \otimes_K T(\mathcal{Q}) \cong \overline{K} \otimes_K T(\mathcal{Q}') \iff \Gamma \text{ is equivalent to } \Gamma'.$$

The answer, unfortunately, is no. However, this idea does yield an interesting result. In [19], Hubery showed that if K is a finite field, then there is a field extension F/K such that $F \otimes_K T(\mathcal{Q})$ is isomorphic to the path algebra of a quiver. Our strategy of tensoring with the algebraic closure allows us to generalize this result for an arbitrary perfect field.

Theorem 5. *Let K be a perfect field and $\mathcal{Q}$ be a K-species with underlying absolute valued quiver Γ containing no oriented cycles such that K_i is a field for each $i \in \Gamma_0$. Then $\overline{K} \otimes_K T(\mathcal{Q})$ is isomorphic to the path algebra of a quiver.*

Proof. Let $A = \overline{K} \otimes_K T(\mathcal{Q})$. Take any $i \in \Gamma_0$. It is a well-known fact that $\overline{K} \otimes_K K_i = \overline{K}^{d_i}$ (see for example [20, Chapter V, Section 6, Proposition 2] or the proof of [21, Theorem 8.46]).

Let $I = \overline{K} \otimes (\sum_{n=1}^{\infty} T^n(M))$. It is clear that I is a two-sided ideal of A. Moreover, since Γ has no cycles, I is also nilpotent. Considering A/I, we see that

$$A/I \cong \overline{K} \otimes_K (\Pi_{i\in\Gamma_0} K_i) \cong \underbrace{\overline{K} \times \cdots \times \overline{K}}_{\sum_{i\in\Gamma_0} d_i \text{ times}}.$$

So, as in Lemma 3, $\operatorname{rad} A = I$ and by [15, Section I, Proposition 6.2], A is a basic finite-dimensional $\overline{K}$-algebra.

We claim that A is also hereditary. It is well-known that a ring is hereditary if and only if it is of global dimension at most 1. According to [22, Theorem 16], if Λ_1 and Λ_2 are K-algebras such that Λ_1 and Λ_2 are semiprimary (recall that a K-algebra Λ is semiprimary if there is a two-sided nilpotent ideal I such that Λ/I is semisimple) and $(\Lambda_1/\operatorname{rad}\Lambda_1)\otimes_K(\Lambda_2/\operatorname{rad}\Lambda_2)$ is semisimple, then $\operatorname{gl.dim}(\Lambda_1\otimes_K\Lambda_2) = \operatorname{gl.dim}\Lambda_1 + \operatorname{gl.dim}\Lambda_2$. The K-algebras $\overline{K}$ and $T(\mathcal{Q})$ satisfy these conditions. Indeed, $\overline{K}$ is simple and thus semiprimary. We know also that $T(\mathcal{Q})/\operatorname{rad} T(\mathcal{Q})$ is semisimple and $\operatorname{rad} T(\mathcal{Q})$ is nilpotent since Γ has no oriented cycles; thus $T(\mathcal{Q})$ is semiprimary. Moreover,

$$\begin{aligned}(\overline{K}/\operatorname{rad}\overline{K})\otimes_K(T(\mathcal{Q})/\operatorname{rad}T(\mathcal{Q})) &\cong \overline{K}\otimes_K(\Pi_{i\in\Gamma_0}K_i)\\ &\cong \underbrace{\overline{K}\times\cdots\times\overline{K}}_{\sum_{i\in\Gamma_0}d_i\text{ times}},\end{aligned}$$

which is semisimple.

Therefore we have that:

$$\operatorname{gl.dim} A = \operatorname{gl.dim}(\overline{K}\otimes_K T(\mathcal{Q})) = \operatorname{gl.dim}\overline{K} + \operatorname{gl.dim} T(\mathcal{Q}).$$

However, $\operatorname{gl.dim}\overline{K} = 0$ (since all $\overline{K}$-modules are free) and $\operatorname{gl.dim} T(\mathcal{Q}) \leq 1$ (since $T(\mathcal{Q})$ is hereditary). Hence, $\operatorname{gl.dim} A \leq 1$ and so A is hereditary. By Theorem 1, A is isomorphic to the path algebra of a quiver. □

Remark 7. *Hubery goes further in [19], constructing an automorphism σ of the quiver Q whose path algebra is isomorphic to $\overline{K}\otimes_K T(\mathcal{Q})$ such that (Q,σ) folds into Γ. It seems likely that this is possible here as well.*

Conjecture 6.8. *Let K be a perfect field, let $\mathcal{Q}$ be a K-species with underlying absolute valued quiver Γ containing no oriented cycles such that K_i is a field for each $i\in\Gamma_0$ and let Q be a quiver such that $\overline{K}\otimes_K T(\mathcal{Q})\cong\overline{K}Q$ (as in Theorem 5). Then there exists an automorphism σ of Q such that (Q,σ) folds into Γ.*

With Theorem 5, we are able to use the methods of [15, Chapter II, Section 3] to construct the quiver, Q, whose path algebra is isomorphic to $A = \overline{K}\otimes_K T(\mathcal{Q})$. That is, the vertices of Q are in one-to-one correspondence with $\{\varepsilon_1,\ldots,\varepsilon_n\}$, a complete set of primitive orthogonal idempotents of A, and the number of arrows from the vertex corresponding to ε_i to the vertex corresponding to ε_j is given by $\dim_{\overline{K}}(\varepsilon_j\cdot(\operatorname{rad}A/\operatorname{rad}A^2)\cdot\varepsilon_i)$. We illustrate this in the next example.

Example 11. *Let Γ be the following absolute valued quiver.*

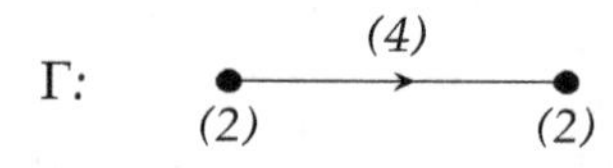

We can construct two $\mathbb{Q}$-species of Γ:

where $\mathbb{Q}(\sqrt{2})^2 = \mathbb{Q}(\sqrt{2})\oplus\mathbb{Q}(\sqrt{2})$. Let $F=\overline{\mathbb{Q}}$, $A = F\otimes_{\mathbb{Q}}T(\mathcal{Q})$ and $B = F\otimes_{\mathbb{Q}}T(\mathcal{Q}')$. We would like to find quivers Q and Q' with $A\cong FQ$ and $B\cong FQ'$.

By direct computation, we see that $\varepsilon_1 = \frac{1}{2}((1 \otimes 1) + (\frac{1}{\sqrt{2}} \otimes \sqrt{2}))$ *and* $\varepsilon_2 = \frac{1}{2}((1 \otimes 1) - (\frac{1}{\sqrt{2}} \otimes \sqrt{2}))$ *form a complete set of primitive orthogonal idempotents of* $F \otimes_{\mathbb{Q}} \mathbb{Q}(\sqrt{2})$*. Thus,* Q *must have 4 vertices. To find the arrows, note that*

$$\begin{aligned}\varepsilon_j \cdot (\operatorname{rad} A / \operatorname{rad} A^2) \cdot \varepsilon_i &= \varepsilon_j \cdot (F \otimes_{\mathbb{Q}} \mathbb{Q}(\sqrt{2})^2) \cdot \varepsilon_i \\ &\cong \varepsilon_j \cdot (F \otimes_{\mathbb{Q}} \mathbb{Q}(\sqrt{2}))^2 \cdot \varepsilon_i \\ &= F(\delta_{ij}\varepsilon_i, 0) \oplus F(0, \delta_{ij}\varepsilon_i)\end{aligned}$$

which has dimension 2 *if* $i = j$ *and* 0 *otherwise. Hence,* A *is isomorphic to the path algebra (over* F*) of*

Q :
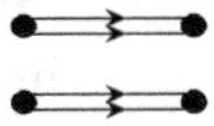

For Q'*, again* $\{\varepsilon_1, \varepsilon_2\}$ *is a complete set of primitive orthogonal idempotents of* $F \otimes_{\mathbb{Q}} \mathbb{Q}(\sqrt{2})$*. Likewise,* $\zeta_1 = \frac{1}{2}((1 \otimes 1) + (\frac{1}{\sqrt{3}} \otimes \sqrt{3}))$ *and* $\zeta_2 = \frac{1}{2}((1 \otimes 1) - (\frac{1}{\sqrt{3}} \otimes \sqrt{3}))$ *form a complete set of primitive orthogonal idempotents of* $F \otimes_{\mathbb{Q}} \mathbb{Q}(\sqrt{3})$*. So* Q' *has 4 vertices. To find the arrows, note that*

$$\zeta_j \cdot (\operatorname{rad} B / \operatorname{rad} B^2) \cdot \varepsilon_i = \zeta_j \cdot (F \otimes_{\mathbb{Q}} \mathbb{Q}(\sqrt{2}, \sqrt{3})) \cdot \varepsilon_i$$

and, since $\zeta_j \cdot \varepsilon_i \neq 0$ *(as elements in* $F \otimes_{\mathbb{Q}} \mathbb{Q}(\sqrt{2}, \sqrt{3})$*), this has dimension* 1 *for all* $i, j \in \{1, 2\}$*. Hence,* B *is isomorphic to the path algebra (over* F*) of*

Q' :
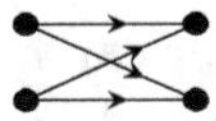

Notice that in Example 11, FQ and FQ' are not isomorphic; this illustrates our earlier point; namely that there is no equivalence on the level of absolute valued quivers such that

$$\overline{K} \otimes_K T(\mathcal{Q}) \cong \overline{K} \otimes_K T(\mathcal{Q}') \iff \Gamma \text{ is equivalent to } \Gamma'$$

Therefore, it seems likely that Theorem 5 is the best that we can hope to achieve.

Note, however, that Theorem 5 fails if all the division rings in our K-modulation are not fields. Consider the following simple example.

Example 12. *View* $\mathbb{H}$*, the quaternions, as an* $\mathbb{R}$*-species (that is,* $\mathbb{H}$ *is an* $\mathbb{R}$*-modulation of the absolute valued quiver with one vertex of value* 4 *and no arrows). Consider the* $\mathbb{C}$*-algebra* $\mathbb{C} \otimes_{\mathbb{R}} \mathbb{H}$*. It is easy to see that* $\mathbb{C} \otimes_{\mathbb{R}} \mathbb{H} \cong M_2(\mathbb{C})$*, the algebra of* 2 *by* 2 *matrices with entries in* $\mathbb{C}$*. This algebra is not basic. Indeed, one can check (by direct computation) that*

$$\left\{ \varepsilon_1 = \begin{pmatrix} 1 & 0 \\ 0 & 0 \end{pmatrix}, \varepsilon_2 = \begin{pmatrix} 0 & 0 \\ 0 & 1 \end{pmatrix} \right\}$$

is a complete set of primitive orthogonal idempotents and that $M_2(\mathbb{C})\varepsilon_1 \cong M_2(\mathbb{C})\varepsilon_2 \cong \mathbb{C}^2$ *as* $M_2(\mathbb{C})$*-modules. Thus,* $\mathbb{C} \otimes_{\mathbb{R}} \mathbb{H}$ *is not isomorphic to the path algebra of a quiver (since all path algebras are basic).*

While we cannot use Theorem 5 for arbitrary K-species, we do have the following.

Corollary 2. *Let* K *be a perfect field and* $\mathcal{Q}$ *be a* K*-species with underlying absolute valued quiver* Γ *containing no oriented cycles. Then* $\overline{K} \otimes_K T(\mathcal{Q})$ *is Morita equivalent to the path algebra of a quiver.*

Proof. It suffices to show that $\overline{K} \otimes_K T(\mathcal{Q})$ is hereditary (and then invoke Part 1 of Corollary 1). In the proof of Theorem 5, all the arguments proving that $\overline{K} \otimes_K T(\mathcal{Q})$ is hereditary go through as before, save for the proof that $(\overline{K} / \operatorname{rad} \overline{K}) \otimes_K (T(\mathcal{Q}) / \operatorname{rad} T(\mathcal{Q})) \cong \overline{K} \otimes_K (\Pi_{i \in \Gamma_0} K_i)$ is semisimple.

To show this in the case that the K_i are not necessarily all fields, pick some $i \in \Gamma_0$ and let Z be the centre of K_i. Then

$$\overline{K} \otimes_K K_i \cong \overline{K} \otimes_K Z \otimes_Z K_i$$

The field Z is a field extension of K and so we may use the same arguments as in the proof of Theorem 5 to show $\overline{K} \otimes_K Z \cong \overline{K} \times \cdots \times \overline{K}$. So

$$\begin{aligned} \overline{K} \otimes_K K_i &\cong (\overline{K} \times \cdots \times \overline{K}) \otimes_Z K_i \\ &\cong (\overline{K} \otimes_Z K_i) \times \cdots \times (\overline{K} \otimes_Z K_i) \end{aligned}$$

One can show that $\overline{K} \otimes_Z K_i \cong M_n(\overline{K})$ for some n, which is a simple ring. Thus, $\overline{K} \otimes_K K_i$ is semisimple, meaning that $\overline{K} \otimes_K (\Pi_{i \in \Gamma_0} K_i)$ is semisimple, completing the proof. □

7. Representations

In this section, we begin by defining representations of quivers and species. We will then see (Proposition 3) that representations of species (resp. quivers) are equivalent to modules of the corresponding tensor ring (resp. path algebra). This fact together with Section 3 (specifically Theorems 1 and 2, and Corollary 1) shows why representations of quivers/species are worth studying; they allow us to understand the representations of any finite-dimensional algebra over a perfect field. We then discuss the root system associated to a valued quiver, which encodes a surprisingly large amount of information about the representation theory of species (see Theorems 6 and 7, and Proposition 4). From Section 1, we know that every valued quiver can be obtained by folding a quiver with automorphism. Thus, we end the section with a discussion on how much of the data of the representation theory of a species is contained in a corresponding quiver with automorphism.

Throughout this section, we make the assumption (unless otherwise specified) that all quivers/species are connected and contain no oriented cycles. Also, whenever there is no need to distinguish between relative or absolute valued quivers, we will simply use the term "valued quiver" and denote it by Ω. We let $\{e_i\}_{i \in \Omega_0}$ be the standard basis of $\mathbb{Z}^{\Omega_0}$ for a valued quiver Ω.

Definition 15 (Representation of a quiver). *A* representation $V = (V_i, f_\rho)_{i \in Q_0, \rho \in Q_1}$ *of a quiver Q over the field K consists of a K-vector space V_i for each $i \in Q_0$ and a K-linear map*

$$f_\rho : V_{t(\rho)} \to V_{h(\rho)}$$

for each $\rho \in Q_1$. If each V_i is finite-dimensional, we call $\underline{\dim}\, V = (\dim_K V_i)_{i \in Q_0} \in \mathbb{N}^{Q_0}$ *the* graded dimension *of V.*

A morphism *of Q representations*

$$\varphi : V = (V_i, f_\rho)_{i \in Q_0, \rho \in Q_1} \to W = (W_i, g_\rho)_{i \in Q_0, \rho \in Q_1}$$

consists of a K-linear map $\varphi_i : V_i \to W_i$ for each $i \in Q_0$ such that $\varphi_{h(\rho)} \circ f_\rho = g_\rho \circ \varphi_{t(\rho)}$ for all $\rho \in Q_1$. That is, the following diagram commutes for all $\rho \in Q_1$.

$$\begin{array}{ccc} V_{t(\rho)} & \xrightarrow{f_\rho} & V_{h(\rho)} \\ {\scriptstyle \varphi_{t(\rho)}} \downarrow & & \downarrow {\scriptstyle \varphi_{h(\rho)}} \\ W_{t(\rho)} & \xrightarrow[g_\rho]{} & W_{h(\rho)} \end{array}$$

We let $\mathfrak{R}_K(Q)$ denote the category of finite-dimensional representations of Q over K.

Definition 16 (Representation of a species). *A representation* $V = (V_i, f_\rho)_{i \in \Delta_0, \rho \in \Delta_1}$ *of a species (or K-species) $\mathcal{Q}$ consists of a K_i-vector space V_i for each $i \in \Delta_0$ and a $K_{h(\rho)}$-linear map*

$$f_\rho : M_\rho \otimes_{K_{t(\rho)}} V_{t(\rho)} \to V_{h(\rho)}$$

for each $\rho \in \Delta_1$. If all V_i are finite-dimensional (over their respective rings), we call $\underline{\dim}\, V = (\dim_{K_i} V_i)_{i \in \Delta_0} \in \mathbb{N}^{\Delta_0}$ *the* graded dimension *of V.*

A morphism *of $\mathcal{Q}$ representations*

$$\varphi : V = (V_i, f_\rho)_{i \in \Delta_0, \rho \in \Delta_1} \to W = (W_i, g_\rho)_{i \in \Delta_0, \rho \in \Delta_1}$$

consists of a K_i-linear map $\varphi_i : V_i \to W_i$ for each $i \in \Delta_0$ such that $\varphi_{h(\rho)} \circ f_\rho = g_\rho \circ (\mathrm{id}_{M_\rho} \otimes \varphi_{t(\rho)})$ for all $\rho \in \Delta_1$. That is, the following diagram commutes for all $\rho \in \Delta_1$.

$$\begin{array}{ccc} M_\rho \otimes_{K_{t(\rho)}} V_{t(\rho)} & \xrightarrow{\ f_\rho\ } & V_{h(\rho)} \\ {\scriptstyle \mathrm{id}_{M_\rho} \otimes \varphi_{t(\rho)}} \downarrow & & \downarrow {\scriptstyle \varphi_{h(\rho)}} \\ M_\rho \otimes_{K_{t(\rho)}} W_{t(\rho)} & \xrightarrow[\ g_\rho\ ]{} & W_{h(\rho)} \end{array}$$

We let $\mathfrak{R}(\mathcal{Q})$ denote the category of finite-dimensional representations of $\mathcal{Q}$. If $\mathcal{Q}$ is a K-species, we use the notation $\mathfrak{R}_K(\mathcal{Q})$.

Note that if $\mathcal{Q}$ is a K-species of a trivially valued absolute valued quiver Γ, then, as before, all $K_i \cong K$ (as K-algebras) and all $M_\rho \cong K$ (as bimodules). Thus, a representation of $\mathcal{Q}$ is a representation of the underlying (non-valued) quiver of Γ. Therefore, by viewing quivers as trivially valued absolute valued quivers, representations of species are a generalization of representations of quivers.

It is well-known that, for a quiver Q, the category $\mathfrak{R}_K(Q)$ is equivalent to KQ-mod, the category of finitely-generated (left) KQ-modules. This fact generalizes nicely for species.

Proposition 3. *Let $\mathcal{Q}$ be a species (possibly with oriented cycles). Then $\mathfrak{R}(\mathcal{Q})$ is equivalent to* $T(\mathcal{Q})$-mod.

Proof. See [17, Proposition 10.1]. While the proof there is given only for K-species, the same arguments hold for species in general. □

Remark 8. *Proposition 3, together with Theorem 4, justifies Remark 1* (i.e., *that we can always assume, without loss of generality, that our valued quivers contain no parallel arrows) since a species with parallel arrows can always be crushed to one with only single arrows and its tensor algebra remains the same. Since $T(\mathcal{Q})$-mod is equivalent to $\mathfrak{R}(\mathcal{Q})$-mod, the representation theory of any species is equivalent to the representation of a species with only single arrows (its crushed species). While allowing parallel arrows in our definition of species is not necessary, there are situations where it may be advantageous as the next example demonstrates.*

Example 13. *Let Δ be the following valued quiver.*

Δ :

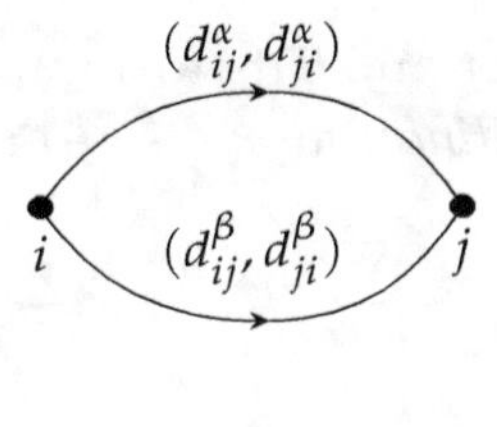

Then

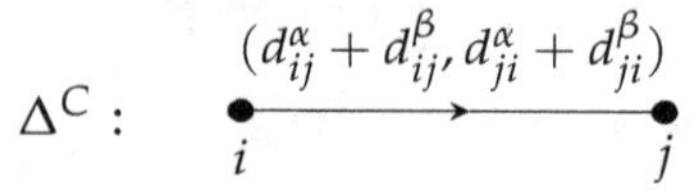

Any modulation of Δ,

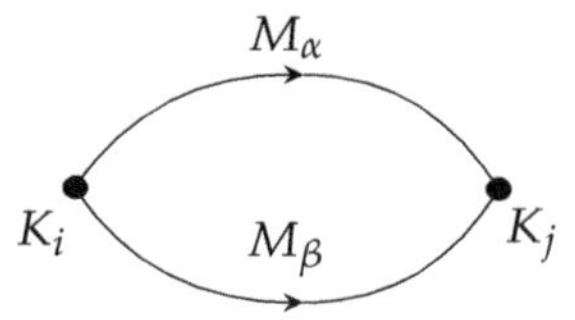

yields a modulation of Δ^C,

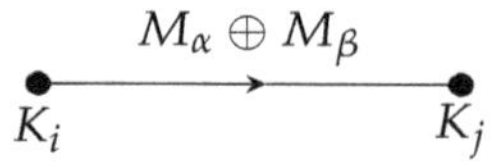

and the representation theory of both of these species is identical. However, the converse is not true. That is, not every modulation of Δ^C *yields a modulation of* Δ. *For example, one can choose a modulation*

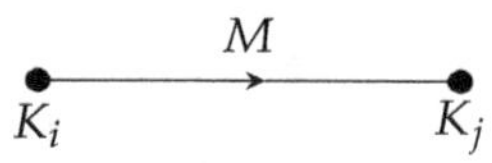

such that M *is indecomposable, and thus cannot be written as* $M = M_1 \oplus M_2$ *(with* $M_1, M_2 \neq 0$*) to yield a modulation of* Δ. *Thus, we can think of modulations of* Δ *as being "special" modulations of* Δ^C *where the bimodule attached to its arrow can be written (nontrivially) as the direct sum of two bimodules.*

Example 13 illustrates why one may wish to allow parallel arrows in the definition of species; they may be used as a way of ensuring that the bimodules in our modulation decompose into a direct sum of proper sub-bimodules.

Definition 17 (Indecomposable representation). *Let* $V = (V_i, f_\rho)_{i\in\Delta_0,\rho\in\Delta_1}$ *and* $W = (W_i, g_\rho)_{i\in\Delta_0,\rho\in\Delta_1}$ *be representations of a species (or a quiver). The* direct sum *of* V *and* W *is*

$$V \oplus W = (V_i \oplus W_i, f_\rho \oplus g_\rho)_{i\in\Delta_0,\rho\in\Delta_1}$$

A representation U *is said to be* indecomposable *if* $U = V \oplus W$ *implies* $V = U$ *or* $W = U$.

Because we restrict ourselves to finite-dimensional representations, the Krull–Schmidt theorem holds. That is, every representation can be written uniquely as a direct sum of indecomposable representations (up to isomorphism and permutation of the components). Thus, the study of all representations of a species (or quiver) reduces to the study of its indecomposable representations.

We say that a species/quiver is of *finite* representation type if it has only finitely many non-isomorphic indecomposable representations. It is of *tame* (or *affine*) representation type if it has infinitely many non-isomorphic indecomposable representations, but they can be divided into finitely many one parameter families. Otherwise, it is of *wild* representation type.

Thus the natural question to ask is: can we classify all species/quivers of finite type, tame type and wild type? The answer, as it turns out, is yes. However, we first need a few additional concepts.

Definition 18 (Euler, symmetric Euler and Tits forms). *The* Euler form *of an absolute valued quiver* Γ *with valuation* $(d_i, m_\rho)_{i\in\Gamma_0,\rho\in\Gamma_1}$ *is the bilinear form* $\langle -, - \rangle : \mathbb{Z}^{\Gamma_0} \times \mathbb{Z}^{\Gamma_0} \to \mathbb{Z}$ *given by:*

$$\langle x, y \rangle = \sum_{i\in\Gamma_0} d_i x_i y_i - \sum_{\rho\in\Gamma_1} m_\rho x_{t(\rho)} y_{h(\rho)}$$

The symmetric Euler form $(-,-) : \mathbb{Z}^{\Gamma_0} \times \mathbb{Z}^{\Gamma_0} \to \mathbb{Z}$ *is given by:*

$$(x,y) = \langle x,y\rangle + \langle y,x\rangle$$

The Tits form $q : \mathbb{Z}^{\Gamma_0} \to \mathbb{Z}$ *is given by:*

$$q(x) = \langle x,x\rangle$$

Remark 9. *If we take* Γ *to be trivially valued* (i.e., *all* $d_i = m_\rho = 1$), *we recover the usual definitions of these forms for quivers (see, for example [23, Definitions 3.6.7, 3.6.8, 3.6.9]).*

Remark 10. *Notice that the symmetric Euler form and the Tits form do not depend on the orientation of our quiver.*

Remark 11. *Given a relative valued quiver* Δ, *we have seen in Lemma 1 that we can choose an absolute valued quiver* Γ *such that* $\mathbf{F}(\Gamma) = \Delta$ *(this is equivalent to making a choice of positive integers* f_i *in Definition 3) with* $d^\rho_{ij} = m_\rho/d_j$ *and* $d^\rho_{ji} = m_\rho/d_i$ *for all* $\rho : i \to j$ *in* Δ_1. *It is easy to see that (as long as the quiver is connected) for any other absolute valued quiver* Γ' *with* $\mathbf{F}(\Gamma') = \Delta$, *there is a* $\lambda \in \mathbb{Q}^+$ *such that* $d'_i = \lambda d_i$ *for all* $i \in \Delta_0$. *Thus, we define the Euler, symmetric Euler and Tits forms on* Δ *to be the corresponding forms on* Γ, *which are well-defined up to positive rational multiple.*

Definition 19 (Generalized Cartan matrix). *Let* $\mathcal{I}$ *be an indexing set.* A generalized Cartan matrix $C = (c_{ij})$, $i,j \in \mathcal{I}$, *is an integer matrix satisfying:*

- $c_{ii} = 2$, *for all* $i \in \mathcal{I}$;
- $c_{ij} \le 0$, *for all* $i \ne j \in \mathcal{I}$;
- $c_{ij} = 0 \iff c_{ji} = 0$, *for all* $i,j \in \mathcal{I}$.

A generalized Cartan matrix C is symmetrizable *if there exists a diagonal matrix D (called the* symmetrizer*) such that DC is symmetric.*

Note that, for any valued quiver Ω, $c_{ij} = 2\dfrac{(e_i,e_j)}{(e_i,e_i)}$ defines a generalized Cartan matrix, since $(e_i,e_i) = 2d_i$ and $(e_i,e_j) = -\sum_\rho m_\rho$ for $i \ne j$, where the sum is taken over all arrows between i and j (regardless of orientation). So,

$$c_{ij} = \begin{cases} 2, & \text{if } i = j, \\ -\sum_\rho m_\rho/d_i = -\sum_\rho d^\rho_{ij}, & \text{if } i \ne j. \end{cases}$$

From this we see that two valued quivers Ω and Ω' have the same generalized Cartan matrix (up to ordering of the rows and columns) if and only if $\Omega^C \cong \Omega'^C$ as relative valued quivers (by this we mean that if Ω and Ω' are relative valued quivers, then $\Omega^C \cong \Omega'^C$ and if they are absolute valued quivers, then $\mathbf{F}(\Omega)^C \cong \mathbf{F}(\Omega')^C$). If all d_i are equal (or alternatively, $\sum_\rho d^\rho_{ij} = \sum_\rho d^\rho_{ji}$ for all adjacent i and j), then the matrix is symmetric, otherwise it is symmetrizable with symmetrizer $D = \operatorname{diag}(d_i)_{i\in\Omega_0}$. Moreover, every symmetrizable Cartan matrix can be obtained in this way. This is one of the motivations for working with species. When working with species we can obtain non-symmetric Cartan matrices, but when restricted to quivers, only symmetric Cartan matrices arise. For every generalized Cartan matrix, we have its associated Kac–Moody Lie algebra.

Definition 20 (Kac–Moody Lie algebra). *Let* $C = (c_{ij})$ *be an* $n \times n$ *generalized Cartan matrix. Then the* Kac–Moody Lie algebra *of C is the complex Lie algebra generated by* e_i, f_i, h_i *for* $1 \le i \le n$, *subject to the following relations.*

- $[h_i,h_j] = 0$ *for all* i,j,

- $[h_i, e_j] = c_{ij}e_j$ *and* $[h_i, f_j] = -c_{ij}f_j$ *for all* i, j,
- $[e_i, f_i] = h_i$ *for each* i *and* $[e_i, f_j] = 0$ *for all* $i \neq j$,
- $(\mathrm{ad}\, e_i)^{1-c_{ij}}(e_j) = 0$ *and* $(\mathrm{ad}\, f_i)^{1-c_{ij}}(f_j)$ *for all* $i \neq j$.

Therefore, to every valued quiver, we can associate a generalized Cartan matrix and its corresponding Kac–Moody Lie algebra. It is only fitting then, that we discuss root systems.

Definition 21 (Root system of a valued quiver). *Let Ω be a valued quiver.*

- *For each $i \in \Omega_0$, define the* simple reflection *through i to be the linear transformation $r_i : \mathbb{Z}^{\Omega_0} \to \mathbb{Z}^{\Omega_0}$ given by:*
$$r_i(x) = x - 2\frac{(x, e_i)}{(e_i, e_i)}e_i$$
- *The* Weyl group, *which we denote by $\mathcal{W}$, is the subgroup of $\mathrm{Aut}(\mathbb{Z}^{\Omega_0})$ generated by the simple reflections r_i, $i \in \Omega_0$.*
- *An element $x \in \mathbb{Z}^{\Omega_0}$ is called a* real root *if $\exists\, w \in \mathcal{W}$ such that $x = w(e_i)$ for some $i \in \Omega_0$.*
- *The* support *of an element $x \in \mathbb{Z}^{\Omega_0}$ is defined as $\mathrm{supp}(x) = \{i \in \Omega_0 \mid x_i \neq 0\}$ and we say $\mathrm{supp}(x)$ is connected if the full subquiver of Ω with vertex set $\mathrm{supp}(x)$ is connected. Then the* fundamental set *is defined as $\mathcal{F} = \{0 \neq x \in \mathbb{N}^{\Omega_0} \mid (x, e_i) \leq 0$ for all $i \in \Omega_0$ and $\mathrm{supp}(x)$ is connected$\}$.*
- *An element $x \in \mathbb{Z}^{\Omega_0}$ is called an* imaginary root *if $x \in \bigcup_{w \in \mathcal{W}} w(\mathcal{F}) \cup w(-\mathcal{F})$.*
- *The* root system *of Ω, denoted $\Phi(\Omega)$ is the set of all real and imaginary roots.*
- *We call a root x* positive *(resp.* negative*) if $x_i \geq 0$ (resp. $x_i \leq 0$) $\forall\, i \in \Omega_0$. We write $\Phi^+(\Omega)$ for the set of positive roots and $\Phi^-(\Omega)$ for the set of negative roots.*

Definition 22 (Stable element). *An element $x \in \mathbb{Z}^{\Omega_0}$ is called* stable *if $w(x) = x$ for all $w \in \mathcal{W}$.*

Remark 12. *It is worth noting that, while a stable element need not be an imaginary root (see [24, Example 6.15]), it is always the sum of imaginary roots (see [24, Lemma 6.16]).*

Definition 23 (Discrete and continuous dimension types). *An indecomposable representation V of a species (or quiver) is of* discrete dimension type *if it is the unique indecomposable representation (up to isomorphism) with graded dimension $\underline{\dim}\, V$. Otherwise, it is of* continuous dimension type.

With all these concepts in mind, we can neatly classify all species of finite and tame representation type. Note that in the case of quivers, this was originally done by Gabriel (see [1]). It was later generalized to species by Dlab and Ringel.

Theorem 6. [3, Main Theorem] *Let $\mathcal{Q}$ be species of a connected relative valued quiver Δ. Then:*

a. *$\mathcal{Q}$ is of finite representation type if and only if the underlying undirected valued graph of Δ is a Dynkin diagram of finite type (see [3] for a list of the Dynkin diagrams). Moreover, $\underline{\dim} : \mathfrak{R}(\mathcal{Q}) \to \mathbb{Z}^{\Delta_0}$ induces a bijection between the isomorphism classes of the indecomposable representations of $\mathcal{Q}$ and the positive real roots of its root system.*
b. *If the underlying undirected valued graph of Δ is an extended Dynkin diagram (see [3] for a list of the extended Dynkin diagrams), then $\underline{\dim} : \mathfrak{R}(\mathcal{Q}) \to \mathbb{Z}^{\Delta_0}$ induces a bijection between the isomorphism classes of the indecomposable representations of $\mathcal{Q}$ of discrete dimension type and the positive real roots of its root system. Moreover, there exists a unique stable element (up to rational multiple) $n \in \Phi(\mathcal{Q})$ and the indecomposable representations of continuous dimension type are those whose graded dimension is a positive multiple of n. If $\mathcal{Q}$ is a K-species, then $\mathcal{Q}$ is of tame representation type if and only if the underlying undirected valued graph of Δ is an extended Dynkin diagram.*

Remark 13. *See [3, p. 57] (and [25]) for a proof that a K-species $\mathcal{Q}$ is tame if and only if Δ is an extended diagram.*

Remark 14. *In the case that the underlying undirected valued graph of Δ is an extended Dynkin diagram, the indecomposable representations of continuous dimension type of $\mathcal{Q}$ can be derived from the indecomposable representations of continuous dimension type of a suitable species with underlying undirected valued graph $\widetilde{\mathrm{A}}_{11}$ or $\widetilde{\mathrm{A}}_{12}$ (see [3, Theorem 5.1]).*

Theorem 6 shows a remarkable connection between the representation theory of species and the theory of root systems of Lie algebras. In the case of quivers, Kac was able to show that this connection is stronger still.

Theorem 7. [26, Theorems 2 and 3] *Let Q be a quiver with no loops (though possibly with oriented cycles) and K an algebraically closed field. Then there is an indecomposable representation of Q of graded dimension α if and only if $\alpha \in \Phi^+(Q)$. Moreover, if α is a real positive root, then there is a unique indecomposable representation of Q (up to isomorphism) of graded dimension α. If α is an imaginary positive root, then there are infinitely many non-isomorphic indecomposable representations of Q of graded dimension α.*

It is not known whether Kac's theorem generalizes fully for species, however, it does for certain classes of species. Indeed, in the case of a species of finite or tame representation type, one can apply Theorem 6. In the case of K-species when K is a finite field, we have the following result by Deng and Xiao.

Proposition 4. [27, Proposition 3.3] *Let $\mathcal{Q}$ be a K-species (K a finite field) containing no oriented cycles. Then there exists an indecomposable representation of $\mathcal{Q}$ of graded dimension α if and only if $\alpha \in \Phi^+(\mathcal{Q})$. Moreover, if α is a real positive root, then there is a unique indecomposable representation of $\mathcal{Q}$ (up to isomorphism) of graded dimension α.*

Based on these results, we see that much of the information about the representation theory of a species is encoded in its underlying valued quiver/graph. Recall from Section 1 that any valued quiver can be obtained by folding a quiver with automorphism. So, one may ask: how much information is encoded in this quiver with automorphism?

We continue our assumption that Q contains no oriented cycles; however for what follows this is more restrictive than we need. It would be enough to assume that Q contains no loops and that no arrow connects two vertices in the same σ-orbit (see [24, Lemma 6.24] for a proof that this is indeed a weaker condition).

Suppose (Q, σ) is a quiver with automorphism and let $V = (V_i, f_\rho)_{i \in Q_0, \rho \in Q_1}$ be a representation of Q. Define a new representation $V^\sigma = (V_i^\sigma, f_i^\sigma)_{i \in Q_0, \rho \in Q_1}$ by $V_i^\sigma = V_{\sigma^{-1}(i)}$ and $f_\rho^\sigma = f_{\sigma^{-1}(\rho)}$.

Definition 24 (Isomorphically invariant representation). *Let (Q, σ) be a quiver with automorphism. A representation $V = (V_i, f_\rho)_{i \in Q_0, \rho \in Q_1}$ is called* isomorphically invariant *(or simply* invariant*) if $V^\sigma \cong V$ as representations of Q.*

We say an invariant representation V is invariant-indecomposable *if $V = W_1 \oplus W_2$ such that W_1 and W_2 are invariant representations implies $W_1 = V$ or $W_2 = V$.*

It is not hard to see that the invariant-indecomposable representations are precisely those of the form

$$V = W \oplus W^\sigma \oplus \cdots \oplus W^{\sigma^{r-1}}$$

where W is an indecomposable representation and r is the least positive integer such that $W^{\sigma^r} \cong W$.

Let $(\mathbb{Z}^{Q_0})^\sigma = \{\alpha \in \mathbb{Z}^{Q_0} \mid \alpha_i = \alpha_j$ for all i and j in the same orbit$\}$. Suppose (Q,σ) folds into Ω and write $\bar{i} \in \Omega_0$ for the orbit of $i \in Q_0$. We then have a well-defined function

$$\mathbf{f} : (\mathbb{Z}^{Q_0})^\sigma \to \mathbb{Z}^{\Omega_0}$$

defined by $\mathbf{f}(\alpha)_{\bar{i}} = \alpha_i$ for any $i \in Q_0$. Notice that if V is an invariant representation of Q, then $\dim V_i = \dim V_{\sigma^{-1}(i)}$ for all $i \in Q_0$. As such, $\dim V_i = \dim V_j$ for all i and j in the same orbit. Thus, $\underline{\dim}\, V \in (\mathbb{Z}^{Q_0})^\sigma$. We have the following result due to Hubery.

Theorem 8. [19, Theorem 1] *Let (Q,σ) be a quiver with automorphism, Ω a valued quiver such that (Q,σ) folds into Ω, and K an algebraically closed field of characteristic not dividing the order of σ.*

a. The images under $\mathbf{f}$ of the graded dimensions of the invariant-indecomposable representations of Q are the positive roots of $\Phi(\Omega)$.
b. If $\mathbf{f}(\alpha)$ is a real positive root, then there is a unique invariant-indecomposable representation of Q with graded dimension α (up to isomorphism).

Theorem 8 tells us that if the indecomposables of $\mathcal{Q}$ are determined by the positive roots of $\Phi(\Omega)$ (such as in the case of species of Dynkin or extended Dynkin type or K-species over finite fields), then finding all the indecomposables of $\mathcal{Q}$ reduces to finding the indecomposables of Q, which, in general, is an easier task.

One may wonder if there is a subcategory of $\mathfrak{R}_K(Q)$, say $\mathfrak{R}_K^\sigma(Q)$, whose objects are the invariant representations of Q, that is equivalent to $\mathfrak{R}(\mathcal{Q})$. One needs to determine what the morphisms of this category should be. The most obvious choice is to let $\mathfrak{R}_K^\sigma(Q)$ be the full subcategory of $\mathfrak{R}_K(Q)$ whose objects are the invariant representations. This, however, does not work. The category $\mathfrak{R}(\mathcal{Q})$ is an Abelian category (this follows from Proposition 3), but $\mathfrak{R}_K^\sigma(Q)$, as we have defined it, is not. As the following example demonstrates, this category does not, in general, have kernels.

Example 14. *Let (Q,σ) be the following quiver with automorphism (where the dotted arrows represent the action of σ).*

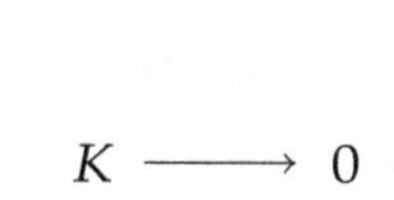

Let V be the following invariant-indecomposable representation of Q.

$$K \longrightarrow 0 \longleftarrow K$$

Let $\varphi : V \to V$ be the morphism defined by

$$\begin{array}{ccccccc} V & & K & \longrightarrow & 0 & \longleftarrow & K \\ \varphi\big\downarrow & & 0\big\downarrow & & 0\big\downarrow & & 1\big\downarrow \\ V & & K & \longrightarrow & 0 & \longleftarrow & K \end{array}$$

Then φ is a morphism of representations since each of the squares in the diagram commutes. However, by a straightforward exercise in category theory, one can show that φ does not have a kernel (in the category $\mathfrak{R}_K^\sigma(Q)$).

Therefore, if we define $\mathfrak{R}_K^\sigma(Q)$ as a full subcategory of $\mathfrak{R}_K(Q)$, it is not equivalent to $\mathfrak{R}(\mathcal{Q})$. It is possible that one could cleverly define the morphisms of $\mathfrak{R}_K^\sigma(Q)$ to avoid this problem, however there are other obstacles to overcome. If $\mathfrak{R}_K^\sigma(Q)$ and $\mathfrak{R}(\mathcal{Q})$ were equivalent, then there should be a bijective correspondence between the (isomorphism classes of) indecomposables in each category. Using the idea of folding, an invariant representation of Q with graded dimension α should be mapped to a representation of $\mathcal{Q}$ with graded dimension $\mathbf{f}(\alpha)$. The following example illustrates the problem with this idea. Note that this example is similar to the example following Proposition 15 in [19], however we approach it in a different fashion.

Example 15. *Let (Q,σ) be the following quiver with automorphism (again, the dotted arrows represent the action of σ).*

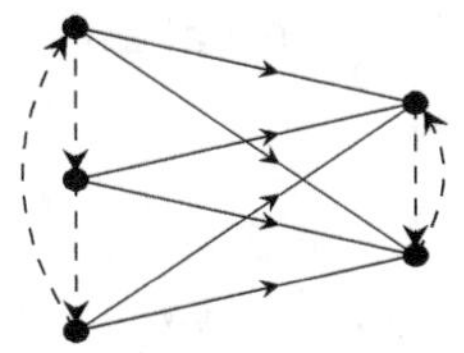

Then (Q,σ) folds into the following absolute valued quiver.

Γ:

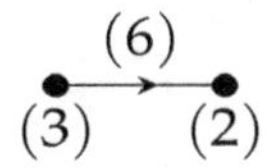

One can easily check that $\beta = (1,1)$ is an imaginary root of $\Phi(\Gamma)$. The only $\alpha \in (\mathbb{Z}^{Q_0})^\sigma$ such that $\mathbf{f}(\alpha) = \beta$ is $\alpha = (1,1,\ldots,1)$. One can show using basic linear algebra that, while there are several non-isomorphic indecomposable representations of Q with graded dimension α (after all, α is an imaginary root of $\Phi(Q)$), all such invariant representations are isomorphc to

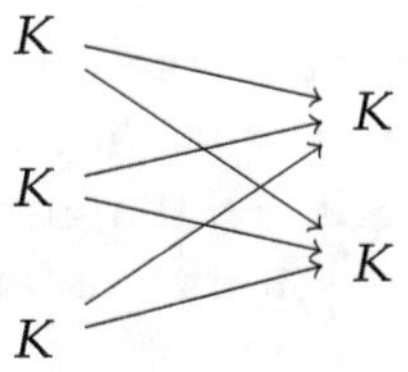

where every arrow represents the identity map id_K. Thus, we have a single isomorphism class of invariant-indecomposables with graded dimension α.

Now, construct a species of Γ. Let $\gamma = 2^{1/6}$ and let $\mathcal{Q}$ be the $\mathbb{Q}$-species given by $\mathbb{Q}(\gamma^2) \xrightarrow{\mathbb{Q}(\gamma)} \mathbb{Q}(\gamma^3)$. Thus the underlying valued quiver of $\mathcal{Q}$ is Γ. There exists an indecomposable representation of $\mathcal{Q}$ with graded dimension β—in fact, there exists more than one.

Let V_1 be the representation $\mathbb{Q}(\gamma^2) \xrightarrow{f_1} \mathbb{Q}(\gamma^3)$ where $f_1 : \mathbb{Q}(\gamma) \otimes_{\mathbb{Q}(\gamma^2)} \mathbb{Q}(\gamma^2) \cong \mathbb{Q}(\gamma) \to \mathbb{Q}(\gamma^3)$ is the $\mathbb{Q}(\gamma^3)$-linear map defined by $1 \mapsto 1$, $\gamma \mapsto 0$ and $\gamma^2 \mapsto 0$.

Let V_2 be the representation $\mathbb{Q}(\gamma^2) \xrightarrow{f_2} \mathbb{Q}(\gamma^3)$ where $f_2 : \mathbb{Q}(\gamma) \otimes_{\mathbb{Q}(\gamma^2)} \mathbb{Q}(\gamma^2) \cong \mathbb{Q}(\gamma) \to \mathbb{Q}(\gamma^3)$ is the $\mathbb{Q}(\gamma^3)$-linear map defined by $1 \mapsto 1$, $\gamma \mapsto 1$ and $\gamma^2 \mapsto 0$.

It is clear that V_1 and V_2 are indecomposable and $\underline{\dim}\, V_1 = \underline{\dim}\, V_2 = \beta$. One can also show that they are not isomorphic as representations of $\mathcal{Q}$. Hence, there are at least two isomorphism classes of indecomposable representations of $\mathcal{Q}$ with graded dimension β.

Therefore, any functor $\mathfrak{R}_K^\sigma(Q) \to \mathfrak{R}(\mathcal{Q})$ mapping invariant representations with graded dimension α to representations with graded dimension $\mathbf{f}(\alpha)$ cannot be essentially surjective, and thus cannot be an equivalence of categories.

While the above example is not enough to conclude that the categories $\mathfrak{R}_K^\sigma(Q)$ and $\mathfrak{R}(\mathcal{Q})$ are not equivalent, it is enough to deduce that one cannot obtain an equivalence via folding.

8. Ringel–Hall Algebras

In this section we define the Ringel–Hall algebra of a species (or quiver). We will construct the generic composition algebra of a species, which is obtained from a subalgebra of the Ringel–Hall algebra, and see that it is isomorphic to the positive part of the quantized enveloping algebra of the corresponding Kac–Moody Lie algebra (see Theorem 9). We then give a similar interpretation of the whole Ringel–Hall algebra (see Theorem 10). For further details, see the expository paper by Schiffmann, [4].

We continue our assumption that all quivers/species have no oriented cycles. Also, we have seen in the last section (Proposition 3) that $\mathfrak{R}(\mathcal{Q})$ is equivalent to $T(\mathcal{Q})$-mod, and so we will simply identify representations of $\mathcal{Q}$ with modules of $T(\mathcal{Q})$.

Definition 25 (Ringel–Hall algebra). *Let $\mathcal{Q}$ be an $\mathbb{F}_q$-species. Let $v = q^{1/2}$ and let $\mathcal{A}$ be an integral domain containing $\mathbb{Z}$ and v, v^{-1}. The* Ringel–Hall algebra, *which we will denote $\mathcal{H}(\mathcal{Q})$, is the free $\mathcal{A}$-module with basis the set of all isomorphism classes of finite-dimensional representations of $\mathcal{Q}$. Multiplication is given by*

$$[A][B] = v^{\langle \underline{\dim}\, A, \underline{\dim}\, B\rangle} \sum_{[C]} g^C_{AB}[C]$$

where g^C_{AB} is the number of subrepresentations (submodules) X of C such that $C/X \cong A$ and $X \cong B$ (as representations/modules) and $\langle -, - \rangle$ is the Euler form (see Definition 18).

Remark 15. *It is well-known (see, for example, [28, Lemma 2.2]) that*

$$\langle \underline{\dim}\, A, \underline{\dim}\, B\rangle = \dim_{\mathbb{F}_q} \mathrm{Hom}_{T(\mathcal{Q})}(A, B) - \dim_{\mathbb{F}_q} \mathrm{Ext}^1{}_{T(\mathcal{Q})}(A, B)$$

In many texts (for example [29] or [8]) this is the way the form $\langle -, - \rangle$ is defined. Also, there does not appear to be a single agreed-upon name for this algebra; depending on the text, it may be called the twisted Hall algebra, *the* Ringel algebra, *the* twisted Ringel–Hall algebra, etc. *Regardless of the name one prefers, it is important not to confuse this algebra with the (untwisted) Hall algebra whose multiplication is given by $[A][B] = \sum_{[C]} g^C_{AB}[C]$.*

Definition 26 (Composition algebra). *Let $\mathcal{Q}$ be an $\mathbb{F}_q$-species with underlying absolute valued quiver Γ and $\mathbb{F}_q$-modulation $(K_i, M_\rho)_{i\in\Gamma_0,\rho\in\Gamma_1}$. The* composition algebra, *$C = C(\mathcal{Q})$, of $\mathcal{Q}$ is the $\mathcal{A}$-subalgebra of $\mathcal{H}(\mathcal{Q})$ generated by the isomorphism classes of the simple representations of $\mathcal{Q}$. Since we assume Γ has no oriented cycles, this means C is generated by the $[S_i]$ for $i \in \Gamma_0$ where $S_i = ((S_i)_j, (S_i)_\rho)_{j\in\Gamma_0,\rho\in\Gamma_1}$ is given by*

$$(S_i)_j = \begin{cases} K_i, & \text{if } i = j, \\ 0, & \text{if } i \neq j, \end{cases} \qquad \text{and} \qquad (S_i)_\rho = 0 \quad \text{for all } \rho \in \Gamma_1$$

Let $\mathcal{S}$ be a set of finite fields K such that $\{|K| \mid K \in \mathcal{S}\}$ is infinite. Let $v_K = |K|^{1/2}$ for each $K \in \mathcal{S}$. Write C_K for the composition algebra of $\mathcal{Q}$ for each finite field K in $\mathcal{S}$ and $[S_i^{(K)}]$ for the corresponding generators. Let C be the subring of $\Pi_{K\in\mathcal{S}} C_K$ generated by $\mathbb{Q}$ and the elements

$$t = (t_K)_{K\in\mathcal{S}},\ t_K = v_K,$$

$$t^{-1} = (t_K^{-1})_{K\in\mathcal{S}},\ t_K^{-1} = v_K^{-1},$$

$$u_i = (u_i^{(K)})_{K\in\mathcal{S}},\ u_i^{(K)} = [S_i^{(K)}].$$

So t lies in the centre of C and, because there are infinitely many v_K, t does not satisfy $p(t) = 0$ for any nonzero polynomial $p(T)$ in $\mathbb{Q}[T]$. Thus, we may view C as the $\mathcal{A}$-algebra generated by the u_i, where $\mathcal{A} = \mathbb{Q}[t, t^{-1}]$ with t viewed as an indeterminate.

Definition 27 (Generic composition algebra). *Using the notation above, the $\mathbb{Q}(t)$-algebra $C^* = \mathbb{Q}(t) \otimes_{\mathcal{A}} C$ is called the* generic composition algebra *of $\mathcal{Q}$. We use the notation $u_i^* = 1 \otimes u_i$.*

Let $\mathcal{Q}$ be an $\mathbb{F}_q$-species with underlying absolute valued quiver Γ. Let (c_{ij}) be the generalized Cartan matrix associated to Γ and let $\mathfrak{g}$ be its associated Kac–Moody Lie algebra (recall Definitions 19 and 20). Let $U_t(\mathfrak{g})$ be the quantized enveloping algebra of $\mathfrak{g}$ and let $U_t(\mathfrak{g}) = U_t^+(\mathfrak{g}) \otimes U_t^0(\mathfrak{g}) \otimes U_t^-(\mathfrak{g})$ be its triangular decomposition (see [2, Chapter 3]). We call $U_t^+(\mathfrak{g})$ the *positive part* of $U_t(\mathfrak{g})$; it is the $\mathbb{Q}(t)$-algebra generated by elements E_i, $i \in \Gamma_0$, modulo the quantum Serre relations

$$\sum_{p=0}^{1-c_{ij}} (-1)^p \begin{bmatrix} 1-c_{ij} \\ p \end{bmatrix} E_i^p E_j E_i^{1-c_{ij}-p} \qquad \text{for all } i \neq j,$$

where

$$\begin{bmatrix} m \\ p \end{bmatrix} = \frac{[m]!}{[p]![m-p]!},$$

$$[n] = \frac{t^n - t^{-n}}{t - t^{-1}}, \qquad [n]! = [1][2]\cdots[n].$$

In [29], Green was able to show that C^* and $U_t^+(\mathfrak{g})$ are canonically isomorphic. Of course, in his paper, Green speaks of modules of hereditary algebras over a finite field K rather than representations of K-species, but as we have seen, these two notions are equivalent.

Theorem 9. *[29, Theorem 3] Let $\mathcal{Q}$ be an $\mathbb{F}_q$-species with underlying absolute valued quiver Γ and let $\mathfrak{g}$ be its associated Kac–Moody Lie algebra. Then, there exists a $\mathbb{Q}(t)$-algebra isomorphism $U_t^+(\mathfrak{g}) \to C^*$ which takes $E_i \mapsto u_i^*$ for all $i \in \Gamma_0$.*

Remark 16. *This result by Green is actually a generalization of an earlier result by Ringel in [30, p. 400] and [31, Theorem 7] who proved Theorem 7.5 in the case that $\mathcal{Q}$ is of finite representation type.*

Theorem 9 gives us an interpretation of the composition algebra in terms of the quantized enveloping algebra of the corresponding Kac–Moody Lie algebra. Later, Sevenhant and Van Den Bergh were able to give a similar interpretation of the whole Ringel–Hall algebra. For this, however, we need the concept of a *generalized Kac–Moody Lie algebra*, which was first defined by Borcherds in [32]. Though some authors have used slightly modified definitions of generalized Kac–Moody Lie algebras over the years, we use here Borcherds' original definition (in accordance with Sevenhant and Van Den Bergh in [8]).

Definition 28 (Generalized Kac–Moody Lie algebra). *Let H be a real vector space with symmetric bilinear product $(-,-) : H \times H \to \mathbb{R}$. Let $\mathcal{I}$ be a countable (but possibly infinite) set and $\{h_i\}_{i\in\mathcal{I}}$ be a subset of H such that $(h_i, h_j) \leq 0$ for all $i \neq j$ and $c_{ij} = 2(h_i, h_j)/(h_i, h_i)$ is an integer if $(h_i, h_i) > 0$. Then, the* generalized Kac–Moody Lie algebra *associated to H, $\{h_i\}_{i\in\mathcal{I}}$ and $(-,-)$ is the Lie algebra (over a field of characteristic 0 containing an isomorphic copy of $\mathbb{R}$) generated by H and elements e_i and f_i for $i \in \mathcal{I}$ whose product is defined by:*

- *$[h, h'] = 0$ for all h and h' in H,*
- *$[h, e_i] = (h, h_i)e_i$ and $[h, f_i] = -(h, h_i)f_i$ for all $h \in H$ and $i \in \mathcal{I}$,*
- *$[e_i, f_i] = h_i$ for each i and $[e_i, f_j] = 0$ for all $i \neq j$,*
- *if $(h_i, h_i) > 0$, then $(\operatorname{ad} e_i)^{1-c_{ij}}(e_j) = 0$ and $(\operatorname{ad} f_i)^{1-c_{ij}}(f_j) = 0$ for all $i \neq j$,*
- *if $(h_i, h_i) = 0$, then $[e_i, e_j] = [f_i, f_j] = 0$.*

Remark 17. *Generalized Kac–Moody Lie algebras are similar to Kac–Moody Lie algebras. The main difference is that generalized Kac–Moody Lie algebras (may) contain simple imaginary roots (corresponding to the h_i with $(h_i, h_i) \leq 0$).*

Let $\mathfrak{g}$ be a generalized Kac–Moody Lie algebra (with the notation of Definition 28) and let $v \neq 0$ be an element of the base field such that v is not a root of unity. Write $d_i = (h_i, h_i)/2$ for the $i \in \mathcal{I}$ such that $(h_i, h_i) > 0$. The *quantized enveloping algebra* $U_v(\mathfrak{g})$ can be defined in the same way as the quantized enveloping algebra of a Kac–Moody Lie algebra (see Section 2 of [8]). The positive part of $U_v^+(\mathfrak{g})$ is the $\mathcal{A}$-algebra generated by elements E_i, $i \in \mathcal{I}$, modulo the quantum Serre relations

$$\sum_{p=0}^{1-c_{ij}} (-1)^p \begin{bmatrix} 1-c_{ij} \\ p \end{bmatrix}_{d_i} E_i^p E_j E_i^{1-c_{ij}-p} \qquad \text{for all } i \neq j, \text{ with } (h_i, h_i) > 0$$

and

$$E_i E_j - E_j E_i \qquad \text{if } (h_i, h_j) = 0,$$

where

$$\begin{bmatrix} m \\ p \end{bmatrix}_{d_i} = \frac{[m]_{d_i}!}{[p]_{d_i}![m-p]_{d_i}!},$$

$$[n]_{d_i} = \frac{(v^{d_i})^n - (v^{d_i})^{-n}}{v^{d_i} - v^{-d_i}}, \qquad [n]_{d_i}! = [1]_{d_i}[2]_{d_i} \cdots [n]_{d_i}.$$

Let n be a positive integer and let $\{e_i\}_{i=1}^n$ be the standard basis of $\mathbb{Z}^n$. Let $v \in \mathbb{R}$ such that $v > 1$ and $\mathcal{A}$ be as in Definition 25. Suppose we have the following:

a. An $\mathbb{N}^n$-graded $\mathcal{A}$-algebra A such that:
   (a) $A_0 = \mathcal{A}$,
   (b) $\dim_{\mathcal{A}} A_\alpha < \infty$ for all $\alpha \in \mathbb{N}^n$,
   (c) $A_{e_i} \neq 0$ for all $1 \leq i \leq n$.
b. A symmetric positive definite bilinear form $[-,-] : A \times A \to \mathcal{A}$ such that $[A_\alpha, A_\beta] = 0$ if $\alpha \neq \beta$ and $[1,1] = 1$ (here we assume $[a,a] \in \mathbb{R}$ for all $a \in A$).
c. A symmetric bilinear form $(-,-) : \mathbb{R}^n \times \mathbb{R}^n \to \mathbb{R}$ such that $(e_i, e_i) > 0$ for all $1 \leq i \leq n$ and $c_{ij} = 2(e_i, e_j)/(e_i, e_i)$ is a generalized Cartan matrix as in Definition 6.9.
d. The tensor product $A \otimes_{\mathcal{A}} A$ can be made into an algebra via the rule

$$(a \otimes b)(c \otimes d) = v^{(\deg(b), \deg(c))}(ac \otimes bd)$$

   for homogeneous a, b, c, d. (here $\deg(x) = \alpha$ if $x \in A_\alpha$). We assume that there is an $\mathcal{A}$-algebra homomorphism $\delta : A \to A \otimes_{\mathcal{A}} A$ which is adjoint under $[-,-]$ to the multiplication (that is, $[\delta(a), b \otimes c]_{A \otimes A} = [a, bc]_A$ where $[a \otimes b, c \otimes d]_{A \otimes A} = [a,c]_A [b,d]_A$).

Proposition 5. [8, Proposition 3.2] *Under the above conditions, A is isomorphic (as an algebra) to the positive part of the quantized enveloping algebra of a generalized Kac–Moody Lie algebra.*

The Ringel–Hall algebra, $\mathcal{H}(\mathcal{Q})$, of a species $\mathcal{Q}$ is $\mathbb{N}^n$-graded by associating to each representation its graded dimension, hence $\mathcal{H}(\mathcal{Q})$ satisfies Condition 1 above. Moreover, the symmetric Euler form satisfies Condition 3 (if we extend it to all $\mathbb{R}^{\Gamma_0}$). Following Green in [29], we define

$$\delta([A]) = \sum_{[B],[C]} v^{\langle \underline{\dim} B, \underline{\dim} C \rangle} g_{BC}^A \frac{|\operatorname{Aut}(B)||\operatorname{Aut}(C)|}{|\operatorname{Aut}(A)|}([B] \otimes [C])$$

and

$$([A], [B])_{\mathcal{H}(\mathcal{Q})} = \frac{\delta_{[A],[B]}}{|\operatorname{Aut}(A)|}$$

In [29, Theorem 1], Green shows that $(-, -)_{\mathcal{H}(\mathcal{Q})}$ satisfies Condition 2 and that δ satisfies Condition 4. Hence, we have the following.

Theorem 10. [8, Theorem 1.1] *Let $\mathcal{Q}$ be an $\mathbb{F}_q$-species. Then, $\mathcal{H}(\mathcal{Q})$ is the positive part of the quantized enveloping algebra of a generalized Kac–Moody algebra.*

Remark 18. *In their paper, Sevenhant and Van Den Bergh state Theorem 10 only for the Ringel–Hall algebra of a quiver, but none of their arguments depend on having a quiver rather than a species. Indeed, many of their arguments are based on those of Green in [29], which are valid for hereditary algebras. Moreover, Sevenhant and Van Den Bergh define the Ringel–Hall algebra to be an algebra opposite to the one we defined in Definition 25 (our definition, which is the one used by Green, seems to be the more standard definition). This does not affect any of the arguments presented.*

References

1. Gabriel, P. Indecomposable Representations. II. In *Symposia Mathematica, Vol. XI (Convegno di Algebra Commutativa, INDAM, Rome, 1971)*; Academic Press: London, UK, 1973; pp. 81–104.
2. Lusztig, G. *Introduction to Quantum Groups*; Modern Birkhäuser Classics, Birkhäuser/Springer: New York, NY, USA, 2010; pp. xiv+346;
3. Dlab, V.; Ringel, C.M. Indecomposable representations of graphs and algebras. *Mem. Amer. Math. Soc.* **1976,** *6*, v+57.
4. Schiffmann, O. Lectures on Hall algebras. arXiv:math/0611617v2. Available online: http://arxiv.org/abs/math/0611617 (accessed on 10 February 2011).
5. Hubery, A. Ringel-Hall Algebras. Available online: http://www1.maths.leeds.ac.uk/~ahubery/RHAlgs.pdf (accessed on 10 February 2011).
6. Drinfel'd, V.G. Hopf algebras and the quantum Yang-Baxter equation. *Dokl. Akad. Nauk SSSR* **1985,** *283*, 1060–1064.
7. Deng, B.; Du, J.; Parshall, B.; Wang, J. Finite Dimensional Algebras and Quantum Groups. In *Mathematical Surveys and Monographs*; American Mathematical Society: Providence, RI, USA, 2008; Volume 150, pp. xxvi+759.
8. Sevenhant, B.; van Den Bergh, M. A relation between a conjecture of Kac and the structure of the hall algebra. *J. Pure Appl. Algebra* **2001,** *160*, 319–332. Available online: http://dx.doi.org/10.1016/S0022-4049(00)00078-5 (accessed on 21 November 2010).
9. Nguefack, B. A non-simply laced version for cluster structures on 2-Calabi-Yau categories. arXiv:0910.5077v1. Avaialble online: http://arxiv.org/abs/0910.5077 (accessed on 26 January 2012).
10. Weil, A. *Basic Number Theory*, 3rd ed.; Springer-Verlag: New York, NY, USA, 1974; pp. xviii+325.
11. Lam, T.Y. A First Course in Noncommutative Rings, Volume 131. In *Graduate Texts in Mathematics*; Springer-Verlag: New York, NY, USA, 1991; pp. xvi+397.
12. Brauer, R. Collected Papers. In *Mathematicians of Our Time*; MIT Press: Cambridge, MA, USA, 1980; Volume 1,17, pp. liv+615.
13. Jans, J.P. On the indecomposable representations of algebras. *Ann. Math.* **1957,** *66*, 418–429.
14. Yoshii, T. On algebras of bounded representation type. *Osaka Math. J.* **1956,** *8*, 51–105.
15. Assem, I.; Simson, D.; Skowroński, A. Elements of the representation theory of associative algebras. In *London Mathematical Society Student Texts*; Cambridge University Press: Cambridge, UK, 2006; Volume 1,65, pp. x+458.
16. Benson, D.J. *Representations and Cohomology I*, 2nd ed.; *Cambridge Studies in Advanced Mathematics*; Cambridge University Press: Cambridge, UK, 1998; Volume 30, pp. xii+246.
17. Dlab, V.; Ringel, C.M. On algebras of finite representation type. *J. Algebra* **1975,** *33*, 306–394.

18. Drozd, Y.A.; Kirichenko, V.V. *Finite-Dimensional Algebras*; Springer-Verlag: Berlin/Heidelberg, Germany, 1994; pp. xiv+249.
19. Hubery, A. Quiver representations respecting a quiver automorphism: A generalisation of a theorem of Kac. *J. Lond. Math. Soc.* **2004**, *69*, 79–96. Available online: http://dx.doi.org/10.1112/S0024610703004988 (accessed on 12 April 2011).
20. Bourbaki, N. *Éléments de mathématique: Algèbre, Chapitre 4 à 7*; Springer: Berlin/Heidelberg, Germany, 2007.
21. Jacobson, N. *Basic Algebra II*, 2nd ed.; W. H. Freeman and Company: New York, NY, USA, 1989; pp. xviii+686.
22. Auslander, M. On the dimension of modules and algebras, III. Global dimension. *Nagoya Math. J.* **1955**, *9*, 67–77.
23. Chen, X.; Nam, K.B.; Pospíchal, T. Quivers and representations. In *Handbook of Algebra. Volume 6*; Elsevier/North-Holland: Amsterdam, The Netherlands, 2009; Volume 6, pp. 507–561. Available online: http://dx.doi.org/10.1016/S1570-7954(08)00209-X (accessed on 18 June 2010).
24. Lemay, J. Valued Graphs and the Representation Theory of Lie Algebras. MSc Thesis, University of Ottawa, Ottawa, ON, Canada, 2011. Available online: http://www.ruor.uottawa.ca/en/handle/10393/20168 (accessed on 3 September 2011).
25. Ringel, C.M. Representations of *K*-species and bimodules. *J. Algebra* **1976**, *41*, 269–302.
26. Kac, V.G. Infinite root systems, representations of graphs and invariant theory. *Invent. Math.* **1980**, *56*, 57–92. Available online: http://dx.doi.org/10.1007/BF01403155 (accessed on 22 November 2010).
27. Deng, B.; Xiao, J. The Ringel-Hall Interpretation to a Conjecture of Kac. Avaialble online: www.mathematik.uni-bielefeld.de/ bruestle/Publications/deng1.ps (accessed on 2 July 2012).
28. Ringel, C.M. Hall algebras and quantum groups. *Invent. Math.* **1990**, *101*, 583–591. Available online: http://dx.doi.org/10.1007/BF01231516 (accessed on 10 February 2011).
29. Green, J.A. Hall algebras, hereditary algebras and quantum groups. *Invent. Math.* **1995**, *120*, 361–377. Available online: http://dx.doi.org/10.1007/BF01241133 (accessed on 12 February 2011).
30. Ringel, C.M. From Representations of Quivers via Hall and Loewy Algebras to Quantum Groups. In *Proceedings of the International Conference on Algebra, Part 2 (Novosibirsk, 1989)*; American Mathematical Society: Providence, RI, USA, 1992; Volume 131, pp. 381–401.
31. Ringel, C.M. The Hall Algebra Approach to Quantum Groups. In *XI Latin American School of Mathematics (Spanish) (Mexico City, 1993)*; Sociedad Matemática Mexicana: México, 1995; Volume 15, pp. 85–114.
32. Borcherds, R. Generalized kac-moody algebras. *J. Algebra* **1988**, *115*, 501–512. Available online: http://dx.doi.org/10.1016/0021-8693(88)90275-X (accessed on 12 February 2011).

6

Frobenius–Schur Indicator for Categories with Duality

Kenichi Shimizu

Graduate School of Mathematics, Nagoya University, Furo-cho, Chikusa-ku, Nagoya 464-8602, Japan;
E-Mail: x12005i@math.nagoya-u.ac.jp

Abstract: We introduce the Frobenius–Schur indicator for categories with duality to give a category-theoretical understanding of various generalizations of the Frobenius–Schur theorem including that for semisimple quasi-Hopf algebras, weak Hopf C^*-algebras and association schemes. Our framework also clarifies a mechanism of how the "twisted" theory arises from the ordinary case. As a demonstration, we establish twisted versions of the Frobenius–Schur theorem for various algebraic objects. We also give several applications to the quantum SL_2.

Keywords: Frobenius–Schur indicator; category with duality; Hopf algebra; quantum groups

1. Introduction

The aim of this paper is to develop a category-theoretical framework to unify various generalizations of the *Frobenius–Schur theorem*. We first recall the Frobenius–Schur theorem for compact groups. Let G be a compact group, and let V be a finite-dimensional continuous representation of G with character χ_V. The *n-th Frobenius–Schur indicator* (or *FS indicator*, for short) of V is defined and denoted by

$$\nu_n(V) = \int_G \chi_V(g^n) d\mu(g) \tag{1.1}$$

where μ is the normalized Haar measure on G. The Frobenius–Schur theorem states that the value of the second FS indicator $\nu_2(V)$ has the following meaning:

Theorem (Frobenius–Schur theorem). *If V is irreducible, then we have*

$$\nu_2(V) = \begin{cases} +1 & \text{if } V \text{ is real} \\ 0 & \text{if } V \text{ is complex} \\ -1 & \text{if } V \text{ is quaternionic} \end{cases} \tag{1.2}$$

Moreover, the following statements are equivalent:

(1) $\nu_2(V) \neq 0$.

(2) *V is isomorphic to its dual representation.*

(3) *There exists a non-degenerate G-invariant bilinear form $b : V \times V \to \mathbb{C}$.*

If one of the above equivalent statements holds, then such a bilinear form b is unique up to scalar multiples and satisfies $b(w,v) = \nu_2(V) \cdot b(v,w)$ for all $v, w \in V$. In other words, b is symmetric if $\nu_2(V) = +1$ and is skew-symmetric if $\nu_2(V) = -1$.

For $n \geq 3$, the representation-theoretic meaning of the n-th FS indicator is less obvious than the second one and there is no such theorem involving the n-th FS indicator. Hence, the second FS indicator could be of special interest. Unless otherwise noted, we simply call ν_2 the *FS indicator* and refer to ν_n for $n \geq 3$ as *higher FS indicators*.

Generalizing those for compact groups, the FS indicator and higher ones have been defined for various algebraic objects, including (quasi-)Hopf algebras, tensor categories and conformal field theories; see [1–9]. Among others, the theory of Ng and Schauenburg [7–9] is especially important since it gives a unified category-theoretical understanding of all of [1–6]. For the case of a semisimple (quasi-)Hopf algebra, a generalization of the Frobenius–Schur theorem is also formulated and proved; see [1–3]. These results have many applications in Hopf algebras and tensor categories; see [10–17].

On the other hand, there are several generalizations in other directions. For example, the earlier result of Linchenko and Montgomery [1] can be thought, in fact, as a generalization of the Frobenius–Schur theorem for a finite-dimensional semisimple algebra with an anti-algebra involution. Based on their result, Hanaki and Terada [18] proved a generalization of the Frobenius–Schur theorem for association schemes and gave some applications to association schemes. Doi [19] reconstructed the results of [1] with an emphasis on the use of the theory of symmetric algebras. Recently, Geck [20] proved a result similar to Doi and gave some applications to finite Coxeter groups.

Unlike Hopf algebras, the representation categories of such algebras do not have a natural structure of a monoidal category and therefore we cannot understand these results in the framework of Ng and Schauenburg. Our first question is:

Question 1.1. *Is there a good category-theoretical framework to understand the FS indicator and the Frobenius–Schur theorem for such algebras?*

The second question is about the twisted versions of some of the above. Given an automorphism τ of a finite group (or a semisimple Hopf algebra), the n-th τ-twisted FS indicator ν_n^τ is defined by twisting the definition of ν_n by τ and the twisted version of the Frobenius–Schur theorem is also formulated and proved; see [21–23]. Our second question is:

Question 1.2. *If there is an answer to Question 1.1, then what is its twisted version?*

In this paper, we give answers to these questions. Following the approaches of [5–7], we see that the duality is what we really need to define the FS indicator. This observation leads us to the notion of a *category with duality*, which has been well-studied in the theory of Witt groups [24,25]. As an answer to Question 1.1, we introduce and study the FS indicator for categories with duality. Considering a suitable category and suitable duality, we can recover various generalizations of the Frobenius–Schur theorem for compact groups. We also introduce a method to "twist" the given duality on a category. This gives an answer to Question 1.2; in fact, the twisted FS indicator can be understood as the "untwisted" FS indicator with respect to the twisted duality.

1.1. Organization and Summary of Results

The present paper is organized as follows: In Section 2, following Mac Lane [26], we recall some basic results on adjoint functors and then introduce a category with duality in terms of adjunctions. By generalizing the definitions of [5–7], we define the FS indicator of an object of a category with duality over a field k (Definition 2.8). We also introduce a general method to "twist" the given duality by an adjunction. Then the twisted FS indicator is defined to be the "untwisted" FS indicator with respect to the twisted duality.

Pivotal Hopf algebras are introduced as a class of Hopf algebras whose representation category is a pivotal monoidal category; see, e.g., [14]. Motivated by this notion, in Section 3, a *pivotal algebra* is defined to be a triple (A, S, g) consisting of an algebra A, an anti-algebra map $S : A \to A$ and an invertible element $g \in A$ satisfying certain conditions (Definition 3.1). The representation category of a pivotal algebra is not monoidal in general but has duality. Therefore the FS indicator of an A-module is defined in the way of Section 2. In Section 3, we study the FS indicator for pivotal algebras and prove some fundamental properties of them.

From our point of view, (1.1) is not the definition but a formula to compute the FS indicator. It is natural to ask when such a formula exists. In Section 3, we also give a formula for a separable pivotal algebra (Theorem 3.8); if a pivotal algebra $A = (A, S, g)$ has a separability idempotent E, then

$$\nu(V) = \sum_i \chi_V\left(S(E_i')gE_i''\right) \quad \left(E = \sum_i E_i' \otimes E_i''\right) \tag{1.3}$$

for all finite-dimensional left A-module V, where χ_V is the character of V. The relation between this formula and the results of [1,19,20] is discussed. By specializing (1.3), we obtain a formula for group-like algebras (Example 3.9) and for finite-dimensional weak Hopf C^*-algebras (Example 3.10). We also obtain the formula of Mason and Ng [2] for finite-dimensional semisimple quasi-Hopf algebras and its twisted version (§3.4).

In Section 4, we introduce a *copivotal coalgebra* as the dual notion of pivotal algebras. Each result of Section 3 has an analogue in the case of copivotal coalgebras. A crucial difference from the case of algebras is that there are infinite-dimensional coseparable coalgebras. For example, the Hopf algebra $R(G)$ of continuous representative functions on a compact group G is coseparable with coseparability idempotent given by the Haar measure on G. The Formula (1.1) is obtained by applying the coalgebraic version of (1.3) to $R(G)$.

In Section 5, we apply our results to the quantum coordinate algebra $\mathcal{O}_q(SL_2)$ and the quantized universal enveloping algebra $U_q(\mathfrak{sl}_2)$. We first determine the FS indicator of all simple right $\mathcal{O}_q(SL_2)$-comodules. In a similar way, we also determine the twisted FS indicator with respect to an involution of $\mathcal{O}_q(SL_2)$ corresponding to the group homomorphism

$$SL_2(k) \to SL_2(k), \quad \begin{pmatrix} a & b \\ c & d \end{pmatrix} \mapsto \begin{pmatrix} a & -b \\ -c & d \end{pmatrix}$$

in the classical limit $q \to 1$. Similar results for $U_q(\mathfrak{sl}_2)$ are also given by using the Hopf pairing between $\mathcal{O}_q(SL_2)$ and $U_q(\mathfrak{sl}_2)$.

Remark 1.3. To answer Question 1.1, we need to work in a "non-monoidal" setting. Since, as we have remarked, the Frobenius–Schur theory has some good applications even in non-monoidal settings, the FS indicator for categories with duality could be interesting. However, to define the higher (twisted) FS indicators, a monoidal structure seems to be necessary. At least, there is a reason why we cannot define higher FS indicators for categories with duality; see Remarks 2.6 and 2.13.

It is interesting to construct a twisted version of [7–9]. One of the referees kindly pointed out to the author that in May 2012, Daniel Sage gave a talk on a category-theoretic definition of the higher twisted FS indicators for Hopf algebras at the Lie Theory Workshop held at University of Southern California. Independently, after the submission of the first version of this paper, the author obtained a description of the higher twisted FS indicator for Hopf algebras by using the crossed product monoidal category. In this paper, we also mention higher twisted FS indicators for pivotal monoidal categories but leave the details for future work; see §2.6.

Remark 1.4. Linchenko and Montgomery [1] established a relation between the FS indicator and invariant bilinear forms on an irreducible representation. Unlike the case of compact groups, a relation between the FS indicator and "reality" of representations is not known in the case of Hopf algebras; in fact, as remarked in [1], the reality of a representation of a Hopf algebra is not defined since, in general, a Hopf algebra does not have a good basis like the group elements of the group algebra. In a forthcoming paper, we will introduce the notions of real, complex and quaternionic representations of a Hopf $*$-algebra and Formulate (1.2) in a Hopf algebraic context. We will also provide an exact quantum analog of the Frobenius–Schur theorem for compact quantum groups.

1.2. Notation

Given a category $\mathcal{C}$ and $X, Y \in \mathcal{C}$, we denote by $\mathrm{Hom}_{\mathcal{C}}(X, Y)$ the set of all morphisms from X to Y. $\mathcal{C}^{\mathrm{op}}$ means the opposite category of $\mathcal{C}$. An object $X \in \mathcal{C}$ is often written as X^{op} when it is regarded an object of $\mathcal{C}^{\mathrm{op}}$. A similar notation is used for morphisms. A functor $F : \mathcal{C} \to \mathcal{D}$ is denoted by F^{op} if it is regarded as a functor $\mathcal{C}^{\mathrm{op}} \to \mathcal{D}^{\mathrm{op}}$.

Throughout, we work over a fixed field k whose characteristic is not two. By an algebra, we mean a unital associative algebra over k. Given a vector space V (over k), we denote by $V^{\vee} = \mathrm{Hom}_k(V, k)$ the dual space of V. For $f \in V^{\vee}$ and $v \in V$, we often write $f(v)$ as $\langle f, v \rangle$. Unless otherwise noted, the unadorned tensor symbol $\otimes$ means the tensor product over k. Given $t \in V^{\otimes n}$, we often write t as

$$t = t^1 \otimes t^2 \otimes \cdots \otimes t^n \in V \otimes V \otimes \cdots \otimes V$$

The comultiplication and the counit of a coalgebra are denoted by Δ and ε, respectively. For an element c of a coalgebra, we use Sweedler's notation

$$\Delta(c) = c_{(1)} \otimes c_{(2)}, \quad \Delta(c_{(1)}) \otimes c_{(2)} = c_{(1)} \otimes c_{(2)} \otimes c_{(3)} = c_{(1)} \otimes \Delta(c_{(2)}), \ \ldots$$

2. Categories with Duality

2.1. Adjunctions

Following Mac Lane [26], we recall basic results on adjunctions. Let $\mathcal{C}$ and $\mathcal{D}$ be categories. An *adjunction* from $\mathcal{C}$ to $\mathcal{D}$ is a triple (F, G, Φ) consisting of functors $F : \mathcal{C} \to \mathcal{D}$ and $G : \mathcal{D} \to \mathcal{C}$ and a natural bijection $\Phi_{X,Y} : \mathrm{Hom}_{\mathcal{D}}(F(X), Y) \to \mathrm{Hom}_{\mathcal{C}}(X, G(Y))$ $(X \in \mathcal{C}, Y \in \mathcal{D})$.

Given an adjunction (F, G, Φ) from $\mathcal{C}$ to $\mathcal{D}$, the *unit* $\eta : \mathrm{id}_{\mathcal{C}} \to GF$ and the *counit* $\varepsilon : FG \to \mathrm{id}_{\mathcal{D}}$ of the adjunction (F, G, Φ) are defined by $\eta_X = \Phi_{X,F(X)}(\mathrm{id}_{F(X)})$ and $\varepsilon_Y = \Phi^{-1}_{G(Y),Y}(\mathrm{id}_{G(Y)})$ for $X \in \mathcal{C}$ and $Y \in \mathcal{D}$, respectively. They satisfy the counit-unit equations

$$\varepsilon_{F(X)} \circ F(\eta_X) = \mathrm{id}_{F(X)} \quad \text{and} \quad G(\varepsilon_Y) \circ \eta_{G(Y)} = \mathrm{id}_{G(Y)} \tag{2.1}$$

for all $X \in \mathcal{C}$ and $Y \in \mathcal{D}$. By using η, the natural bijection Φ is expressed as

$$\Phi_{X,Y}(f) = G(f) \circ \eta_X \quad (f \in \mathrm{Hom}_{\mathcal{C}}(F(X), Y) \tag{2.2}$$

Similarly, by using ε, the inverse of Φ is expressed as

$$\Phi^{-1}_{X,Y}(g) = \varepsilon_Y \circ F(g) \quad (g \in \mathrm{Hom}_{\mathcal{D}}(X, G(Y)) \tag{2.3}$$

Note that $\circ$ at the right-hand side stands for the composition in $\mathcal{D}$. We will deal with the case where $\mathcal{D} = \mathcal{C}^{\mathrm{op}}$, the opposite category of $\mathcal{C}$.

Each adjunction is determined by its unit and counit; indeed, let $F : \mathcal{C} \to \mathcal{D}$ and $G : \mathcal{D} \to \mathcal{C}$ be functors, and let $\eta : \mathrm{id}_{\mathcal{C}} \to GF$ and $\varepsilon : FG \to \mathrm{id}_{\mathcal{D}}$ be natural transformations satisfying (2.1). If we define Φ by (2.2), then the triple (F, G, Φ) is an adjunction from $\mathcal{C}$ to $\mathcal{D}$ whose unit and counit are η and ε. From this reason, we abuse terminology and refer to such a quadruple $(F, G, \eta, \varepsilon)$ as an adjunction from $\mathcal{C}$ to $\mathcal{D}$.

2.2. Categories with Duality

The following terminologies are taken from Balmer [24] and Calmés-Hornbostel [25].

Definition 2.1. A *category with duality* is a triple $\mathcal{C} = (\mathcal{C}, (-)^{\vee}, j)$ consisting of a category $\mathcal{C}$, a contravariant functor $(-)^{\vee} : \mathcal{C} \to \mathcal{C}$ and a natural transformation $j : \mathrm{id}_{\mathcal{C}} \to (-)^{\vee\vee}$ satisfying

$$(j_X)^{\vee} \circ j_{X^{\vee}} = \mathrm{id}_{X^{\vee}} \tag{2.4}$$

for all $X \in \mathcal{C}$. If, moreover, j is a natural isomorphism, then we say that $\mathcal{C}$ is a *category with strong duality*, or, simply, $\mathcal{C}$ is *strong*.

Let $\mathcal{C}$ be a category with duality. We call the functor $(-)^\vee : \mathcal{C} \to \mathcal{C}$ the *duality functor* of $\mathcal{C}$. A pivotal monoidal category is an example of categories with duality; see [3, Appendix]. Thus we call the natural transformation $j : \mathrm{id}_{\mathcal{C}} \to (-)^{\vee\vee}$ the *pivotal morphism* of $\mathcal{C}$.

Example 2.2. Let H be a Hopf algebra with antipode S. If H is *involutory, i.e.*, $S^2 = \mathrm{id}_H$, then the category $\mathrm{mod}(H)$ of left H-modules is a category with duality; the duality functor is given by taking the dual H-module and the pivotal morphism is given by the canonical map

$$\iota_V : V \to V^{\vee\vee} = \mathrm{Hom}_k(\mathrm{Hom}_k(V,k),k), \quad \langle \iota(v), f\rangle = \langle f, v\rangle \quad (v \in V, f \in V^\vee) \tag{2.5}$$

The full subcategory $\mathrm{mod}_{fd}(H)$ of $\mathrm{mod}(H)$ of finite-dimensional left H-modules is a category with strong duality since ι_V is an isomorphism if (and only if) $\dim_k V < \infty$.

Let D denote the duality functor of $\mathcal{C}$ regarded as a covariant functor from $\mathcal{C}$ to $\mathcal{C}^{\mathrm{op}}$. Definition 2.1 says that the quadruple $(D, D^{\mathrm{op}}, j, j^{\mathrm{op}}) : \mathcal{C} \to \mathcal{C}^{\mathrm{op}}$ is an adjunction. Hence we obtain a natural bijection

$$\begin{aligned} \mathrm{T}_{X,Y} : \mathrm{Hom}_{\mathcal{C}}(X, Y^\vee) &= \mathrm{Hom}_{\mathcal{C}^{\mathrm{op}}}(D(Y), X^{\mathrm{op}}) \\ &\longrightarrow \mathrm{Hom}_{\mathcal{C}}(Y, D^{\mathrm{op}}(X^{\mathrm{op}})) = \mathrm{Hom}_{\mathcal{C}}(Y, X^\vee) \quad (X, Y \in \mathcal{C}) \end{aligned} \tag{2.6}$$

which we call the *transposition map*. By (2.2), $\mathrm{T}_{X,Y}$ is expressed as

$$\mathrm{T}_{X,Y}(f) = f^\vee \circ j_Y \quad (f \in \mathrm{Hom}_{\mathcal{C}}(X, Y^\vee)) \tag{2.7}$$

By (2.3), we have $\mathrm{T}_{X,Y}^{-1}(g) = g^\vee \circ j_X = \mathrm{T}_{Y,X}(g)$ for $g \in \mathrm{Hom}_{\mathcal{C}}(Y, X^\vee)$. Hence we have

$$\mathrm{T}_{Y,X} \circ \mathrm{T}_{X,Y} = \mathrm{id}_{\mathrm{Hom}_{\mathcal{C}}(X, Y^\vee)} \tag{2.8}$$

Note that j is not necessarily an isomorphism. By understanding a category with duality as a kind of adjunction, we obtain the following characterization of categories with strong duality.

Lemma 2.3. *For a category $\mathcal{C}$ with duality, the following are equivalent:*
(1) *$\mathcal{C}$ is a category with strong duality.*
(2) *The duality functor $(-)^\vee : \mathcal{C} \to \mathcal{C}^{\mathrm{op}}$ is an equivalence.*

Proof. Let, in general, $(F, G, \eta, \varepsilon)$ be an adjunction between some categories. Then F is fully faithful if and only if η is an isomorphism [26, IV.3]. Now we apply this result to the above quadruple $(D, D^{\mathrm{op}}, j, j^{\mathrm{op}})$ as follows: If $\mathcal{C}$ is strong, then D is fully faithful. Since $X \cong X^{\vee\vee} = D(X^\vee)$ $(X \in \mathcal{C})$, D is essentially surjective. Hence, D is an equivalence. The converse is clear, since an equivalence of categories is fully faithful. □

Following [25], we introduce duality preserving functors and related notions:

Definition 2.4. Let $\mathcal{C}$ and $\mathcal{D}$ be categories with duality. A *duality preserving functor* from $\mathcal{C}$ to $\mathcal{D}$ is a pair (F, ξ) consisting of a functor $F : \mathcal{C} \to \mathcal{D}$ and a natural transformation $\xi : F(X^\vee) \to F(X)^\vee$ $(X \in \mathcal{C})$ making

$$\begin{array}{ccc} F(X) & \xrightarrow{F(j_X)} & F(X^{\vee\vee}) \\ {\scriptstyle j_{F(X)}}\downarrow & & \downarrow{\scriptstyle \xi_{X^\vee}} \\ F(X)^{\vee\vee} & \xrightarrow[\xi_X^\vee]{} & F(X^\vee)^\vee \end{array} \tag{2.9}$$

commute for all $X \in \mathcal{C}$. If, moreover, ξ is an isomorphism, then (F, ξ) is said to be a *strong*. If ξ is the identity, then (F, ξ) is said to be *strict*.

Now let $(F,\xi),(G,\zeta) : \mathcal{C} \to \mathcal{D}$ be such functors. A *morphism of duality preserving functors* from (F,ξ) to (G,ζ) is a natural transformation $h : F \to G$ making

$$\begin{array}{ccc} F(X^\vee) & \xrightarrow{\xi_X} & F(X)^\vee \\ {\scriptstyle h_{X^\vee}}\downarrow & & \uparrow{\scriptstyle h_X^\vee} \\ G(X^\vee) & \xrightarrow[\zeta_X]{} & G(X)^\vee \end{array}$$

commute for all $X \in \mathcal{C}$.

If $(F,\xi) : \mathcal{C} \to \mathcal{D}$ and $(G,\zeta) : \mathcal{D} \to \mathcal{E}$ are duality preserving functors between categories with duality, then the composition $G \circ F : \mathcal{C} \to \mathcal{E}$ becomes a duality preserving functor with

$$G(F(X^\vee)) \xrightarrow{G(\xi_X)} G(F(X)^\vee) \xrightarrow{\zeta_{F(X)}} G(F(X))^\vee \quad (X \in \mathcal{C})$$

One can check that categories with duality form a 2-category; 1-arrows are duality preserving functors and 2-arrows are morphisms of duality preserving functors. Hence we can define an *isomorphism* and an *equivalence* of categories with duality in the usual way.

Given a duality preserving functor $(F,\xi) : \mathcal{C} \to \mathcal{D}$, we define

$$\tilde{F}_{X,Y} : \mathrm{Hom}_{\mathcal{C}}(X, Y^\vee) \to \mathrm{Hom}_{\mathcal{D}}(F(X), F(Y)^\vee) \quad (X, Y \in \mathcal{C})$$

by $\tilde{F}_{X,Y}(f) = \xi_Y \circ F(f)$ for $f : X \to Y^\vee$. $\tilde{F}$ is compatible with the transposition map in the sense that the diagram

$$\begin{array}{ccc} \mathrm{Hom}_{\mathcal{C}}(X, Y^\vee) & \xrightarrow{\tilde{F}_{X,Y}} & \mathrm{Hom}_{\mathcal{D}}(F(X), F(Y)^\vee) \\ {\scriptstyle \mathsf{T}_{X,Y}}\downarrow & & \downarrow{\scriptstyle \mathsf{T}_{F(X),F(Y)}} \\ \mathrm{Hom}_{\mathcal{C}}(Y, X^\vee) & \xrightarrow[\tilde{F}_{X,Y}]{} & \mathrm{Hom}_{\mathcal{D}}(F(Y), F(X)^\vee) \end{array} \tag{2.10}$$

commutes for all $X, Y \in \mathcal{C}$. Indeed, we have

$$\begin{aligned} \mathsf{T}_{F(X),F(Y)}(\tilde{F}_{X,Y}(f)) &= (\xi_Y \circ F(f))^\vee \circ j_{F(Y)} = F(f)^\vee \circ \xi_Y^\vee \circ j_{F(Y)} \\ &= F(f)^\vee \circ \xi_{X^\vee} \circ F(j_X) = \xi_Y \circ F(f^\vee) \circ F(j_X) = \tilde{F}_{X,Y}(\mathsf{T}_{X,Y}(f)) \end{aligned}$$

Now suppose that $\mathcal{C}$ is a category with strong duality. Then:

Lemma 2.5. *$\mathcal{C}^{\mathrm{op}}$ is a category with duality with the same duality functor as $\mathcal{C}$ and pivotal morphism $(j^{-1})^{\mathrm{op}}$. The duality functor on $\mathcal{C}$ is an equivalence of categories with duality between $\mathcal{C}$ and $\mathcal{C}^{\mathrm{op}}$.*

Hence, from (2.10) with $(F,\xi) = ((-)^\vee, \mathrm{id}_{(-)^\vee}) : \mathcal{C}^{\mathrm{op}} \to \mathcal{C}$, we see that

$$\begin{array}{ccccc} \mathrm{Hom}_{\mathcal{C}}(X^\vee, Y) & = & \mathrm{Hom}_{\mathcal{C}^{\mathrm{op}}}(Y^{\mathrm{op}}, (X^\vee)^{\mathrm{op}}) & \xrightarrow{(-)^\vee} & \mathrm{Hom}_{\mathcal{C}}(Y^\vee, X^{\vee\vee}) \\ {\scriptstyle \mathsf{T}^{\mathrm{op}}_{X,Y}}\downarrow & & & & \downarrow{\scriptstyle \mathsf{T}_{Y^\vee,X^\vee}} \\ \mathrm{Hom}_{\mathcal{C}}(Y^\vee, X) & = & \mathrm{Hom}_{\mathcal{C}^{\mathrm{op}}}(X^{\mathrm{op}}, (Y^\vee)^{\mathrm{op}}) & \xrightarrow[(-)^\vee]{} & \mathrm{Hom}_{\mathcal{C}}(X^\vee, Y^{\vee\vee}) \end{array} \tag{2.11}$$

commutes for all $X, Y \in \mathcal{C}$, where $\mathsf{T}^{\mathrm{op}}_{X,Y}$ is the transposition map for $\mathcal{C}^{\mathrm{op}}$ regarded as a map $\mathrm{Hom}_{\mathcal{C}}(X^\vee, Y) \to \mathrm{Hom}_{\mathcal{C}}(Y^\vee, X)$. Explicitly, it is given by $\mathsf{T}^{\mathrm{op}}_{X,Y}(f) = j_X^{-1} \circ f^\vee$.

Remark 2.6. If $\mathcal{C}$ is a pivotal monoidal category, then there is a natural bijection

$$\mathrm{Hom}_{\mathcal{C}}(X^\vee, Y) \cong \mathrm{Hom}_{\mathcal{C}}(\mathbf{1}, X \otimes Y) \quad (X, Y \in \mathcal{C}),$$

where $\otimes$ is the tensor product of $\mathcal{C}$ and $\mathbf{1} \in \mathcal{C}$ is the unit object. The diagram

$$\begin{array}{ccc} \mathrm{Hom}_{\mathcal{C}}(X^\vee, X) & \xrightarrow{\cong} & \mathrm{Hom}_{\mathcal{C}}(\mathbf{1}, X \otimes X) \\ {\scriptstyle \mathsf{T}^{\mathrm{op}}_{X,X}} \downarrow & & \downarrow {\scriptstyle E^{(2)}_X} \\ \mathrm{Hom}_{\mathcal{C}}(X^\vee, X) & \xrightarrow[\cong]{} & \mathrm{Hom}_{\mathcal{C}}(\mathbf{1}, X \otimes X) \end{array}$$

commutes, where the horizontal arrows are the canonical bijections and $E^{(2)}_X$ is the map used in [8] to define the FS indicator.

2.3. Frobenius–Schur Indicator

Recall that a category $\mathcal{C}$ is said to be *k-linear* if each hom-set is a vector space over k and the composition of morphisms is k-bilinear. A functor $F : \mathcal{C} \to \mathcal{D}$ between k-linear categories is said to be *k-linear* if the map $\mathrm{Hom}_{\mathcal{C}}(X, Y) \to \mathrm{Hom}_{\mathcal{D}}(F(X), F(Y))$, $f \mapsto F(f)$ is k-linear for all $X, Y \in \mathcal{C}$. Note that $\mathcal{C}^{\mathrm{op}}$ is k-linear if $\mathcal{C}$ is. Thus the k-linearity of a contravariant functor makes sense.

Definition 2.7. By a *category with duality over k*, we mean a k-linear category with duality whose duality functor is k-linear.

For simplicity, in this section, we always assume that a category $\mathcal{C}$ with duality over k satisfies the following finiteness condition:

$$\dim_k \mathrm{Hom}_{\mathcal{C}}(X, Y) < \infty \quad \text{for all } X, Y \in \mathcal{C} \tag{2.12}$$

Definition 2.8. Let $\mathcal{C}$ be a category with duality over k. The *Frobenius–Schur indicator* (or FS indicator, for short) of $X \in \mathcal{C}$ is defined and denoted by $\nu(X) = \mathrm{Tr}(\mathsf{T}_{X,X})$, where Tr means the trace of a linear map.

The following is a list of basic properties of the FS indicator:

Proposition 2.9. *Let $\mathcal{C}$ be a category with duality over k and let $X \in \mathcal{C}$.*
(a) *$\nu(X)$ depends on the isomorphism class of $X \in \mathcal{C}$.*
(b) *$\nu(X) = \dim_k B^+_{\mathcal{C}}(X) - \dim_k B^-_{\mathcal{C}}(X)$, where $B^{\pm}_{\mathcal{C}}(X) = \{b : X \to X^\vee \mid \mathsf{T}_{X,X}(b) = \pm b\}$.*
(c) *Let $X_1, X_2 \in \mathcal{C}$. If their biproduct $X_1 \oplus X_2$ exists, then we have $\nu(X_1 \oplus X_2) = \nu(X_1) + \nu(X_2)$.*

Proof. (a) Let $p : X \to Y$ be an isomorphism in $\mathcal{C}$. Then

$$\mathrm{Hom}_{\mathcal{C}}(p, p^\vee) : \mathrm{Hom}_{\mathcal{C}}(Y, Y^\vee) \to \mathrm{Hom}_{\mathcal{C}}(X, X^\vee), \quad f \mapsto p^\vee \circ f \circ p$$

is an isomorphism. By the naturality of the transposition map, the diagram

$$\begin{array}{ccc} \mathrm{Hom}_{\mathcal{C}}(Y, Y^\vee) & \xrightarrow{\mathsf{T}_{Y,Y}} & \mathrm{Hom}_{\mathcal{C}}(Y, Y^\vee) \\ {\scriptstyle \mathrm{Hom}_{\mathcal{C}}(p,p^\vee)} \downarrow & & \downarrow {\scriptstyle \mathrm{Hom}_{\mathcal{C}}(p,p^\vee)} \\ \mathrm{Hom}_{\mathcal{C}}(X, X^\vee) & \xrightarrow[\mathsf{T}_{X,X}]{} & \mathrm{Hom}_{\mathcal{C}}(X, X^\vee) \end{array}$$

commutes. Hence, we have $\nu(X) = \mathrm{Tr}(\mathsf{T}_{X,X}) = \mathrm{Tr}(\mathsf{T}_{Y,Y}) = \nu(Y)$.

(b) The result follows from (2.8) and the fact that the trace of an operator is the sum of its eigenvalues.

(c) For $a = 1, 2$, let $i_a : X_a \to X_1 \oplus X_2$ and $p_a : X_1 \oplus X_2 \to X_a$ be the inclusion and the projection, respectively. For $a, b, c, d = 1, 2$, we set

$$\mathsf{T}^{cd}_{ab} = p_{cd} \circ \mathsf{T}_{X_1 \oplus X_2, X_1 \oplus X_2} \circ i_{ab} : \mathrm{Hom}_{\mathcal{C}}(X_a, X_b^\vee) \to \mathrm{Hom}_{\mathcal{C}}(X_c, X_d^\vee)$$

where $i_{ab} = \mathrm{Hom}_{\mathcal{C}}(p_a, p_b^\vee)$ and $p_{cd} = \mathrm{Hom}_{\mathcal{C}}(i_c, i_d^\vee)$. By linear algebra, we have

$$\mathrm{Tr}(\mathsf{T}_{X_1 \oplus X_2, X_1 \oplus X_2}) = \mathrm{Tr}(\mathsf{T}_{11}^{11}) + \mathrm{Tr}(\mathsf{T}_{12}^{12}) + \mathrm{Tr}(\mathsf{T}_{21}^{21}) + \mathrm{Tr}(\mathsf{T}_{22}^{22}) \tag{2.13}$$

Now, by the naturality of the transposition map, we compute

$$\begin{aligned} \mathsf{T}_{ab}^{ab} &= \mathrm{Hom}_{\mathcal{C}}(i_a, i_b^\vee) \circ \mathsf{T}_{X_1 \oplus X_2, X_1 \oplus X_2} \circ \mathrm{Hom}_{\mathcal{C}}(p_a, p_b^\vee) \\ &= \mathrm{Hom}_{\mathcal{C}}(i_a, i_b^\vee) \circ \mathrm{Hom}_{\mathcal{C}}(p_b, p_a^\vee) \circ \mathsf{T}_{X_a, X_b} \\ &= \mathrm{Hom}_{\mathcal{C}}(p_b \circ i_a, p_a^\vee \circ i_b^\vee) \circ \mathsf{T}_{X_a, X_b} \end{aligned}$$

Hence T_{ab}^{ab} is equal to T_{X_a, X_a} if $a = b$ and is zero if otherwise. Combining this result with (2.13), we obtain $\nu(X_1 \oplus X_2) = \nu(X_1) + \nu(X_2)$. □

The FS indicator is an invariant of categories with duality over k. Indeed, the commutativity of (2.10) yields the following proposition:

Proposition 2.10. *Let $(F, \xi) : \mathcal{C} \to \mathcal{D}$ be a strong duality preserving functor. If F is k-linear and fully faithful, then we have $\nu(F(X)) = \nu(X)$ for all $X \in \mathcal{C}$.*

Similarly, we obtain the following proposition from (2.11):

Proposition 2.11. *Suppose that $\mathcal{C}$ is a category with strong duality over k. Then, for all $X \in \mathcal{C}$, we have $\nu(X^\vee) = \mathrm{Tr}(\mathsf{T}_{X,X}^{\mathrm{op}}) = \nu(X^{\mathrm{op}})$.*

Let $\mathcal{A}$ be a k-linear Abelian category. Recall that a nonzero object of $\mathcal{A}$ is said to be *simple* if it has no proper subobjects. We say that a simple object $V \in \mathcal{A}$ is *absolutely simple* if $\mathrm{End}_{\mathcal{A}}(V) \cong k$. Note that the opposite category $\mathcal{A}^{\mathrm{op}}$ is also k-linear and Abelian. It is easy to see that an object of $\mathcal{A}$ is (absolutely) simple if and only if it is (absolutely) simple as an object of $\mathcal{A}^{\mathrm{op}}$.

Proposition 2.12. *Let $\mathcal{C}$ be an Abelian category with strong duality over k, and let $X \in \mathcal{C}$.*

(a) If X is a finite biproduct of simple objects, then $\nu(X) = \nu(X^\vee)$.

(b) If X is absolutely simple, then $\nu(X) \in \{0, \pm 1\}$. $\nu(X) \neq 0$ if and only if X is self-dual, that is, X is isomorphic to $X^\vee$.

Proof. (a) We first claim that if $V \in \mathcal{C}$ is simple, then $\nu(V) = \nu(V^\vee)$. Let $V \in \mathcal{C}$ be a simple object. Since $(-)^\vee : \mathcal{C} \to \mathcal{C}^{\mathrm{op}}$ is an equivalence, $V^\vee$ is simple as an object of $\mathcal{C}^{\mathrm{op}}$ and hence it is simple as an object of $\mathcal{C}$. If V is isomorphic to $V^\vee$, then our claim is obvious. Otherwise, we have

$$\mathrm{Hom}_{\mathcal{C}}(V, V^\vee) = 0 \quad \text{and} \quad \mathrm{Hom}_{\mathcal{C}}(V^\vee, V^{\vee\vee}) \cong \mathrm{Hom}_{\mathcal{C}}(V^\vee, V) = 0$$

by Schur's lemma. Therefore $\nu(V) = 0 = \nu(V^\vee)$ follows.

Now write X as $X = V_1 \oplus \cdots \oplus V_m$ for some simple objects $V_1, \ldots, V_m \in \mathcal{C}$. By the above arguments and the additivity of the FS indicator, we have

$$\nu(X^\vee) = \nu(V_1^\vee) + \cdots + \nu(V_m^\vee) = \nu(V_1) + \cdots + \nu(V_m) = \nu(X)$$

(b) Suppose that $X \in \mathcal{C}$ is absolutely simple. If X is isomorphic to $X^\vee$, then

$$\dim_k \mathrm{Hom}_{\mathcal{C}}(X, X^\vee) = \dim_k \mathrm{End}_{\mathcal{C}}(X) = 1$$

and hence $\nu(X)$ is either $+1$ or -1 by Proposition 2.9 (b). Otherwise, $\nu(X) = 0$ as we have seen in the proof of (a). □

Remark 2.13. If $\mathcal{C}$ is a pivotal monoidal category over k, then the n-th FS indicator $\nu_n(X)$ of $X \in \mathcal{C}$ is defined for each integer $n \geq 2$; see [8]. The commutativity of (2.10) implies $\nu_2(X) = \nu(X^\vee)$

(see also Remark 2.6). However, in view of Proposition 2.12, $\nu(X) = \nu_2(X)$ always holds in the case where $\mathcal{C}$ is strong, Abelian, and semisimple. We prefer our Definition 2.8 since it is more convenient when we discuss the relation between the FS indicator and invariant bilinear forms.

One would like to define higher FS indicators for an object of a category with duality over k by extending that for an object of a pivotal monoidal category over k. This is impossible because of the following example: For a group G, we denote by Vec_{fd}^G the k-linear pivotal monoidal category of finite-dimensional G-graded vector spaces over k. The n-th FS indicator of $V = \bigoplus_{x \in G} V_x \in \mathsf{Vec}_{fd}^G$ is given by

$$\nu_n(V) = \sum_{x \in G[n]} \dim_k(V_x), \quad \text{where } G[n] = \{x \in G \mid x^n = 1\} \tag{2.14}$$

Now we put

$$\mathcal{C} = \mathsf{Vec}_{fd}^{\mathbb{Z}_4 \times \mathbb{Z}_4} \quad \text{and} \quad \mathcal{D} = \mathsf{Vec}_{fd}^{\mathbb{Z}_2 \times \mathbb{Z}_8}$$

There exists a bijection $f : \mathbb{Z}_4 \times \mathbb{Z}_4 \to \mathbb{Z}_2 \times \mathbb{Z}_8$ such that $f(x^{-1}) = f(x)^{-1}$ for all $x \in \mathbb{Z}_4 \times \mathbb{Z}_4$. f induces an equivalence $\mathcal{C} \approx \mathcal{D}$ of categories with duality over k. If we could define the n-th FS indicator for categories with duality over k, then there would exist at least one equivalence $F : \mathcal{C} \to \mathcal{D}$ such that $\nu_n(F(X)) = \nu_n(X)$ for all $X \in \mathcal{C}$ and $n \geq 2$. However, by (2.14), there is no such equivalence.

2.4. *Separable Functors*

Let H be an involutory Hopf algebra, e.g., the group algebra of a group G. We consider the category $\mathcal{C} = \mathsf{mod}_{fd}(H)$ of Example 2.2. The FS indicator $\nu(V)$ of $V \in \mathcal{C}$ is interpreted as follows: Let $\mathrm{Bil}_H(V)$ denote the set of all H-invariant bilinear forms on V. The transposition map induces

$$\Sigma_V : \mathrm{Bil}_H(V) \to \mathrm{Bil}_H(V), \quad \Sigma_V(b)(v, w) = b(w, v) \quad (b \in \mathrm{Bil}_H(V), v, w \in V)$$

via the canonical isomorphism $\mathrm{Bil}_H(V) \cong \mathrm{Hom}_H(V, V^\vee)$. Now let $\mathrm{Bil}_H^{\pm}(V)$ denote the eigenspace of Σ_V with eigenvalue ± 1. Then we have

$$\nu(V) = \mathrm{Tr}(\mathsf{T}_{V,V}) = \mathrm{Tr}(\Sigma_V) = \dim_k \mathrm{Bil}_H^+(V) - \dim_k \mathrm{Bil}_H^-(V)$$

Hence, from our definition, the relation between $\nu(V)$ and H-invariant bilinear forms on V is clear. On the other hand, it is not obvious that $\nu(V)$ is expressed by a formula like (1.1). It should be emphasized that, from our point of view, (1.1) is not the definition of $\nu(V)$ but rather a formula to compute $\nu(V)$. We note that a similar point of view is effectively used to derive a formula of the FS indicator for semisimple finite-dimensional quasi-Hopf algebras in [3].

A key notion to derive (1.1) is a *separable functor* [27]; a functor $U : \mathcal{C} \to \mathcal{V}$ is said to be *separable* if there exists a natural transformation

$$\Pi_{X,Y} : \mathrm{Hom}_{\mathcal{V}}(U(X), U(Y)) \to \mathrm{Hom}_{\mathcal{C}}(X, Y) \quad (X, Y \in \mathcal{C})$$

such that $\Pi_{X,Y}(U(f)) = f$ for all $f \in \mathrm{Hom}_{\mathcal{C}}(X, Y)$. Such a natural transformation Π is called a *section* of U. Suppose that $\mathcal{C}$ and $\mathcal{V}$ be k-linear. We say that a section Π of U is *k-linear* if $\Pi_{X,Y}$ is k-linear for all $X, Y \in \mathcal{C}$.

Now let $\mathcal{C}$ be a category with duality over k, and let $\mathcal{V}$ be a k-linear category satisfying (2.12). A k-linear functor $U : \mathcal{C} \to \mathcal{V}$ induces a k-linear map

$$U_{X,Y} : \mathrm{Hom}_{\mathcal{C}}(X, Y) \to \mathrm{Hom}_{\mathcal{V}}(U(X), U(Y)) \quad f \mapsto U(f) \quad (X, Y \in \mathcal{C})$$

If U has a k-linear section Π, we define a linear map

$$\widetilde{\mathsf{T}}_{X,Y} : \mathrm{Hom}_{\mathcal{V}}(U(X), U(Y^\vee)) \to \mathrm{Hom}_{\mathcal{V}}(U(Y), U(X^\vee)) \quad (X, Y \in \mathcal{C}) \tag{2.15}$$

so that the diagram

$$\begin{array}{ccc}
\mathrm{Hom}_{\mathcal{V}}(\mathcal{U}(X),\mathcal{U}(Y^\vee)) & \xrightarrow{\widetilde{\mathrm{T}}_{X,Y}} & \mathrm{Hom}_{\mathcal{V}}(\mathcal{U}(Y),\mathcal{U}(X^\vee)) \\
\Pi_{X,Y^\vee}\big\downarrow & & \big\uparrow \mathcal{U}_{Y,X^\vee} \\
\mathrm{Hom}_{\mathcal{C}}(X,Y^\vee) & \xrightarrow[\mathrm{T}_{X,Y}]{} & \mathrm{Hom}_{\mathcal{C}}(Y,X^\vee)
\end{array}$$

commutes. By using the well-known identity $\mathrm{Tr}(AB) = \mathrm{Tr}(BA)$, we have

$$\begin{aligned}
\mathrm{Tr}(\widetilde{\mathrm{T}}_{X,X}) &= \mathrm{Tr}(\mathcal{U}_{X,X^\vee} \circ \mathrm{T}_{X,X} \circ \Pi_{X,X^\vee}) \\
&= \mathrm{Tr}(\Pi_{X,X^\vee} \circ \mathcal{U}_{X,X^\vee} \circ \mathrm{T}_{X,X}) = \mathrm{Tr}(\mathrm{T}_{X,X}) = \nu(X)
\end{aligned} \tag{2.16}$$

Before we explain how (1.1) is derived from (2.16), we recall the following lemma in linear algebra: Let $f : V \to W$ and $g : W \to V$ be linear maps between finite-dimensional vector spaces. We define

$$T : V \otimes W \to V \otimes W, \quad T(v \otimes w) = g(w) \otimes f(v). \quad (v \in V, w \in W)$$

Lemma 2.14. $\mathrm{Tr}(T) = \mathrm{Tr}(fg)$.

The dual of this lemma is also useful: Let $B(V,W)$ be the set of bilinear maps $V \times W \to k$ and consider the map $T^\vee : B(V,W) \to B(V,W)$, $T^\vee(b)(v,w) = b(g(w), f(v))$. Since $T^\vee$ is the dual map of T under the identification $B(V,W) \cong (V \otimes W)^\vee$, we have $\mathrm{Tr}(T^\vee) = \mathrm{Tr}(fg)$.

Proof. Let $\{v_i\}$ and $\{w_j\}$ be a basis of V and W, and let $\{v^i\}$ and $\{w^j\}$ be the dual basis to $\{v_i\}$ and $\{w_j\}$, respectively. Then we have

$$\mathrm{Tr}(T) = \sum_{i,j} \langle v^i, g(w_j) \rangle \langle w^j, f(v_i) \rangle = \sum_j \langle w^j, f(g(w_j)) \rangle = \mathrm{Tr}(fg)$$

Here, the first and the last equality follow from $\mathrm{Tr}(f) = \sum_i \langle v^i, f(v_i) \rangle$ and the second follows from $x = \sum_i \langle v^i, x \rangle v_i$. □

Now, for simplicity, we assume k to be an algebraically closed field of characteristic zero. Let H be a finite-dimensional semisimple Hopf algebra over k, e.g., the group algebra of a finite group G. Then, by the theorem of Larson and Radford [28], H is involutory. Let $\mathsf{Vec}_{fd}(k)$ denote the category of finite-dimensional vector spaces over k. The forgetful functor $\mathsf{mod}_{fd}(H) \to \mathsf{Vec}_{fd}(k)$ is a separable functor with k-linear section

$$\begin{aligned}
&\Pi_{V,W} : \mathrm{Hom}_k(V,W) \to \mathrm{Hom}_H(V,W) \quad (V, W \in \mathsf{mod}(H)) \\
&\Pi_{V,W}(f)(v) = S(\Lambda_{(1)}) f(\Lambda_{(2)} v) \quad (f \in \mathrm{Hom}_k(V,W), v \in V)
\end{aligned}$$

where $\Lambda \in H$ is the *Haar integral* (i.e., the two-sided integral such that $\varepsilon(\Lambda) = 1$). Let $\mathrm{Bil}(V)$ denote the set of all bilinear forms on V. Instead of the map (2.15), we prefer to consider the map

$$\widetilde{\Sigma}_V : \mathrm{Bil}(V) \to \mathrm{Bil}(V), \quad \widetilde{\Sigma}_V(b)(v,w) = b(\Lambda_{(1)} v, \Lambda_{(2)} w), \quad (b \in \mathrm{Bil}(V), v, w \in V)$$

which makes the diagram

$$\begin{array}{ccccccc}
\mathrm{Bil}(V) & \cong & \mathrm{Hom}_k(V,V^\vee) & \xrightarrow{\Pi_{V,V^\vee}} & \mathrm{Hom}_H(V,V^\vee) & \cong & \mathrm{Bil}_H(V) \\
\widetilde{\Sigma}_V\big\downarrow & & \widetilde{\mathrm{T}}_{V,V}\big\downarrow & & \big\downarrow \mathrm{T}_{V,V} & & \big\downarrow \Sigma_V \\
\mathrm{Bil}(V) & \cong & \mathrm{Hom}_k(V,V^\vee) & \xleftarrow[\text{inclusion}]{} & \mathrm{Hom}_H(V,V^\vee) & \cong & \mathrm{Bil}_H(V)
\end{array}$$

commutes; see §3, especially Theorem 3.8, for the details. Let $\rho : H \to \mathrm{End}_k(V)$ be the algebra map corresponding to the action $H \otimes V \to V$. By Lemma 2.14, we have

$$\nu(V) = \mathrm{Tr}(\widetilde{\Sigma}_V) = \mathrm{Tr}\left(\rho(\Lambda_{(1)}) \circ \rho(\Lambda_{(2)})\right) = \chi_V(\Lambda_{(1)}\Lambda_{(2)})$$

where $\chi_V = \mathrm{Tr} \circ \rho$. This is the FS indicator for H introduced by Linchenko and Montgomery [1]. In the case where $H = kG$, we have $\nu(V) = |G|^{-1} \sum_{g \in G} \chi_V(g^2)$ since $\Lambda = |G|^{-1} \sum_{g \in G} g$. Formula (1.1) for compact groups is obtained in a similar way; see §4 for details.

2.5. *Twisted Duality*

To deal with some twisted versions of the Frobenius–Schur theorem, we introduce a *twisting adjunction* of a category with duality and give a method to twist the original duality functor by using a twisting adjunction. Our method can be thought as a generalization of the arguments in [23, §4].

Let $\mathcal{C}$ be a category with duality. Suppose that we are given an adjunction $(F, G, \eta, \varepsilon) : \mathcal{C} \to \mathcal{C}$ and a natural transformation $\xi_X : F(X^\vee) \to G(X)^\vee$. Define ζ_X by

$$\zeta_X : G(X^\vee) \xrightarrow{j_{G(X^\vee)}} G(X^\vee)^{\vee\vee} \xrightarrow{\xi^\vee_{X^\vee}} F(X^{\vee\vee})^\vee \xrightarrow{F(j_X)^\vee} F(X)^\vee \tag{2.17}$$

ζ_X is a natural transformation making the following diagrams commute:

$$\begin{array}{ccc} F(X) & \xrightarrow{F(j_X)} & F(X^{\vee\vee}) \\ {\scriptstyle j_{F(X)}}\downarrow & & \downarrow{\scriptstyle \xi_{X^\vee}} \\ F(X)^{\vee\vee} & \xrightarrow[\zeta^\vee_X]{} & G(X^\vee)^\vee, \end{array} \qquad \begin{array}{ccc} G(X) & \xrightarrow{G(j_X)} & G(X^{\vee\vee}) \\ {\scriptstyle j_{G(X)}}\downarrow & & \downarrow{\scriptstyle \zeta_{X^\vee}} \\ G(X)^{\vee\vee} & \xrightarrow[\xi^\vee_X]{} & F(X^\vee)^\vee \end{array} \tag{2.18}$$

Indeed, the commutativity of the first diagram is checked as follows:

$$\zeta^\vee_X \circ j_{F(X)} = j^\vee_{G(X^\vee)} \circ \xi^{\vee\vee}_{X^\vee} \circ F(j_X)^{\vee\vee} \circ j_{F(X)} = j^\vee_{G(X^\vee)} \circ j_{G(X^\vee)^\vee} \circ \xi_{X^\vee} \circ F(j_X) = \xi_{X^\vee} \circ F(j_X)$$

The commutativity of the second one is checked in a similar way as follows:

$$\begin{aligned} \zeta_{X^\vee} \circ G(j_X) &= F(j_{X^\vee})^\vee \circ \xi^\vee_{X^{\vee\vee}} \circ j_{G(X^{\vee\vee})} \circ G(j_X) \\ &= F(j_{X^\vee})^\vee \circ \xi^\vee_{X^{\vee\vee}} \circ G(j_X)^{\vee\vee} \circ j_{G(X)} = F(j_{X^\vee})^\vee \circ F(j^\vee_X)^\vee \circ \xi^\vee_X \circ j_{G(X)} = \xi^\vee_X \circ j_{G(X)} \end{aligned}$$

Now we define the *twisted duality functor* by $X^\sharp = G(X^\vee)$. The problem is when $\mathcal{C}$ is a category with duality with this new duality functor $(-)^\sharp$.

Lemma 2.15. *Define* $\omega : \mathrm{id}_{\mathcal{C}} \to (-)^\sharp \circ (-)^\sharp$ *by*

$$\omega_X : X \xrightarrow{\eta_X} GF(X) \xrightarrow{GF(j)} GF(X^{\vee\vee}) \xrightarrow{G(\xi_{X^\vee})} G(G(X^\vee)^\vee) = X^{\sharp\sharp} \tag{2.19}$$

The triple $(\mathcal{C}, (-)^\sharp, \omega)$ *is a category with duality if*

$$\left(FG(X^\vee) \xrightarrow{F(\zeta_X)} F(F(X)^\vee) \xrightarrow{\xi_{F(X)}} (GF(X))^\vee \xrightarrow{\eta^\vee_X} X^\vee\right) = \varepsilon_{X^\vee} \tag{2.20}$$

holds for all $X \in \mathcal{C}$. *If, moreover,* $\mathcal{C}$ *is strong, then the following are equivalent:*

(1) *F is an equivalence.*
(2) *G is an equivalence.*
(3) *$(\mathcal{C}, (-)^\sharp, \omega)$ is strong.*

Proof. Let $X \in \mathcal{C}$. By (2.17), ω_X can be expressed in two ways as follows:

$$\omega_X = G(\xi_{X^\vee} \circ F(j_X)) \circ \eta_X = G(\zeta_X \circ j_{F(X)}) \circ \eta_X$$

For each $X \in \mathcal{C}$, we compute

$$\begin{aligned} \omega_X^\sharp \circ \omega_{X^\sharp} &= G(\omega_X^\vee) \circ \omega_{G(X^\vee)} = G(\omega_X^\vee) \circ G\big(\xi_{G(X^\vee)^\vee} \circ F(j_{G(X^\vee)})\big) \circ \eta_{G(X^\vee)} \\ &= G\big(\eta_X^\vee \circ G(\zeta_X^\vee \circ j_{F(X)})^\vee \circ \xi_{G(X^\vee)^\vee} \circ F(j_{G(X^\vee)})\big) \circ \eta_{G(X^\vee)} \\ &= G\big(\eta_X^\vee \circ \xi_{F(X)} \circ F(j_{F(X)}^\vee \circ j_{F(X)^\vee} \circ \zeta_X)\big) \circ \eta_{G(X^\vee)} \\ &= G\big(\eta_X^\vee \circ \xi_{F(X)} \circ F(\zeta_X)\big) \circ \eta_{G(X^\vee)} = G(\varepsilon_{X^\vee}) \circ \eta_{G(X^\vee)} = \mathrm{id}_{G(X^\vee)} = \mathrm{id}_{X^\sharp} \end{aligned}$$

Here, the fourth equality follows from the naturality of ξ, the fifth from (2.4), the sixth from the Assumption (2.20), and the seventh from (2.1). Now we have shown that the triple $(\mathcal{C}, (-)^\sharp, \omega_X)$ is a category with duality.

It is easy to prove the rest of the statement; (1) $\Leftrightarrow$ (2) follows from basic properties of adjunctions. To show (2) $\Leftrightarrow$ (3), recall Lemma 2.3. □

In view of Lemma 2.15, we call a quintuple $\mathbf{t} = (F, G, \eta, \varepsilon, \xi)$ satisfying (2.20) a *twisting adjunction* for $\mathcal{C}$. Given such a quintuple $\mathbf{t}$, we denote by $\mathcal{C}^{\mathbf{t}}$ the triple $(\mathcal{C}, (-)^\sharp, \omega)$ constructed in Lemma 2.15. Now we introduce an involution of a category with duality:

Definition 2.16. An *involution* of $\mathcal{C}$ is a triple $\mathbf{t} = (F, \xi, \eta)$ such that (F, ξ) is a strong duality preserving functor on $\mathcal{C}$ and η is an isomorphism

$$\eta : (\mathrm{id}_{\mathcal{C}}, \mathrm{id}_{(-)^\vee}) \to (F, \xi) \circ (F, \xi) \tag{2.21}$$

of duality preserving functors satisfying

$$\eta_{F(X)} = F(\eta_X) \tag{2.22}$$

for all $X \in \mathcal{C}$. We say that $\mathbf{t}$ is *strict* if ξ and η are identities.

Note that (2.21) is an isomorphism of such functors if and only if

$$\eta_X^\vee \circ \xi_{F(X)} \circ F(\xi_X) \circ \eta_{X^\vee} = \mathrm{id}_{X^\vee} \tag{2.23}$$

holds for all $X \in \mathcal{C}$.

An involution (F, ξ, η) of $\mathcal{C}$ is a special type of twisting adjunction; indeed, by (2.22), the quadruple (F, F, η, η^{-1}) is an adjunction. By the definition of duality preserving functors, the natural transformation (2.17) is given by

$$\zeta_X = F(j_X)^\vee \circ \xi_{X^\vee}^\vee \circ j_{F(X^\vee)} = F(j_X)^\vee \circ \xi_{X^{\vee\vee}} \circ F(j_{X^\vee}) = \xi_{X^{\vee\vee}} \circ F(j_X^\vee) \circ F(j_{X^\vee}) = \xi_X$$

Since the counit is η^{-1}, (2.20) is equivalent to (2.23). Hence $(F, F, \eta, \eta^{-1}, \xi)$ is a twisting adjunction for $\mathcal{C}$. From now, we identify an involution of $\mathcal{C}$ with the corresponding twisting adjunction.

Suppose that $\mathcal{C}$ is a category with duality over k. Let $\mathbf{t} = (F, G, \dots)$ be a twisting adjunction for $\mathcal{C}$. $\mathbf{t}$ is said to be *k-linear* if the functor $F : \mathcal{C} \to \mathcal{C}$ is k-linear. If this is the case, G is also k-linear as the right adjoint of F (see, e.g., [26, IV.1]) and hence $\mathcal{C}^{\mathbf{t}}$ is a category with duality over k.

Definition 2.17. Let $\mathcal{C}$ be a category with duality over k, and let $\mathbf{t}$ be a k-linear twisting adjunction for $\mathcal{C}$. The ($\mathbf{t}$-)*twisted FS indicator* $\nu^{\mathbf{t}}(X)$ of $X \in \mathcal{C}$ is defined by

$$\nu^{\mathbf{t}}(X) = \nu(X^{\mathbf{t}})$$

where $X^{\mathbf{t}} \in \mathcal{C}^{\mathbf{t}}$ is the object X regarded as an object of $\mathcal{C}^{\mathbf{t}}$.

To study the twisted FS indicator, it is useful to introduce the twisted transposition map. Let $\mathcal{C}$ be a category with duality (not necessarily over k). Given a twisting adjunction $\mathbf{t} = (F, G, \eta, \varepsilon, \xi)$ for $\mathcal{C}$, the ($\mathbf{t}$-)*twisted transposition map*

$$\mathsf{T}^{\mathbf{t}}_{X,Y} : \mathrm{Hom}_{\mathcal{C}}(F(X), Y^{\vee}) \to \mathrm{Hom}_{\mathcal{C}}(F(Y), X^{\vee}) \quad (X, Y \in \mathcal{C}) \tag{2.24}$$

is defined for $f : F(X) \to Y^{\vee}$ by

$$\mathsf{T}^{\mathbf{t}}_{X,Y}(f) : F(Y) \xrightarrow{F(f^{\vee} \circ j_Y)} F(F(X)^{\vee}) \xrightarrow{\xi_{F(X)}} (GF(X))^{\vee} \xrightarrow{\eta_X^{\vee}} X^{\vee}$$

For a while, let $\mathsf{T}^{\sharp}_{X,Y} : \mathrm{Hom}_{\mathcal{C}}(X, Y^{\sharp}) \to \mathrm{Hom}_{\mathcal{C}}(Y, X^{\sharp})$ denote the transposition map for $\mathcal{C}^{\mathbf{t}}$. One can easily check that the diagram

$$\begin{array}{ccc} \mathrm{Hom}_{\mathcal{C}}(F(X), Y^{\vee}) & \xrightarrow{\cong} & \mathrm{Hom}_{\mathcal{C}}(X, G(Y^{\vee})) = \mathrm{Hom}_{\mathcal{C}}(X, Y^{\sharp}) \\ {\scriptstyle \mathsf{T}^{\mathbf{t}}_{X,Y}} \downarrow & & \downarrow {\scriptstyle \mathsf{T}^{\sharp}_{X,Y}} \\ \mathrm{Hom}_{\mathcal{C}}(F(Y), X^{\vee}) & \xrightarrow[\cong]{} & \mathrm{Hom}_{\mathcal{C}}(Y, G(X^{\vee})) = \mathrm{Hom}_{\mathcal{C}}(Y, X^{\sharp}) \end{array} \tag{2.25}$$

commutes for all $X, Y \in \mathcal{C}$, where the horizontal arrows are the natural bijection given by (2.2). Hence, under the condition of Definition 2.17, we have

$$\nu^{\mathbf{t}}(X) = \mathrm{Tr}(\mathsf{T}^{\sharp}_{X,X}) = \mathrm{Tr}(\mathsf{T}^{\mathbf{t}}_{X,X})$$

The properties of the twisted FS indicator can be obtained as follows: Apply previous results to $\mathcal{C}^{\mathbf{t}}$ and then interpret the results in terms of $\mathcal{C}$ and $\mathbf{t}$ by using the commutative Diagram (2.25). Following this scheme, a twisted version of Proposition 2.12 is established as follows:

Proposition 2.18. *Let $\mathcal{C}$ be an Abelian category with strong duality over k, let $X \in \mathcal{C}$, and let $\mathbf{t} = (F, G, \dots)$ be a k-linear twisting adjunction for $\mathcal{C}$ such that F is an equivalence of categories. Then:*

(a) *If X is a finite biproduct of simple objects, then $\nu^{\mathbf{t}}(F(X)) = \nu^{\mathbf{t}}(X^{\vee})$.*

(b) *If X is absolutely simple, then $\nu^{\mathbf{t}}(X) \in \{0, \pm 1\}$. $\nu^{\mathbf{t}}(X) \neq 0$ if and only if $F(X) \cong X^{\vee}$.*

Now let H be a finite-dimensional semisimple Hopf algebra over an algebraically closed field k of characteristic zero. We give two examples of k-linear twisting adjunctions for $\mathrm{mod}_{fd}(H)$.

Example 2.19. An automorphism τ of H induces a strict monoidal autoequivalence on $\mathrm{mod}_{fd}(H)$. If $\tau^2 = \mathrm{id}_H$, then it gives rise to a strict involution of $\mathrm{mod}_{fd}(H)$, which we denote by the same symbol τ. The study of the τ-twisted FS indicator leads us to the results of Sage and Vega [23]; see §3.

Example 2.20. Let $L \in \mathrm{mod}_{fd}(H)$ be a left H-module such that $h_{(1)}\ell \otimes h_{(2)} = h_{(2)}\ell \otimes h_{(1)}$ holds for all $h \in H$ and $\ell \in L$. This condition implies that, for each $X \in \mathrm{mod}_{fd}(H)$, the map

$$\mathrm{flip} : L \otimes X \to X \otimes L, \quad \ell \otimes x \mapsto x \otimes \ell \quad (\ell \in L, x \in X)$$

is an isomorphism of left H-modules. Fix a basis $\{\ell_i\}$ of L and define

$$\begin{aligned} \eta_X : X \to L \otimes L^{\vee} \otimes X, &\quad \eta_X(x) = \textstyle\sum \ell_i \otimes \ell_i^{\vee} \otimes x \\ \varepsilon_X : L^{\vee} \otimes L \otimes X \to X, &\quad \varepsilon_X(\ell_i^{\vee} \otimes \ell_j \otimes x) = \delta_{ij} x \end{aligned}$$

for $x \in X$, where $\{\ell_i^{\vee}\}$ is the dual basis of $\{\ell_i\}$. The quadruple $(F = L^{\vee} \otimes (-), G = L \otimes (-), \eta, \varepsilon)$ is an adjunction on $\mathrm{mod}_{fd}(H)$. Moreover, if we define ξ_X by

$$\xi_X : L^{\vee} \otimes X^{\vee} \xrightarrow{\cong} (X \otimes L)^{\vee} \xrightarrow{\mathrm{flip}^{\vee}} (L \otimes X)^{\vee}$$

then the quintuple $\mathbf{t}(L) := (F, G, \eta, \varepsilon, \xi)$ is a twisting adjunction for $\mathsf{mod}_{fd}(H)$. Since, in general, F and G are not monoidal, $\mathbf{t}(L)$ is of different type of twisting adjunctions from Example 2.19.

Following the above notation, we write $X^\sharp$ for $G(X^\vee)$. To interpret the $\mathbf{t}(L)$-twisted FS indicator $\nu(V; L) := \nu^{\mathbf{t}(L)}(V)$, we recall that there is an isomorphism

$$\mathrm{Hom}_H(X, Y^\sharp) = \mathrm{Hom}_H(X, L \otimes Y^\vee) \cong \mathrm{Hom}_H(X \otimes Y, L)$$

natural in $X, Y \in \mathsf{mod}_{fd}(H)$. The transposition map for $\mathsf{mod}_{fd}(H)^{\mathbf{t}(L)}$ induces

$$\Sigma_V^L : \mathrm{Hom}_H(V \otimes V, L) \to \mathrm{Hom}_H(V \otimes V, L), \quad \Sigma_V^L(b)(v, w) = b(w, v)$$

via the above isomorphism. Hence we have $\nu(V; L) = \dim_k B_H^+(V; L) - \dim_k B_H^-(V; L)$, where $B_H^\pm(V; L)$ is the eigenspace of Σ_V^L with eigenvalue ± 1.

Let $B(V; L)$ denote the set of all bilinear maps $V \times V \to L$. To express $\nu(V; L)$ by using the characters of V and L, we use the map

$$\widetilde{\Sigma}_V^L : B(V; L) \to B(V; L), \quad \widetilde{\Sigma}_V^L(b)(v, w) = S(\Lambda_{(1)}) \cdot b(\Lambda_{(2)} w, \Lambda_{(3)} v)$$

which makes the following diagrams commute:

$$\begin{array}{ccccc}
B(V; L) \cong \mathrm{Hom}_k(V \otimes V, L) & \xrightarrow{\Pi_{V \otimes V, L}} & \mathrm{Hom}_H(V \otimes V, L) & \cong & \mathrm{Hom}_H(V, V^\sharp) \\
\widetilde{\Sigma}_V^L \downarrow & & \downarrow \Sigma_V^L & & \downarrow \mathsf{T}_{V,V}^\sharp \\
B(V; L) \cong \mathrm{Hom}_k(V \otimes V, L) & \xleftarrow[\text{inclusion}]{} & \mathrm{Hom}_H(V \otimes V, L) & \cong & \mathrm{Hom}_H(V, V^\sharp)
\end{array}$$

Now, let, in general, $f : A \to B$, $g : B \to A$ and $h : M \to M$ be linear maps between finite-dimensional vector spaces. Then, in a similar way as Lemma 2.14, one can show that the trace of

$$T : \mathrm{Hom}_k(A \otimes B, M) \to \mathrm{Hom}_k(A \otimes B, M), \quad T(\mu)(a \otimes b) = h\mu(g(b) \otimes f(a))$$

is given by $\mathrm{Tr}(T) = \mathrm{Tr}(h)\,\mathrm{Tr}(fg)$. By using this formula, we have

$$\nu(V; L) = \mathrm{Tr}(\widetilde{\Sigma}_V^L) = \chi_L(S(\Lambda_{(1)}))\chi_V(\Lambda_{(2)}\Lambda_{(3)})$$

If $\dim_k L = 1$, then $L \otimes (-)$ is an equivalence and hence Proposition 2.18 can be applied to the above example. By the above arguments, we now obtain in the following another type of twisted version of the Frobenius–Schur theorem for semisimple Hopf algebras.

Theorem 2.21. *Let $\alpha : H \to k$ be an algebra map such that $\alpha(h_{(1)})h_{(2)} = \alpha(h_{(2)})h_{(1)}$ holds for all $h \in H$ (or, equivalently, let α be a central grouplike element of the dual Hopf algebra $H^\vee$), and let L be the left H-module corresponding to α. Then, for all simple module $V \in \mathsf{mod}_{fd}(H)$, we have*

$$\nu(V; L) = \alpha(S(\Lambda_{(1)}))\chi_V(\Lambda_{(2)}\Lambda_{(3)}) \in \{0, \pm 1\}$$

Moreover, for a simple module $V \in \mathsf{mod}_{fd}(H)$, the following statements are equivalent:

(1) $\nu(V; L) \neq 0$.

(2) $V \cong L \otimes V^\vee$.

(3) *There exists a non-degenerate bilinear form $b : V \otimes V \to k$ satisfying*

$$b(h_{(1)}v, h_{(2)}w) = \alpha(h)b(v, w) \quad \text{for all } h \in H \text{ and } v, w \in V$$

If one of the above equivalent statements holds, then such a bilinear form b is unique up to scalar multiples and satisfies $b(w, v) = \nu(V; L)b(v, w)$ for all $v, w \in V$.

For the case where $H = kG$ is the group algebra of a finite group G, the above theorem has been obtained by Mizukawa [29, Theorem 3.5].

2.6. Group Action on a Pivotal Monoidal Categories

We have concentrated on studying generalizations of the second FS indicator. Here we briefly explain how to define the higher twisted FS indicators for k-linear pivotal monoidal categories by generalizing those for semisimple Hopf algebras due to Sage and Vega [23]. As we have remarked in Section 1, the details are left for future work.

For a set S, we denote by $\underline{S}$ the category whose objects are the elements of S and whose morphisms are the identity morphisms. If G is a group, then $\underline{G}$ is a strict monoidal category with tensor product given by $x \otimes y = xy$ $(x, y \in G)$.

Let $\mathcal{C}$ be a k-linear pivotal monoidal category with pivotal structure j. We denote by $\underline{\mathrm{Aut}}_{\mathrm{piv}}(\mathcal{C})$ the category of k-linear monoidal autoequivalences of $\mathcal{C}$ that *preserve the pivotal structure* in the sense of [8]. This is a strict monoidal category with respect to the composition of monoidal functors. By an *action* of G on $\mathcal{C}$, we mean a strong monoidal functor

$$\underline{G} \to \underline{\mathrm{Aut}}_{\mathrm{piv}}(\mathcal{C}), \quad g \mapsto F_g \quad (g \in G)$$

Note that, by definition, there are natural isomorphisms $\mathrm{id}_{\mathcal{C}} \cong F_1$ and $F_x \circ F_y \cong F_{xy}$ of monoidal functors. We say that an action $\underline{G} \to \underline{\mathrm{Aut}}_{\mathrm{piv}}(\mathcal{C})$ is *strict* if it is strict as a monoidal functor and, moreover, $F_g : \mathcal{C} \to \mathcal{C}$ is strict as a monoidal functor for all $g \in G$.

Now suppose that an action of G on $\mathcal{C}$ is given. The *crossed product* $\mathcal{C} \rtimes G$ is a monoidal category defined as follows: As a k-linear category, $\mathcal{C} \rtimes G = \bigoplus_{g \in G} \mathcal{C} \rtimes g$, where $\mathcal{C} \rtimes g = \mathcal{C}$ is a copy of $\mathcal{C}$. Given an object $X \in \mathcal{C}$, we denote by (X, g) the object X regarded as an object of $\mathcal{C} \rtimes g \subset \mathcal{C} \rtimes G$. The tensor product of $\mathcal{C} \rtimes G$ is given by

$$(X, g) \otimes (Y, g') = (X \otimes F_g(Y), gg')$$

see [30] and [31] for details. We now claim:

Lemma 2.22. *$\mathcal{C} \rtimes G$ is rigid. The dual object of $(X, g) \in \mathcal{C} \rtimes g$ is given by*

$$(X, g)^\vee = (F_{g^{-1}}(X^\vee), g^{-1})$$

Moreover, $\mathcal{C} \rtimes G$ is a pivotal monoidal category with pivotal structure given by

$$(X, g) = X \xrightarrow{j_X} X^{\vee\vee} \cong F_g F_{g^{-1}}(X^{\vee\vee}) \xrightarrow{F_g(\xi_{g^{-1};X^\vee})} F_g(F_{g^{-1}}(X^\vee)^\vee) = (X, g)^{\vee\vee}$$

where $\xi_{g^{-1};V} : F_{g^{-1}}(V^\vee) \to F_{g^{-1}}(V)^\vee$ is the duality transform [8, §1] of $F_{g^{-1}} : \mathcal{C} \to \mathcal{C}$.

In the most important case for us where the action of G is strict, this lemma is easy to prove. The proof for general cases is tedious and omitted for brevity.

Definition 2.23. Let $\mathcal{C}$ be a k-linear pivotal monoidal category satisfying (2.12) and suppose that an action of G on $\mathcal{C}$ is given. For a positive integer n and an element $g \in G$, we define the *g-twisted n-th FS indicator $\nu_n^g(V)$* of $V \in \mathcal{C}$ by $\nu_n^g(V) = \nu_n((V, g))$, where ν_n in the right-hand side stands for the n-th FS indicator of Ng and Schauenburg [8] for the k-linear pivotal monoidal category $\mathcal{C} \rtimes G$.

For a pair of positive integers (n, r), the (n, r)-th FS indicator $\nu_{n,r}$ is also defined in [8]. It is clear how to define the *g-twisted (n, r)-th FS indicator $\nu_{n,r}^g$* of $V \in \mathcal{C}$.

We explain that our definition agrees with that of Sage and Vega [23]. For simplicity, we treat all monoidal categories as if they were strict. Note that, as an object of $\mathcal{C}$, we have

$$(V, g)^{\otimes n} = V \otimes F_g(V) \otimes F_g^2(V) \otimes \cdots \otimes F_g^{n-1}(V) \quad (:= \widetilde{V^{\otimes n}})$$

Now let H be a finite-dimensional semisimple Hopf algebra over an algebraically closed field k of characteristic zero. Then the group $G = \mathrm{Aut}_{\mathrm{Hopf}}(H)^{\mathrm{op}}$ naturally acts on $\mathcal{C} = \mathrm{mod}_{fd}(H)$. If $g : H \to H$ is an automorphism such that $g^n = \mathrm{id}_H$, then the map

$$E^{(n)}_{(V,g)} : \mathrm{Hom}_{\mathcal{C}\rtimes G}(\mathbf{1}_{\mathcal{C}\rtimes G}, (V,g)^{\otimes n}) \to \mathrm{Hom}_{\mathcal{C}\rtimes G}(\mathbf{1}_{\mathcal{C}\rtimes G}, (V,g)^{\otimes n})$$

used to define the n-th FS indicator in [8] coincides with the map

$$\alpha : (\widetilde{V^{\otimes n}})^H \to (\widetilde{V^{\otimes n}})^H, \quad \sum_{i_1,\dots,i_n} v^1_{i_1} \otimes v^2_{i_2} \cdots \otimes v^n_{i_n} \mapsto \sum_{i_1,\dots,i_n} v^2_{i_1} \otimes \cdots \otimes v^n_{i_n} \otimes v^1_{i_1}$$

under the canonical identification $(\widetilde{V^{\otimes n}})^H = \mathrm{Hom}_{\mathcal{C}\rtimes G}(\mathbf{1}_{\mathcal{C}\rtimes G}, (V,g)^{\otimes n})$. Since the twisted FS indicator of [23] is equal to the trace of the map α [23, Theorem 3.5], our definition agrees with that [23] in the case where both are defined.

Recall that a pivotal monoidal category is a category with duality. If $G = \langle a \mid a^2 = 1\rangle$ acts on a k-linear pivotal monoidal category $\mathcal{C}$, then the functor $F_a : \mathcal{C} \to \mathcal{C}$ is naturally an involution of $\mathcal{C}$ in the sense of Definition 2.16. Let $\mathbf{t}$ denote this involution. By using the crossed product $\mathcal{C} \rtimes G$, the category $\mathcal{C}^{\mathbf{t}}$ constructed in Lemma 2.15 can be described as follows:

Proposition 2.24. *$\mathcal{C}^{\mathbf{t}} \to \mathcal{C} \rtimes G$, $X \mapsto (X, a)$ is a k-linear fully faithful duality preserving functor.*

This is clear from Lemma 2.22 and the definition of $\mathcal{C}^{\mathbf{t}}$.

3. Pivotal Algebras

3.1. Pivotal Algebras

In this section, we introduce and study a class of algebras such that its representation category is a category with strong duality. We first recall that $\mathrm{mod}_{fd}(H)$ is a pivotal monoidal category if H is a pivotal Hopf algebra [14]. Note that the monoidal structure of $\mathrm{mod}_{fd}(H)$ is defined by using the comultiplication of H. Since we do not need a monoidal structure, it seems to be a good way to consider "pivotal Hopf algebras with no comultiplication". This is the notion of *pivotal algebras*, which is formally defined as follows:

Definition 3.1. A *pivotal algebra* is a triple (A, S, g) consisting of an algebra A, an anti-algebra map $S : A \to A$, and an invertible element $g \in A$ satisfying $S(g) = g^{-1}$ and $S^2(a) = gag^{-1}$ for all $a \in A$.

Let $A = (A, S, g)$ be a pivotal algebra. We denote by $\mathrm{mod}(A)$ the category of left A-modules and by $\mathrm{mod}_{fd}(A)$ its full subcategory of finite-dimensional modules. Given $V \in \mathrm{mod}(A)$, we can make its dual space $V^\vee$ into a left A-module by

$$\langle af, v\rangle := \langle f, S(a)v\rangle \quad (a \in A, f \in V^\vee, v \in V) \tag{3.1}$$

The assignment $V \mapsto V^\vee$ extends to a contravariant endofunctor on $\mathrm{mod}(A)$. Now, for each $V \in \mathrm{mod}(A)$, we define $j_V : V \to V^{\vee\vee}$ by

$$\langle j_V(v), f\rangle = \langle f, gv\rangle \quad (v \in V, f \in V^\vee) \tag{3.2}$$

The following computation shows that $j_V : V \to V^{\vee\vee}$ is A-linear:

$$\langle a \cdot j_V(v), f\rangle = \langle S(a)f, gv\rangle = \langle f, S^2(a)gv\rangle = \langle f, gav\rangle = \langle j_V(av), f\rangle$$

It is obvious that j_V is natural in $V \in \mathrm{mod}(A)$. Now we verify (2.4) as follows:

$$\langle (j_V)^\vee j_{V^\vee}(f), v\rangle = \langle j_{V^\vee}(f), j_V(v)\rangle = \langle gf, gv\rangle = \langle f, S(g)gv\rangle = \langle f, v\rangle$$

Note that j_V is an isomorphism if and only if $\dim_k V < \infty$. We conclude:

Proposition 3.2. *Let A be a pivotal algebra. Then $\mathsf{mod}(A) = (\mathsf{mod}(A), (-)^\vee, j)$ is an Abelian category with duality over k. The full subcategory $\mathsf{mod}_{fd}(A)$ is an Abelian category with strong duality over k.*

Let A and B be algebras. Given an algebra map $f : A \to B$, we can make each left B-module into a left A-module by defining $a \cdot v = f(a)v$ $(a \in A, v \in V)$. We denote by $f^\natural(V)$ the left A-module obtained in this way from V. The assignment $V \mapsto f^\natural(V)$ extends to a functor

$$f^\natural : \mathsf{mod}(B) \to \mathsf{mod}(A) \tag{3.3}$$

By restriction, we also obtain a functor

$$f^\natural|_{fd} : \mathsf{mod}_{fd}(B) \to \mathsf{mod}_{fd}(A) \tag{3.4}$$

It is easy to see that these functors are k-linear, exact and faithful. If, moreover, f is surjective, then they are full.

Suppose that $A = (A, S, g)$ and $B = (B, S', g')$ are pivotal algebras. A *morphism of pivotal algebras* from A to B is an algebra map $f : A \to B$ satisfying $f(g) = g'$ and $S'(f(a)) = f(S(a))$ for all $a \in A$. If f is such a morphism, then the Functors (3.3) and (3.4) are strict duality preserving functors.

An *involution* of A is a morphism $\tau : A \to A$ of pivotal algebras such that $\tau^2 = \mathrm{id}_A$. Such a τ gives rise to a strict involution of $\mathsf{mod}(A)$, which is usually denoted by the same symbol τ. The proof of the following proposition is straightforward and omitted.

Proposition 3.3. *Let $A = (A, S, g)$ be a pivotal algebra, and let τ be an involution of A. Put $S^\tau = S \circ \tau (= \tau \circ S)$. Then:*

(a) *The triple $A^\tau = (A, S^\tau, g)$ is a pivotal algebra.*

(b) $\mathrm{id}_{\mathsf{mod}(A)}$ *is a strict duality preserving functor* $\mathsf{mod}(A^\tau) \to \mathsf{mod}(A)^\tau$.

This implies that the τ-twisted FS indicator of $V \in \mathsf{mod}_{fd}(A)$ is equal to the untwisted FS indicator of V regarded as a left A^τ-module. Thus, in principle, the theory of the τ-twisted FS indicator reduces to that of the untwisted FS indicator of A^τ, which is again a pivotal algebra.

3.2. FS Indicator for Pivotal Algebras

Let $A = (A, S, g)$ be a pivotal algebra, and let $V \in \mathsf{mod}_{fd}(A)$. Since $\mathsf{mod}_{fd}(A)$ is a category with duality over k satisfying (2.12), we can define $\nu(V)$ in the way of Section 2.

The FS indicator $\nu(V)$ is interpreted as follows: Let $\mathrm{Bil}(V)$ be the set of all bilinear forms on V. Recall that there is a canonical isomorphism

$$B_V : \mathrm{Hom}_k(V, V^\vee) \to \mathrm{Bil}(V), \quad B_V(f)(v, w) = \langle f(v), w \rangle \tag{3.5}$$

Let $\mathrm{Bil}_A(V)$ be the subset of $\mathrm{Bil}(V)$ consisting of those $b \in \mathrm{Bil}(V)$ such that

$$b(av, w) = b(v, S(a)w) \quad (a \in A, v, w \in V) \tag{3.6}$$

The set $\mathrm{Bil}_A(V)$ is in fact the image of $\mathrm{Hom}_A(V, V^\vee) \subset \mathrm{Hom}_k(V, V^\vee)$ under the canonical isomorphism (3.5). Now we define $\Sigma_V : \mathrm{Bil}_A(V) \to \mathrm{Bil}_A(V)$ so that

$$\begin{array}{ccc} \mathrm{Hom}_A(V, V^\vee) & \xrightarrow{B_V} & \mathrm{Bil}_A(V) \\ {\scriptstyle \mathsf{T}_{V,V}} \downarrow & & \downarrow {\scriptstyle \Sigma_V} \\ \mathrm{Hom}_A(V, V^\vee) & \xrightarrow[B_V]{} & \mathrm{Bil}_A(V) \end{array}$$

commutes. If $b = B_V(f)$ for some $f \in \mathrm{Hom}_A(V, V^\vee)$, then we have

$$\Sigma_V(b)(v,w) = B_V(f^\vee j_V)(v,w) = \langle f^\vee j_V(v), w\rangle = \langle f(w), gv\rangle = b(w, gv) \tag{3.7}$$

for all $v, w \in V$. In view of (3.7), we set

$$\mathrm{Bil}_A^{\pm}(V) = \{b \in \mathrm{Bil}_A(V) \mid b(w, gv) = \pm b(v,w) \text{ for all } v, w \in V\}$$

Then, as a counterpart of Proposition 2.9 (b), we have a formula

$$\nu(V) = \dim_k \mathrm{Bil}_A^+(V) - \dim_k \mathrm{Bil}_A^-(V) \tag{3.8}$$

Rephrasing the results of Section 2 by using these notations, we immediately obtain the following theorem:

Theorem 3.4. *If $V \in \mathsf{mod}_{fd}(A)$ is absolutely simple, then we have $\nu(V) \in \{0, \pm 1\}$. Moreover, the following are equivalent:*

(1) $\nu(V) \neq 0$.

(2) *V is isomorphic to $V^\vee$ as a left A-module.*

(3) *There exists a non-degenerate bilinear form b on V satisfying* (3.6).

If one of the above statements holds, then such a bilinear form b is unique up to scalar multiples and satisfies $b(w, gv) = \nu(V) \cdot b(v,w)$ for all $v, w \in V$.

We denote by R_A the left regular representation of A. If A is finite-dimensional, then the FS indicator of R_A is defined. In the case where $A = kG$ is the group algebra of a finite group G, there is a well-known formula $\nu(\mathsf{R}_{kG}) = \#\{x \in G \mid x^2 = 1\}$. This formula is generalized to finite-dimensional pivotal algebras as follows:

Theorem 3.5. *Let $A = (A, S, g)$ be a finite-dimensional pivotal algebra.*

(a) $\nu(\mathsf{R}_A) = \mathrm{Tr}(Q)$, *where* $Q : A \to A$, $Q(a) = S(a)g$.

(b) *If A is Frobenius, then $\mathsf{R}_A \cong \mathsf{R}_A^\vee$ as left A-modules. Hence, $\nu(\mathsf{R}_A) = \nu(\mathsf{R}_A^\vee)$.*

The part (b) is motivated by Remark 2.13; as we have mentioned, our definition of the FS indicator is different from that of [8]. Therefore, if, for example, A is a finite-dimensional pivotal Hopf algebra, then there are two definitions of the FS indicator of the regular representation of A. Nevertheless they are equal since a finite-dimensional Hopf algebra is Frobenius.

Proof. (a) Recall that there is an isomorphism $\Phi : \mathrm{Hom}_A(\mathsf{R}_A, \mathsf{R}_A^\vee) \to \mathsf{R}_A^\vee$ given by $\Phi(f) = f(1)$. For $f \in \mathrm{Hom}_A(\mathsf{R}_A, \mathsf{R}_A^\vee)$ and $a \in A$, we have

$$\langle \Phi \mathsf{T}_{A,A}(f), a\rangle = \langle (f^\vee j_V)(1), a\rangle = \langle j_V(1), f(a)\rangle = \langle f(a), g\rangle$$

Recalling that $f : \mathsf{R}_A \to \mathsf{R}_A^\vee$ is A-linear, we compute

$$\langle f(a), g\rangle = \langle f(a \cdot 1), g\rangle = \langle a \cdot f(1), g\rangle = \langle \Phi(f), S(a)g\rangle, = \langle Q^\vee \Phi(f), a\rangle$$

Hence we have $\mathsf{T}_{A,A} = \Phi^{-1} \circ Q^\vee \circ \Phi$ and therefore

$$\nu(A) = \mathrm{Tr}(\mathsf{T}_{A,A}) = \mathrm{Tr}(Q^\vee) = \mathrm{Tr}(Q)$$

(b) Suppose that A is Frobenius. By definition, there exists $\phi \in A^\vee$ such that the bilinear map $A \times A \to k$, $(a,b) \mapsto \phi(ab)$ $(a, b \in A)$ is non-degenerate. By using ϕ, we define a linear map $f : \mathsf{R}_A \to \mathsf{R}_A^\vee$ by $\langle f(a), b\rangle = \phi(S(a)b)$ $(a, b \in A)$. It is obvious that f is bijective. For $a, b, c \in A$, we compute

$$\langle f(ab), c\rangle = \phi(S(ab)c) = \phi(S(b)S(a)c) = \langle f(b), S(a)c\rangle = \langle a \cdot f(b), c\rangle$$

Thus f is A-linear and therefore $\mathsf{R}_A \cong \mathsf{R}_A^\vee$ as left A-modules. □

The following Theorem 3.6 is motivated by the *trace-like invariant* of Hopf algebras studied in [32] and [33]. Given $V \in \mathsf{mod}(A)$, we denote by $\rho_V : A \to \mathrm{End}_k(V)$ the algebra map induced by the action of A. Let $I_V := \mathrm{Ker}(\rho_V)$ denote the annihilator of V. By the definition of the dual module $V^\vee$, we have $I_{V^\vee} = S(I_V)$. Hence, if V is self-dual, then $\mathrm{Im}(\rho_V) \to \mathrm{Im}(\rho_V)$, $\rho_V(a) \mapsto \rho_V(S(a))$ $(a \in A)$ is well-defined. We also note that if $V \in \mathsf{mod}_{fd}(A)$ is absolutely simple, then:

$$\text{The algebra map } \rho_V : A \to \mathrm{End}_k(V) \text{ is surjective} \tag{3.9}$$

Theorem 3.6. *Let $V \in \mathsf{mod}_{fd}(A)$ be an absolutely simple module. If V is self-dual, then, by the above arguments, the map*

$$S_V : \mathrm{End}_k(V) \to \mathrm{End}_k(V), \quad \rho_V(a) \mapsto \rho_V(S(a)) \quad (a \in A)$$

is well-defined. By using S_V, we also define

$$Q_V : \mathrm{End}_k(V) \to \mathrm{End}_k(V), \quad Q_V(f) = S_V(f) \circ \rho_V(g) \quad (f \in \mathrm{End}_k(V))$$

Then we have:

$$\text{(a) } \mathrm{Tr}(S_V) = \nu(V) \cdot \chi_V(g), \qquad \text{(b) } \mathrm{Tr}(Q_V) = \nu(V) \cdot \dim_k(V)$$

Proof. (a) This can be proved in the same way as [14, Proposition 4.5]. Here we give another proof: Fix an isomorphism $p : V \to V^\vee$ of left A-modules and define $q : V \to V^\vee$ by $q = p^\vee \iota$, where $\iota = \iota_V$ is the canonical isomorphism (2.5). Our first claim is

$$S_V(f) = q^{-1} f^\vee q \quad (f \in \mathrm{End}_k(V))$$

Indeed, (3.9), there exists $a\ in A$ such that $f = \rho_V(a)$ for some $a \in A$. Hence, we compute

$$f^\vee q = (p\rho_V(a))^\vee \iota = (\rho_V(Sa)^\vee p)^\vee \iota = p^\vee \rho_V(Sa)^{\vee\vee} \iota = q\rho_V(Sa) = qS_V(f)$$

Next, we determine the map $V \otimes V^\vee \to V \otimes V^\vee$ induced by S_V via

$$V \otimes V^\vee \to \mathrm{End}_k(V), \quad v \otimes \lambda \mapsto (x \mapsto \lambda(x)v) \quad (\lambda \in V^\vee, v, x \in V)$$

If $f \in \mathrm{End}_k(V)$ is the element corresponding to $v \otimes \lambda \in V \otimes V^\vee$, then we have

$$\langle f^\vee q(x), y \rangle = \langle p^\vee \iota(x), f(y) \rangle = \langle p(v), x \rangle \, \lambda(y) \quad (x, y \in V)$$

and therefore $S_V(f)(x) = \langle p(v), x \rangle q^{-1}(\lambda)$. This means that $S_V(f)$ corresponds to the element $q^{-1}(\lambda) \otimes p(v) \in V \otimes V^\vee$ via the above isomorphism.

By the above observation, we have that the trace of S_V is equal to that of

$$V \otimes V^\vee \to V \otimes V^\vee, \quad v \otimes \lambda \mapsto q^{-1}(\lambda) \otimes p(v) \quad (v \in V, \lambda \in V^\vee)$$

Applying Lemma 2.14, we have $\mathrm{Tr}(S_V) = \mathrm{Tr}(q^{-1}p)$. Now we recall the definition of the transposition map and compute $q = p^\vee \iota = p^\vee j_V \rho_V(g)^{-1} = \mathrm{T}_{V,V}(p)\rho_V(g)^{-1} = \nu(V) \cdot p\rho_V(g)^{-1}$. Hence, we conclude $\mathrm{Tr}(S_V) = \mathrm{Tr}(q^{-1}p) = \nu(V)\,\mathrm{Tr}(\rho_V(g)) = \nu(V)\chi_V(g)$.

(b) The triple $E = (\mathrm{End}_k(V), S_V, \rho_V(g))$ is a pivotal algebra and $\rho_V : A \to E$ is a morphism of pivotal algebras. Let V_0 denote the vector space V regarded as a left E-module. By Proposition 2.10, the functor $\rho_V^\natural : \mathsf{mod}(E) \to \mathsf{mod}(A)$ preserves the FS indicator. Since $V = \rho_V^\natural(V_0)$, we have $\nu(V_0) = \nu(V)$.

Now let $d = \dim_k(V)$. Then we have $\mathsf{R}_E \cong V_0^{\oplus d}$ as left E-modules and therefore $\nu(E) = \nu(V_0)d = \nu(V)d$ by Proposition 2.9. On the other hand, $\nu(E) = \mathrm{Tr}(Q_V)$ by Proposition 3.5. Thus $\mathrm{Tr}(Q_V) = \nu(V)d$ follows. □

The following is a generalization of [2, Theorem 8.8 (iii)].

Corollary 3.7. *Suppose that k is algebraically closed and that $A = (A, S, g)$ is a finite-dimensional semisimple pivotal algebra. Let $\{V_i\}_{i=1,\dots,n}$ be a complete set of representatives of the isomorphism classes of simple left A-modules. Then*

$$\mathrm{Tr}(S) = \sum_{i=1}^{n} \nu(V_i)\chi_i(g)$$

where $\chi_i = \chi_{V_i}$ is the character of V_i.

Proof. Put $I = \{1, \dots, n\}$. For $i \in I$, let $\rho_i : A \to \mathrm{End}_k(V_i)$ denote the action of A on V_i. By the Artin–Wedderburn theorem, we have an isomorphism

$$A \to \mathrm{End}_k(V_1) \oplus \cdots \oplus \mathrm{End}_k(V_n), \quad a \mapsto (\rho_1(a), \dots, \rho_n(a))$$

of algebras. $S : A \to A$ induces an anti-algebra map

$$\tilde{S} : \mathrm{End}_k(V_1) \oplus \cdots \oplus \mathrm{End}_k(V_n) \to \mathrm{End}_k(V_1) \oplus \cdots \oplus \mathrm{End}_k(V_n)$$

via the isomorphism. For each $i \in I$, we have $\tilde{S}(\mathrm{End}_k(V_i)) \subset \mathrm{End}_k(V_{i^*})$, where $i^* \in I$ is the element such that $V_i^\vee \cong V_{i^*}$. Hence we obtain

$$\mathrm{Tr}(S) = \mathrm{Tr}(\tilde{S}) = \sum_{i \in I, i^* = i} \mathrm{Tr}(\widetilde{S}|_{\mathrm{End}_k(V_i)}) = \sum_{i \in I, i^* = i} \nu(V_i)\chi_i(g)$$

by Theorem 3.6. The sum in the right-hand side is equal to $\sum_{i=1}^{n} \nu(V_i)\chi_i(g)$ since $\nu(V_i) = 0$ unless $i = i^*$. The proof is done. □

3.3. Separable Pivotal Algebras

Recall that an algebra A is said to be *separable* if it has a *separability idempotent*, *i.e.*, an element $E \in A \otimes A$ such that $E^1 E^2 = 1$ and $aE^1 \otimes E^2 = E^1 \otimes E^2 a$ for all $a \in A$. If such an element exists, then the forgetful functor $\mathsf{mod}(A) \to \mathsf{Vec}_{fd}$ is separable with section $\Pi_{V,W} : \mathrm{Hom}_k(V, W) \to \mathrm{Hom}_A(V, W)$ $(V, W \in \mathsf{mod}(A))$ given by

$$\Pi_{V,W}(f)(v) = E^1 f(E^2 v) \quad (f \in \mathrm{Hom}_k(V, W))$$

Hence, if a pivotal algebra $A = (A, S, g)$ is separable (as an algebra), then we can apply the arguments of §2.4. This is a rationale for the following theorem:

Theorem 3.8. *Let $A = (A, S, g)$ be a separable pivotal algebra with separability idempotent $E \in A \otimes A$. Then, for all $V \in \mathsf{mod}_{fd}(A)$, we have*

$$\nu(V) = \chi_V(S(E^1) g E^2)$$

Proof. Define $\widetilde{\Sigma}_V : \mathrm{Bil}(V) \to \mathrm{Bil}(V)$ so that the following diagrams commute:

$$\begin{array}{ccccccc}
\mathrm{Bil}(V) & \xrightarrow{B_V^{-1}} & \mathrm{Hom}_k(V, V^\vee) & \xrightarrow{\Pi_{V,V^\vee}} & \mathrm{Hom}_A(V, V^\vee) & \xrightarrow{B_V} & \mathrm{Bil}_A(V) \\
{\scriptstyle \widetilde{\Sigma}_V}\downarrow & & {\scriptstyle \widetilde{\mathrm{T}}_{V,V}}\downarrow & & \downarrow{\scriptstyle \mathrm{T}_{V,V}} & & \downarrow{\scriptstyle \Sigma_V} \\
\mathrm{Bil}(V) & \xleftarrow[B_V]{} & \mathrm{Hom}_k(V, V^\vee) & \xleftarrow[\text{inclusion}]{} & \mathrm{Hom}_A(V, V^\vee) & \xleftarrow[B_V^{-1}]{} & \mathrm{Bil}_A(V)
\end{array}$$

By the arguments in §2.4, $\nu(V)$ is equal to $\mathrm{Tr}(\widetilde{\mathrm{T}}_{V,V})$. However, to make the computation easier, we prefer to compute $\mathrm{Tr}(\widetilde{\Sigma}_V)$, which is also equal to $\nu(V)$.

Let $\Pi'_V : \mathrm{Bil}(V) \to \mathrm{Bil}_A(V)$ be the composition of the arrows of the first row of the above diagram. If $b = B_V(f)$ for some $f \in \mathrm{Hom}_k(V, V^\vee)$, we have

$$\Pi'_V(b)(v,w) = \left\langle \Pi_{V,V^\vee}(f)(v), w \right\rangle = \left\langle f(E^2 v), S(E^1)w \right\rangle = b\left(E^2 v, S(E^1)w\right)$$

Hence, by (3.7), we have

$$\widetilde{\Sigma}_V(b)(v,w) = \Sigma_V\left(\Pi'_V(b)\right)(v,w) = b\left(E^2 w, S(E^1)gv\right)$$

Applying Lemma 2.14, we obtain $\nu(V) = \chi_V(S(E^1)gE^2)$. □

We discuss the relation between Theorem 3.8 and the results of [1,19,20]. Let $A = (A, S, g)$ be a pivotal algebra such that the algebra A is symmetric with trace form $\phi : A \to k$. By definition, the map

$$A \times A \to k, \quad (a,b) \mapsto \phi(ab) \quad (a, b \in A) \tag{3.10}$$

is a non-degenerate bilinear symmetric form. Fix a basis $\{b_i\}_{i\in I}$ of A and let $\{b_i^\vee\}_{i\in I}$ be the dual basis of $\{b_i\}$ with respect to (3.10). As remarked in [19], we have

$$\sum_{i\in I} ab_i \otimes b_i^\vee = \sum_{i\in I} b_i \otimes b_i^\vee a$$

for all $a \in A$. Hence $v_A = \sum_{i\in I} b_i b_i^\vee \in A$ is a central element, called the *volume* of (A, ϕ).

Now we suppose that the base field k is of characteristic zero and A is split semisimple over k. Then, as Doi showed in [19], the volume v_A is invertible and hence $E = \sum_{i\in I} b_i \otimes b_i^\vee v_A^{-1} \in A \otimes A$ is a separability idempotent of A. By Theorem 3.8, the FS indicator of $V \in \mathrm{Rep}(A)$ is given by

$$\nu(V) = \sum_{i\in I} \chi_V(S(b_i)g b_i^\vee v_A^{-1})$$

If V is simple, then, by Schur's lemma, v_A acts on V as $\chi_V(v_A)\dim_k(V)^{-1} \cdot \mathrm{id}_V$. Hence, we have

$$\nu(V) = \frac{\dim_k(V)}{\chi_V(v_A)} \sum_{i\in I} \chi_V(S(b_i)g b_i^\vee) \tag{3.11}$$

The *Schur element* of a simple module $V \in \mathrm{Rep}(A)$ is given by $c_V = \chi_V(v_A)\dim_k(V)^{-2}$ (see Remark 1.6 of [19]). By using the Schur element, $\nu(V)$ is expressed as

$$\nu(V) = \frac{1}{c_V \dim_k(V)} \sum_{i\in I} \chi_V(S(b_i)g b_i^\vee) \tag{3.12}$$

Letting $g = 1$, we recover the results of [1,19]. Geck [20] assumed that $g = 1$ and A has a basis $\{b_i\}_{i\in I}$ such that $b_i^\vee = S(b_i)$. If this is the case, then we have

$$\nu(V) = \frac{1}{c_V \dim_k(V)} \sum_{i\in I} \chi_V(b_i^2)$$

Example 3.9 (Group-like algebras). As a generalization of the group algebra of a finite group and the adjacency algebra of an association scheme, Doi [19,34] introduced a *group-like algebra*; it is defined to be a quadruple $(A, \varepsilon, \mathcal{B}, *)$ consisting of a finite-dimensional algebra A, an algebra map $\varepsilon : A \to k$, a basis $\mathcal{B} = \{b_i\}_{i\in I}$ of A indexed by a set I, and an involutive map $* : I \to I$, $i \mapsto i^*$ satisfying the following conditions:

(G0) There is a special element $0 \in I$ such that $b_0 = 1$ is the unit of A.
(G1) $\varepsilon(b_i) = \varepsilon(b_{i^*}) \neq 0$ for all $i \in I$.
(G2) $p_{ij}^k = p_{j^*i^*}^{k^*}$ for all $i, j, k \in I$, where p_{ij}^k is given by $b_i \cdot b_j = \sum_{k \in I} p_{ij}^k b_k$ $(i, j, k \in I)$.
(G3) $p_{ij}^0 = \delta_{ij^*}\varepsilon(b_i)$ for all $i \in I$.

Now let $A = (A, \varepsilon, \mathcal{B}, *)$ be a group-like algebra. Define a linear map $S : A \to A$ by $S(b_i) = b_{i^*}$ for $i \in I$. One can check that the triple $A = (A, S, 1)$ is a pivotal algebra. By an *involution* of A, we mean an involutive map $\tau : I \to I$ satisfying

$$\tau(i^*) = \tau(i)^* \quad \text{and} \quad p_{\tau(i),\tau(j)}^{\tau(k)} = p_{ij}^k$$

for all $i, j, k \in I$. Such a map gives rise to an involution of the pivotal algebra $(A, S, 1)$.

In what follows, we compute the τ-twisted FS indicator $\nu^\tau(V)$ of a simple module $V \in \mathrm{Rep}(A)$ under the assumption that the base field is $\mathbb{C}$ and $\varepsilon(b_i) > 0$ for all $i \in I$. Then, by the results of Doi [19], A is semisimple. Note that A is a symmetric algebra with trace form given by $\phi(b_i) = \delta_{i0}$ and the dual basis of $\{b_i\}$ with respect to (3.10) is given by $b_i^\vee = \varepsilon(b_i)^{-1} b_{i^*}$. Applying (3.11) and (3.12) to $A^\tau = (A, S \circ \tau, 1)$, we obtain the following formula:

$$\nu^\tau(V) = \frac{\dim_{\mathbb{C}}(V)}{\chi_V(v_A)} \sum_{i \in I} \frac{1}{\varepsilon(b_i)} \chi_V(b_{\tau(i)} b_i) = \frac{1}{c_V \dim_{\mathbb{C}}(V)} \sum_{i \in I} \frac{1}{\varepsilon(b_i)} \chi_V(b_{\tau(i)} b_i)$$

In particular, applying this formula to the adjacency algebra of an association scheme, we recover the formula of Hanaki and Terada [18].

To obtain the twisted Frobenius–Schur theorem for this class of algebras, combine the above formula with Theorem 3.4; we then have $\nu^\tau(V) \in \{0, \pm 1\}$. Moreover, $\nu^\tau(V) \neq 0$ if and only if there exists a non-degenerate bilinear form β on V such that $\beta(b_{\tau(i)} v, w) = \beta(v, b_{i^*} w)$ for all $i \in I$ and $v, w \in V$. Such a bilinear form β is symmetric if $\nu^\tau(V) = +1$ and skew-symmetric if $\nu^\tau(V) = -1$.

Example 3.10 (Weak Hopf C^*-algebras). We assume that the base field is $\mathbb{C}$. A *weak Hopf algebra* is an algebra H which is a coalgebra at the same time such that there exists a special map $S : H \to H$ called the antipode; see [35] and [36] for the precise definition. We note that the antipode of a weak Hopf algebra is known to be an anti-algebra map.

Let H be a finite-dimensional weak Hopf C^*-algebra; see [35, §4] for the precise definition. There exists an element $g \in H$, called the *canonical grouplike element* [35, §4], satisfying $S(g) = g^{-1}$, $S^2(x) = gxg^{-1}$ for all $x \in H$, and some other good properties. In particular, the triple (H, S, g) is a pivotal algebra and therefore the FS indicator $\nu(V)$ is defined for each $V \in \mathsf{mod}_{fd}(H)$.

We can express $\nu(V)$ by using the *Haar integral* [35, §3]; if $\Lambda \in H$ is the Haar integral in H, then $E = S(\Lambda_{(1)}) \otimes \Lambda_{(2)}$ is a separability idempotent of H (*cf.* the proof of Theorem 3.13 of [35]). Applying Theorem 3.8, we have

$$\nu(V) = \chi_V(S^2(\Lambda_{(1)}) g \Lambda_{(2)}) = \chi_V(g \Lambda_{(1)} \Lambda_{(2)}) \tag{3.13}$$

Combining the above formula with Theorem 3.4, we obtain the Frobenius–Schur theorem for semisimple weak Hopf algebras. We finally give some remarks concerning this example:

(1) Takahiro Hayashi (in private communication with the author) has proved (3.13) and analogous formulas of the higher FS indicators for weak Hopf algebras in the case where $S^2 = \mathrm{id}_H$.

(2) The formula of Linchenko and Montgomery [1] is the case where H is an ordinary Hopf algebra. If this is the case, then the C^*-condition for H is not needed since we have $S^2 = \mathrm{id}_H$ by the theorem of Larson and Radford [28]. It is not known whether every semisimple weak Hopf algebra H has a grouplike element g such that $S^2(h) = ghg^{-1}$ for all $h \in H$. An affirmative answer to this question proves Conjecture 2.8 of [37], which states that every fusion category admits a pivotal structure.

(3) We do not know whether our Formula (3.13) is equivalent to [5, (4.3)] or [36, (3.70)]. In [2,5,36], formulas are proved by finding a central element e such that $\nu(V) = \chi_V(e)$ for all V. On the other hand, the element $S(E^1) g E^2$ of our Theorem 3.8 is not central in general. In §3.4, we give a formula

of the FS indicator for quasi-Hopf algebras and its twisted version. For the above reason, it is not straightforward to derive the formula of Mason and Ng [2] from our formula.

(4) By an involution of H, we mean an involutive algebra map $\tau : H \to H$ which is also a coalgebra map. Such a map τ is in fact an involution of the pivotal algebra (H, S, g) and the τ-twisted FS indicator of $V \in \mathrm{Rep}(H)$ is given by $\nu^{\tau}(V) = \chi_V(g\tau(\Lambda_{(1)})\Lambda_{(2)})$. We omit the details since these results can be proved in a similar way as the case of quasi-Hopf algebras; see §3.4.

Example 3.11 (Twisting by $L \otimes (-)$). By using separable pivotal algebras, Theorem 2.21 can be proved as follows: Let H, α and L be as in that theorem and define $T_\alpha : H \to H$ by $T_\alpha(h) = \alpha(h_{(1)})S(h_{(2)})$ $(h \in H)$. It is easy to see that the triple $H' = (H, T_\alpha, 1)$ is a pivotal algebra and the identity functor is a strict duality preserving functor between $\mathsf{mod}_{fd}(H)^{\mathbf{t}(L)}$ and $\mathsf{mod}_{fd}(H')$. Hence, by Proposition 2.10 and Theorem 3.8, we have

$$\nu(V; L) = \chi_V(T_\alpha S(\Lambda_{(1)})\Lambda_{(2)}) = \alpha(S(\Lambda_{(1)}))\chi_V(\Lambda_{(2)}\Lambda_{(3)})$$

for all $V \in \mathsf{mod}_{fd}(H)$. The meaning of $\nu(V; L)$ is obtained by applying Theorem 3.4 to H'.

3.4. *Quasi-Hopf Algebras*

We derive a formula of Mason and Ng [2] and its twisted version from our results.

Recall that a *quasi-Hopf algebra* [38] is a data $H = (H, \Delta, \varepsilon, \Phi, S, \alpha, \beta)$ consisting of an algebra H, algebra maps $\Delta : H \to H \otimes H$ and $\varepsilon : H \to k$, an anti-algebra automorphism $S : H \to H$, elements $\alpha, \beta \in H$ and an invertible element $\Phi \in H^{\otimes 3}$ with inverse $\overline{\Phi}$ satisfying numerous conditions. Let $H_i = (H_i, \Delta_i, \varepsilon_i, \Phi_i, S_i, \alpha_i, \beta_i)$ be quasi-Hopf algebras $(i = 1, 2)$. A *morphism of quasi-Hopf algebras* from H_1 to H_2 is an algebra map $f : H_1 \to H_2$ satisfying

$$\Delta_2 f = (f \otimes f)\Delta_1, \quad \varepsilon_2 f = \varepsilon_1, \quad \Phi_2 = (f \otimes f \otimes f)(\Phi_1)$$
$$S_2 f = f S_1, \quad \alpha_2 = f(\alpha_1), \quad \beta_2 = f(\beta_2)$$

Hence, by an *involution* of a quasi-Hopf algebra H, we shall mean a morphism $\tau : H \to H$ of quasi-Hopf algebras such that $\tau^2 = \mathrm{id}_H$.

If H is a quasi-Hopf algebra, then $\mathsf{mod}_{fd}(H)$ is a rigid monoidal category. Given $V \in \mathsf{mod}_{fd}(H)$, we denote by $e_V : V^\vee \otimes V \to k$ and $c_V : k \to V \otimes V^\vee$ the evaluation and the coevaluation, respectively. Here we need to recall that the dual module of V is defined by the same way as (3.1) and the maps e_V and c_V are given by

$$e_V(\lambda \otimes v) = \langle \lambda, \alpha v \rangle \quad (\lambda \in V^\vee, v \in V); \quad c_V(1) = \sum_{i=1}^{n} \beta v_i \otimes v^i$$

where $\{v_i\}_{i=1,\dots,n}$ is a basis of V and $\{v^i\}$ is the dual basis.

We need additional assumptions on H so that H is a pivotal algebra. In what follows, we suppose that k is an algebraically closed field of characteristic zero and H is a finite-dimensional semisimple quasi-Hopf algebra. Then $\mathsf{mod}_{fd}(H)$ is a fusion category [37] such that each its object has an integral Frobenius–Perron dimension. Therefore, by the results of [37], $\mathsf{mod}_{fd}(H)$ has a *canonical pivotal structure, i.e.*, an isomorphism $j : \mathrm{id}_{\mathsf{mod}_{fd}(H)} \to (-)^{\vee\vee}$ of k-linear monoidal functors such that, for all $V \in \mathsf{mod}_{fd}(H)$, the composition

$$k \xrightarrow{c_V} V \otimes V^\vee \xrightarrow{j_V \otimes \mathrm{id}_{V^\vee}} V^{\vee\vee} \otimes V^\vee \xrightarrow{e_{V^\vee}} k$$

maps $1 \in k$ to $\dim_k(V) \in k$. Now let $g \in H$ be the image of $1 \in H$ under

$$H \xrightarrow{j_H} H^{\vee\vee} \xrightarrow{\iota_H^{-1}} H$$

where ι_H is the Isomorphism (2.5). We call g the *canonical pivotal element* of H. By definition, g is invertible and satisfies $S(g)g = 1$ and $S^2(h) = gag^{-1}$ for all $a \in H$; see [2] and [3] for details. Hence (H, S, g) is a pivotal algebra. Now we remark:

Lemma 3.12. *Let $f : H_1 \to H_2$ be an isomorphism between finite-dimensional semisimple quasi-Hopf algebras. Then we have $f(g_1) = g_2$, where $g_i \in H_i$ is the canonical pivotal element of H_i.*

Proof. f induces a functor $f^\natural : \mathrm{mod}_{fd}(H_1) \to \mathrm{mod}_{fd}(H_2)$. By the definition of morphisms of quasi-Hopf algebras, the functor $f^\natural$ is a k-linear strict monoidal equivalence. The result follows from the fact that such a functor preserves the canonical pivotal structure [8, Corollary 6.2]. □

From this lemma, we see that an involution τ of the quasi-Hopf algebra H is an involution of the pivotal algebra (H, S, g). As we have observed in §3.2, τ gives rise to an involution of $\mathrm{mod}_{fd}(H)$ and hence the τ-twisted FS indicator $\nu^\tau(V)$ is defined for $V \in \mathrm{mod}_{fd}(H)$. By Theorem 3.4 applied to $A = (H, S \circ \tau, g)$, we have the following property of ν^τ:

Theorem 3.13. *Let $V \in \mathrm{mod}_{fd}(H)$ be a simple module. Then $\nu^\tau(V) \in \{0, \pm 1\}$ and the following statements are equivalent:*

(1) $\nu^\tau(V) \neq 0$.

(2) *$\tau^\natural(V)$ is isomorphic to the dual module $V^\vee$ as a H-module.*

(3) *There exists a non-degenerate bilinear form b on V satisfying*

$$b(\tau(h)v, w) = b(v, S(h)w) \quad \text{for all } v, w \in V$$

If one of the above statements holds, then such a bilinear form b is unique up to scalar multiples and satisfies $b(w, gv) = \nu^\tau(V) \cdot b(v, w)$ for all $v, w \in V$.

Next we express the number $\nu^\tau(V)$ by using the character of V. To that end, it is sufficient to find a separability idempotent of H. Let $\Lambda \in H$ be the Haar integral of H (see [39] and [40]). We set

$$p_L = \Phi^2 S^{-1}(\Phi^1 \beta) \otimes \Phi^3, \qquad q_L = S(\overline{\Phi}^1)\alpha\overline{\Phi}^2 \otimes \overline{\Phi}^3$$
$$p_R = \overline{\Phi}^1 \otimes \overline{\Phi}^2 \beta S(\overline{\Phi}^1), \qquad q_R = \Phi^1 \otimes S^{-1}(\alpha\Phi^3)\Phi^2$$

and fix $p \in \{p_L, p_R\}$ and $q \in \{q_L, q_R\}$. Following [2, Lemma 3.1], we have

$$\Lambda_{(1)} p^1 a \otimes \Lambda_{(2)} p^2 = \Lambda_{(1)} p^1 \otimes \Lambda_{(2)} p^2 S(a) \tag{3.14}$$
$$S(a) q^1 \Lambda_{(1)} \otimes q^2 \Lambda_{(2)} = q^1 \Lambda_{(1)} \otimes a q^2 \Lambda_{(2)} \tag{3.15}$$

for all $a \in A$. From these identities, we see that both $S(\Lambda_{(1)} p^1) \otimes \alpha \Lambda_{(2)} p^2$ and $q^1 \Lambda_{(1)} \beta \otimes S(q^2 \Lambda_{(2)})$ are separability idempotents. Applying Theorem 3.8 to $(H, S\tau, g)$, we have

$$\nu^\tau(V) = \chi_V\left(S\tau S(\Lambda_{(1)} p^1) g \alpha \Lambda_{(2)} p^2\right) = \chi_V\left(S\tau(q^1 \Lambda_{(1)} \beta) g S(q^2 \Lambda_{(2)})\right)$$

for all $V \in \mathrm{mod}_{fd}(H)$. Hence, by using the former expression, we compute

$$\begin{aligned}
\nu^\tau(V) &= \chi_V(S^2(\tau(\Lambda_{(1)} p^1)) g \cdot \alpha \Lambda_{(2)} p^2) = \chi_V(g \cdot \tau(\Lambda_{(1)} p^1) \cdot \alpha \Lambda_{(2)} p^2) \\
&= \chi_V(g \cdot \tau(\Lambda_{(1)} p^1 \tau(\alpha)) \cdot \Lambda_{(2)} p^2) \overset{(3.14)}{=} \chi_V(g \cdot \tau(\Lambda_{(1)} p^1) \Lambda_{(2)} p^2 \cdot S\tau(\alpha))) \\
&= \chi_V(S\tau(\alpha) g \cdot \tau(\Lambda_{(1)} p^1) \Lambda_{(2)} p^2)
\end{aligned}$$

Note that the formula of Mason and Ng in [2] does not involve g. To exclude g from the above formula of $\nu^\tau(V)$, we require:

Lemma 3.14. *Fix $p \in \{p_L, p_R\}$ and $q \in \{q_L, q_R\}$. Then we have*

$$g^{-1} S(\beta) = S(\Lambda_{(1)} p^1) \Lambda_{(2)} p^2, \quad S(\alpha) g = S(q^2 \Lambda_{(2)}) q^1 \Lambda_{(1)} \tag{3.16}$$

Proof. The first identity is proved in [3] (where our g appears as g^{-1}) and the second can be proved in a similar way. For the sake of completeness, we give a detailed proof of the second identity.

Let V be a simple H-module and set $c = c_V(1)$. The map

$$e_1 : V \otimes V^\vee \to k, \quad e_2(v \otimes f) = \langle q^2\Lambda_{(2)} f, q^1\Lambda_{(1)} v\rangle \quad (v \in V, f \in V^\vee)$$

is an H-linear map such that $e_1(c) = \dim_k(V)$. On the other hand, by the definition of the canonical pivotal structure, we see that

$$e_2 : V \otimes V^\vee \xrightarrow{j_V \otimes \mathrm{id}_{V^\vee}} V^{\vee\vee} \otimes V^\vee \xrightarrow{ev} k$$

has the same property. Since $\mathrm{Hom}_H(V \otimes V^\vee, k) \cong \mathrm{Hom}_H(V, V) \cong k$, we have $e_1 = e_2$. This implies that $\langle f, S(\alpha)gv\rangle = \langle f, S(q^2\Lambda_{(2)})t^1\Lambda_{(1)}v\rangle$ holds for all $f \in V^\vee$ and $v \in V$. In conclusion, $S(\alpha)g = S(q^2\Lambda_{(2)})q^1\Lambda_{(1)}$ holds on each simple module V. Since H is semisimple, the identity holds in H. $\square$

By Lemmas 3.12 and 3.14, we have $S\tau(\alpha)g = \tau(S(\alpha)g) = S(\tau(q^2\Lambda'_{(2)})) \cdot \tau(q^1\Lambda'_{(1)})$, where $\Lambda' = \Lambda$ is a copy of Λ. Hence we compute:

$$\begin{aligned}
\nu^\tau(V) &= \chi_V(S(\tau(q^2\Lambda'_{(2)})) \cdot \tau(q^1\Lambda'_{(1)}) \cdot \tau(\Lambda_{(1)}p^1)\Lambda_{(2)}p^2) \\
&= \chi_V(\tau(q^1\Lambda'_{(1)}\Lambda_{(1)}p^1) \cdot \Lambda_{(2)}p^2 S(\tau(q^2\Lambda'_{(2)}))) \\
&= \chi_V(\tau(q^1\Lambda'_{(1)}\Lambda_{(1)}p^1\tau(q^2\Lambda'_{(2)})) \cdot \Lambda_{(2)}p^2) \\
&= \chi_V(\tau(q^1\Lambda'_{(1)}\Lambda_{(1)}p^1)q^2\Lambda'_{(2)}\Lambda_{(2)}p^2)
\end{aligned}$$

Since $\Delta : H \to H \otimes H$ is an algebra map, we have $\Lambda'_{(1)}\Lambda_{(1)} \otimes \Lambda'_{(2)}\Lambda_{(2)} = \Delta(\Lambda'\Lambda) = \varepsilon(\Lambda')\Delta(\Lambda) = \Delta(\Lambda)$. Hence, we finally obtain $\nu^\tau(V) = \chi_V(\tau(q^1\Lambda_{(1)}p^1)q^2\Lambda_{(2)}p^2)$. Letting $\tau = \mathrm{id}_H$, we recover the results of Mason and Ng [2]. Assuming H to be a Hopf algebra, we recover the results of Sage and Vega [23].

4. Coalgebras

4.1. Copivotal Coalgebras

In this section, we introduce the dual notion of pivotal algebras and study the Frobenius–Schur theory for them. For the reader's convenience, we briefly recall some basic results on coalgebras.

Given a coalgebra C, we denote by $\mathrm{com}(C)$ the category of *right* C-comodules and by $\mathrm{com}_{fd}(C)$ its full subcategory of finite-dimensional objects. We express the coaction of $V \in \mathrm{com}(C)$ as

$$\rho_V : V \to V \otimes C, \quad v \mapsto v_{(0)} \otimes v_{(1)} \quad (v \in V)$$

The *convolution product* of $\lambda, \mu \in C^\vee$ is defined by $\langle \lambda \star \mu, c\rangle = \langle \lambda, c_{(1)}\rangle\langle \mu, c_{(2)}\rangle$ for all $c \in C$. $C^\vee$ is an algebra, called the *dual algebra*, with multiplication $\star$ and unit ε. The algebra $C^\vee$ acts from the left on each $V \in \mathrm{com}(C)$ by $\rightharpoonup : C^\vee \otimes V \to V$, $\lambda \rightharpoonup v = v_{(0)}\langle \lambda, v_{(1)}\rangle$ $(\lambda \in C^\vee, v \in V)$. This defines a k-linear fully faithful functors $\mathrm{com}(C) \to \mathrm{mod}(C^\vee)$ and

$$\mathrm{com}_{fd}(C) \to \mathrm{mod}_{fd}(C^\vee) \tag{4.1}$$

which are not equivalences in general. If C is finite-dimensional, then these functors are isomorphisms of categories. See, e.g., [41] for details.

Fix a basis $\{v_i\}_{i=1,\dots,n}$ of $V \in \mathrm{com}_{fd}(C)$. Then we can define $c_{ij} \in C$ by $\rho_V(v_j) = \sum_{i=1}^n v_i \otimes c_{ij}$ $(j = 1,\dots,n)$. The matrix (c_{ij}) is called the *matrix corepresentation* of V with respect to the basis $\{v_i\}$. By the definition of comodules, we have

$$\Delta(c_{ij}) = \sum_{s=1}^{n} c_{is} \otimes c_{sj}, \quad \varepsilon(c_{ij}) = \delta_{ij} \tag{4.2}$$

for all $i, j = 1,\dots,n$. Hence $C_V = \mathrm{span}_k\{c_{ij} \mid i,j = 1,\dots,n\}$ is a subcoalgebra of C. We call C_V the *coefficient subcoalgebra* of C. C_V has a special element $t_V = \sum_i c_{ii}$, called the *character* of V. If we regard V as a left $C^\vee$-module via (4.1) and denote its character by χ_V, then we have

$$\chi_V(\lambda) = \langle \lambda, c_{11}\rangle + \cdots + \langle \lambda, c_{nn}\rangle = \lambda(t_V) \tag{4.3}$$

for all $\lambda \in C^\vee$.

Now let V be a finite-dimensional vector space with basis $\{v_i\}_{i=1,\dots,n}$ and let $\{v^i\}$ denote the dual basis. Then $\mathrm{End}^c(V) = V^\vee \otimes V$ has a basis $e_{ij} = v^i \otimes v_j$ $(i,j = 1,\dots,n)$ and turns into a coalgebra with $\Delta(e_{ij}) = \sum_{s=1}^n e_{is} \otimes e_{sj}$, $\varepsilon(e_{ij}) = \delta_{ij}$. $\mathrm{End}^c(V)$ coacts on V from the right by

$$V \to V \otimes \mathrm{End}^c(V), \quad v_j \mapsto \sum_{j=1}^{n} v_i \otimes e_{ij} \quad (j = 1,\dots,n)$$

Suppose that C coacts on V. Let (c_{ij}) be the matrix corepresentation of V with respect to $\{v_i\}$. By (4.2), the linear map $\phi : \mathrm{End}^c(V) \to C$, $\phi(e_{ij}) = c_{ij}$ is a coalgebra map. Conversely, if a coalgebra map $\phi : \mathrm{End}^c(V) \to C$ is given, V is a right C-comodule by

$$\rho : V \to V \otimes C, \quad v_j \mapsto \sum_{i=1}^{n} v_i \otimes \phi(e_{ij}) \quad (j = 1,\dots,n)$$

These constructions give a bijection between the set of linear maps $\rho : V \to V \otimes C$ making V into a right C-comodule and the set of coalgebra maps $\phi : \mathrm{End}^c(V) \to C$.

Suppose that $V \in \mathrm{com}_{fd}(C)$ is absolutely simple. As the dual of (3.9), we have that the corresponding coalgebra map $\phi : \mathrm{End}^c(V) \to C$ is injective. Let (c_{ij}) be the matrix corepresentation of V with respect to some basis of V. The injectivity of ϕ implies that the set $\{c_{ij}\}$ is linearly independent.

Now we introduce *copivotal coalgebras* as the dual notion of pivotal algebras:

Definition 4.1. A *copivotal coalgebra* is a triple (C, S, γ) consisting of a coalgebra C, an anti-coalgebra map $S : C \to C$ and a linear map $\gamma : C \to k$ satisfying

$$S^2(c) = \langle \gamma, c_{(1)}\rangle c_{(2)} \langle \overline{\gamma}, c_{(3)}\rangle \quad \text{and} \quad \overline{\gamma} = \gamma \circ S$$

for all $c \in C$, where $\overline{\gamma} : C \to k$ is the inverse of γ with respect to $\star$.

Let $C = (C, S, \gamma)$ is a copivotal coalgebra. If $V \in \mathrm{com}_{fd}(C)$, then we can make $V^\vee$ into a right C-comodule as follows: First fix a basis $\{v_i\}$ of V and let (c_{ij}) be the matrix corepresentation of V with respect to the basis $\{v_i\}$. Then we define the coaction of C on $V^\vee$ by

$$\rho_{V^\vee} : V^\vee \to V^\vee \otimes C, \quad \rho_{V^\vee}(v^j) = \sum_{i=1}^{n} v^i \otimes S(c_{ji}) \quad (i = 1,\dots,n) \tag{4.4}$$

where $\{v^i\}$ is the dual basis of $\{v_i\}$. This coaction does not depend on the choice of the basis and has the following characterization, which is rather useful than the above explicit formula:

$$\langle f_{(0)}, v\rangle f_{(1)} = \langle f, v_{(0)}\rangle S(v_{(1)}) \quad (f \in V^\vee, v \in V)$$

For each $V \in \mathsf{com}_{fd}(C)$, we define $j_V : V \to V^{\vee\vee}$ by

$$\langle j_V(v), f\rangle = \langle f, \gamma \rightharpoonup v\rangle \quad (= \langle f, v_{(0)}\rangle\langle \gamma, v_{(1)}\rangle) \quad (f \in V^\vee, v \in V)$$

In a similar way as Proposition 3.2, we prove:

Proposition 4.2. $\mathsf{mod}_{fd}(C)$ *is a category with strong duality over k.*

The triple $C^\vee = (C^\vee, S^\vee, \gamma)$ is a pivotal algebra, which we call the *dual pivotal algebra* of C. Let $V \in \mathsf{mod}_{fd}(C)$. For all $\lambda \in C^\vee$, $f \in V^\vee$ and $v \in V$, we have

$$\langle \lambda \rightharpoonup f, v\rangle = \langle f_{(0)}, v\rangle\langle \lambda, f_{(1)}\rangle = \langle f, v_{(0)}\rangle\langle \lambda, S(v_{(1)})\rangle = \langle f, S^\vee(\lambda) \rightharpoonup v\rangle$$

This implies that the Functor (4.1) is in fact a strict duality preserving functor. In what follows, we often regard $\mathsf{com}_{fd}(C)$ as a full subcategory of $\mathsf{mod}_{fd}(C^\vee)$.

4.2. FS Indicator for Copivotal Coalgebras

Let $C = (C, S, \gamma)$ be a copivotal coalgebra, and let $V \in \mathsf{com}_{fd}(V)$. We denote by $\mathrm{Bil}(V)$ the set of all bilinear forms on V and by $\mathrm{Bil}_C(V)$ its subset consisting of those $b \in \mathrm{Bil}(V)$ satisfying

$$b(v_{(0)}, w)v_{(1)} = b(v, w_{(0)})S(w_{(1)}) \quad \text{for all } v, w \in V \tag{4.5}$$

$\mathrm{Bil}_C(V)$ is the image of $\mathrm{Hom}_C(V, V^\vee) \subset \mathrm{Hom}_k(V, V^\vee)$ under the canonical Isomorphism (3.5). Define $\Sigma_V : \mathrm{Bil}_C(V) \to \mathrm{Bil}_C(V)$ in the same way as before. Then, for all $b \in \mathrm{Bil}_C(V), v, w \in V$, we have

$$\Sigma_V(b)(v, w) = b(w, \gamma \rightharpoonup v)$$

Now let $\mathrm{Bil}_C^\pm(V)$ be the eigenspace of Σ_V with eigenvalue ± 1:

$$\mathrm{Bil}_C^\pm(V) = \{b \in \mathrm{Bil}_C(V) \mid b(w, \gamma \rightharpoonup v) = \pm b(v, w) \text{ for all } v, w \in V\}$$

Then we have $\nu(V) = \dim_k \mathrm{Bil}_C^+(V) - \dim_k \mathrm{Bil}_C^-(V)$ as a counterpart of Proposition 2.9 (b). Now we immediately obtain the following coalgebraic version of Theorem 3.4:

Theorem 4.3. *If* $V \in \mathsf{com}_{fd}(C)$ *is absolutely simple, then we have* $\nu(V) \in \{0, \pm 1\}$. *Moreover, the following are equivalent:*

(1) $\nu(V) \neq 0$.

(2) *V is isomorphic to* $V^\vee$ *as a right C-comodule.*

(3) *There exists a non-degenerate bilinear form b on V satisfying* (4.5).

If one of the above statements holds, then such a bilinear form b is unique up to scalar multiples and satisfies $b(w, \gamma \rightharpoonup v) = \nu(V) \cdot b(v, w)$ *for all* $v, w \in V$.

We prove several statements concerning the FS indicator of $V \in \mathsf{com}_{fd}(C)$. The proof will be done by reducing to the case of pivotal algebras in the following way: First fix a subcoalgebra $D \subset C$ satisfying

$$\dim_k(D) < \infty, \quad C_V \subset D \quad \text{and} \quad S(D) \subset D \tag{4.6}$$

Note that such a subcoalgebra D always exists. Indeed, by (4.4), we have $C_{X^\vee} = S(C_X)$ for all $X \in \mathsf{com}_{fd}(C)$. If V is a subcomodule of X, then C_V is a subcoalgebra of C_X. Therefore, since $X = V \oplus V^\vee$ is self-dual and has V as a subcomodule, $D = C_{V \oplus V^\vee}$ satisfies (4.6).

It is obvious that the triple $D = (D, S|_D, \gamma|_D)$ is a copivotal coalgebra and hence $D^\vee$ is a pivotal algebra. As we remarked in the above, the functor

$$F_D : \mathsf{mod}_{fd}(D^\vee) \xrightarrow[(4.1)]{\cong} \mathsf{com}_{fd}(D) \xrightarrow{\text{inclusion}} \mathsf{com}_{fd}(C)$$

is a k-linear fully faithful strict duality preserving functor. Hence, by Proposition 2.10, F_D preserves the FS indicator. By *regarding V as a left $D^\vee$-module,* we mean taking $V_0 \in \mathrm{mod}_{fd}(D^\vee)$ such that $F_D(V_0) = V$ and then identifying V_0 with V.

Now we prove an analogue of Theorem 3.5. Let R_C denote the coalgebra C regarded as a right C-comodule by the comultiplication.

Theorem 4.4. *Let $C = (C, S, \gamma)$ be a finite-dimensional copivotal coalgebra.*
(a) $\nu(\mathsf{R}_C^\vee) = \mathrm{Tr}(Q)$*, where* $Q : C \to C, c \mapsto S(c_{(1)})\langle \gamma, c_{(2)} \rangle$.
(b) *If C is co-Frobenius, then $\mathsf{R}_C \cong \mathsf{R}_C^\vee$ as right C-comodules. Hence, $\nu(\mathsf{R}_C) = \nu(\mathsf{R}_C^\vee)$.*

Proof. (a) Write $A = C^\vee$ and regard the right C-comodule $\mathsf{R}_C^\vee$ as a left A-module via (4.1). To avoid confusion, we denote by $\rightharpoonup$ the action of A on R_A and by $\rightharpoondown$ that on $\mathsf{R}_C^\vee$. Since the coaction of $\mu \in \mathsf{R}_C^\vee$ is characterized as $\langle \mu_{(0)}, c \rangle \mu_{(1)} = \langle \mu, c_{(1)} \rangle S(c_{(2)})$, the action $\rightharpoondown : A \times \mathsf{R}_C^\vee \to \mathsf{R}_C^\vee$ is given by

$$\langle f \rightharpoondown \mu, c \rangle = \langle f, S(c_{(2)}) \rangle \langle \mu, c_{(1)} \rangle \quad (f \in A, \mu \in \mathsf{R}_C^\vee, c \in C)$$

Consider the map $S^\vee : \mathsf{R}_A \to \mathsf{R}_C^\vee$. For $\lambda \in A$, $f \in \mathsf{R}_A$ and $c \in C$, we have

$$\begin{aligned}\langle S^\vee(\lambda \rightharpoonup f), c \rangle &= \langle \lambda \star f, S(c) \rangle = \langle \lambda, S(c_{(2)}) \rangle \langle f, S(c_{(1)}) \rangle \\ &= \langle \lambda, S(c_{(2)}) \rangle \langle S^\vee(f), c_{(1)} \rangle = \langle \lambda \rightharpoondown S^\vee(f), c \rangle\end{aligned}$$

and therefore $S^\vee : \mathsf{R}_A \to \mathsf{R}_C^\vee$ is an isomorphism of A-modules. Applying Theorem 3.5 to $A = C^\vee$, we see that $\nu(\mathsf{R}_C^\vee) = \nu(\mathsf{R}_A)$ is equal to the trace of the map $Q' : A \to A, \lambda \mapsto S^\vee(\lambda) \star \gamma$ $(\lambda \in C^\vee)$. Since $\langle Q'(\lambda), c \rangle = \langle \lambda, S(c_{(1)}) \rangle \langle \gamma, c_{(2)} \rangle = \langle \lambda, Q(c) \rangle$, Q' is the dual map of Q. Hence, we have $\nu(\mathsf{R}_C^\vee) = \mathrm{Tr}(Q') = \mathrm{Tr}(Q)$.

(b) If C is co-Frobenius, then A is Frobenius. Thus, by Theorem 3.5, there is an isomorphism $\varphi : \mathsf{R}_A \to \mathsf{R}_A^\vee$ of A-modules. In the proof of (1), we see that $S^\vee : \mathsf{R}_A \to \mathsf{R}_C^\vee$ is an isomorphism of A-modules. Regarding them as isomorphisms in the category $\mathrm{com}_{fd}(C)$, we obtain an isomorphism

$$\mathsf{R}_C \xrightarrow{\ j\ } \mathsf{R}_C^{\vee\vee} \xrightarrow{\ S^{\vee\vee}\ } \mathsf{R}_A^\vee \xrightarrow{\ \varphi\ } \mathsf{R}_A \xrightarrow{\ S^\vee\ } \mathsf{R}_C^\vee$$

of right C-comodules. □

The following is a coalgebraic version of Theorem 3.6.

Theorem 4.5. *Let $V \in \mathrm{com}_{fd}(C)$ be an absolutely simple comodule and suppose that V is self-dual. Then, by* (4.4), *the map*

$$S_V : C_V \to C_V, \quad S_V(c) = S(c) \quad (c \in C)$$

is well-defined. We also define

$$Q_V : C_V \to C_V, \quad Q_V(c) = S(\gamma \rightharpoonup c) \quad (= S(c_{(1)})\gamma(c_{(2)})) \quad (c \in C)$$

Then we have:

$$(1)\ \mathrm{Tr}(S_V) = \nu(V) \cdot \gamma(t_V) \qquad (2)\ \mathrm{Tr}(Q_V) = \nu(V) \cdot \dim_k(V)$$

Proof. (1) We regard V as a left $(C_V)^\vee$-module and denote its character by χ_V. Applying Theorem 3.6 to the dual pivotal algebra $(C_V)^\vee$ and by using (4.3), we have $\mathrm{Tr}(S_V) = \nu(V) \cdot \chi_V(\gamma) = \nu(V) \cdot \gamma(t_V)$.

(2) We regard C_V as a right C_V-comodule. Since $C_V \cong \mathrm{End}^c(V)$ is co-Frobenius, by Theorem 4.4, we have $\nu(C_V) = \mathrm{Tr}(Q_V)$. Let $d = \dim_k(V)$. Since $C_V \cong V^{\oplus d}$ as a right C-comodule, we have $\nu(C_V) = \nu(V)d$. Hence, $\mathrm{Tr}(Q_V) = \nu(V)d$. □

Applying Corollary 3.7 to the dual pivotal algebra of C, we have:

Corollary 4.6. *Suppose that k is algebraically closed and that $C = (C, S, \gamma)$ is a finite-dimensional cosemisimple copivotal coalgebra. Let $\{V_i\}_{i=1,\dots,n}$ be a complete set of representatives of the isomorphism classes of simple right C-comodules. Then*

$$\mathrm{Tr}(S) = \sum_{i=1}^{n} \nu(V_i)\gamma(t_i)$$

where $t_i = t_{V_i}$ is the character of V_i.

4.3. Coseparable Copivotal Coalgebras

A coalgebra C is said to be *coseparable* if it has a *coseparability idempotent, i.e.*, a bilinear form $\lambda : C \times C \to k$ satisfying $c_{(1)}\lambda(c_{(2)}, d) = \lambda(c, d_{(1)})d_{(2)}$ and $\lambda(c_{(1)}, c_{(2)}) = \varepsilon(c)$ for all $c, d \in C$. If such a form exists, then the forgetful functor $\mathsf{com}(C) \to \mathsf{Vec}(k)$ is separable with section $\Pi_{V,W}$: $\mathrm{Hom}_k(V, W) \to \mathrm{Hom}_C(V, W)$ given by

$$\Pi_{V,W}(f) : V \xrightarrow{\rho_V} V \otimes C \xrightarrow{f \otimes \mathrm{id}_V} W \otimes C \xrightarrow{\rho_W} W \otimes C \otimes C \xrightarrow{\mathrm{id}_W \otimes \lambda} W$$

for $f \in \mathrm{Hom}_k(V, W)$. The following theorem can be proved by the arguments of §2.4. Nevertheless, to avoid notational difficulties, we do not use Π and prove the theorem by reducing to Theorem 3.8.

Theorem 4.7. *If $C = (C, S, \gamma)$ is a coseparable copivotal coalgebra with coseparability idempotent λ, then, for all $V \in \mathsf{com}_{fd}(C)$, we have*

$$\nu(V) = \lambda(S(\gamma \rightharpoonup t_{V(1)}), t_{V(2)}) \quad \left(= \lambda(S(t_{V(1)}), t_{V(3)})\gamma(t_{V(2)})\right)$$

Proof. Fix a subcoalgebra D of C satisfying (4.6). D is coseparable with $\lambda_D = \lambda|_{D \times D}$. Since D is finite-dimensional, there exist finite number of linear maps $\lambda'_i, \lambda''_i : D \to k$ such that

$$\lambda_D(x, y) = \sum_i \lambda'_i(x)\lambda''_i(y)$$

for all $x, y \in D$. It is easy to see that $E = \sum_i \lambda'_i \otimes \lambda''_i$ is a separability idempotent for the dual pivotal algebra $D^\vee$. Now we regard V as a left $D^\vee$-module and denote its character by χ_V. Applying Theorem 3.8 to D, we obtain

$$\nu(V) = \sum_i \chi_V(S^\vee_X(\lambda'_i) \star \gamma \star \lambda''_i)$$

Now the desired formula is obtained by using (4.3). □

A *copivotal Hopf algebra* is a Hopf algebra $H = (H, \Delta, \varepsilon, S)$ equipped with an algebra map γ : $H \to k$ satisfying $S^2(x) = \langle \gamma, x_{(1)} \rangle x_{(2)} \langle \gamma, S(x_{(3)}) \rangle$ for all $x \in H$. Since γ is an algebra map, $\gamma \circ S$ is the inverse of γ with respect to the convolution product. Therefore a copivotal Hopf algebra is a copivotal coalgebra.

A *Haar functional* of a Hopf algebra H is a linear map $\lambda : H \to k$ satisfying $\langle \lambda, 1 \rangle = 1$ and $\langle \lambda, x_{(1)} \rangle x_{(2)} = \varepsilon(x)1 = x_{(1)} \langle \lambda, x_{(2)} \rangle$ for all $x \in H$. If λ is a Haar functional of H, then the map

$$\tilde{\lambda} : H \times H \to k, \quad \tilde{\lambda}(x, y) = \langle \lambda, S(x)y \rangle \quad (x, y \in H)$$

is a coseparability idempotent of the coalgebra H. Note that we have

$$S^2(x_{(1)})\langle \gamma, x_{(2)} \rangle = \langle \gamma, x_{(1)} \rangle x_{(2)} \langle \gamma, S(x_{(3)}) \rangle \langle \gamma, x_{(4)} \rangle = \langle \gamma, x_{(1)} \rangle x_{(2)}$$

for all $x \in H$. The following corollary is a direct consequence of Theorem 4.7:

Corollary 4.8. *Regard a copivotal Hopf algebra $H = (H, \Delta, \varepsilon, S; \gamma)$ as a copivotal coalgebra. If there exists a Haar functional $\lambda : H \to k$ on H, then we have*

$$\nu(V) = \langle \gamma, t_{V(1)} \rangle \langle \lambda, t_{V(2)} t_{V(3)} \rangle$$

for all $V \in \mathsf{com}_{fd}(H)$.

We shall explain how can we obtain (1.1) from Corollary 4.8.

Example 4.9. We work over $\mathbb{C}$. Let G be a compact group. A function $f : G \to \mathbb{C}$ is said to be *representative* if there exist finite number of functions $f_i, g_i : G \to \mathbb{C}$ such that $f(xy) = \sum_i f_i(x) g_i(y)$ for all $x, y \in G$. We denote by $R(G)$ the algebra of continuous representative functions on G. $R(G)$ is in fact a Hopf algebra; the comultiplication, the counit and the antipode are given by

$$f_{(1)}(x) f_{(2)}(y) = f(xy), \quad \varepsilon(f) = f(1), \quad S(f)(x) = f(x^{-1})$$

for $f \in R(G)$, $x, y \in G$. Define $\lambda : R(G) \to \mathbb{C}$ by $\lambda(f) = \int_G f(x) d\mu(x)$, where μ is the normalized Haar measure on G. We see that λ is a Haar functional of $R(G)$ (in fact, this is the origin of this term).

The group G acts continuously from the left on each $V \in \mathsf{com}_{fd}(R(G))$ by $x \cdot v = v_{(1)}(x) \cdot v_{(0)}$ ($x \in G, v \in V$). Conversely, if V is a finite-dimensional continuous representation of G, then $R(G)$ coacts from the right on V. If we fix a basis $\{v_i\}_{i=1,\dots,n}$ of V, the coaction of $R(G)$ is described as follows: Define $f_{ij} : G \to \mathbb{C}$ by

$$x \cdot v_i = \sum_{i=1}^{n} f_{ij}(x) v_j \quad (x \in G, i = 1, \dots, n) \tag{4.7}$$

Then each f_{ij} is an element of $R(G)$. The coaction of $R(G)$ on V is defined by

$$V \to V \otimes R(G), \quad v_i \mapsto \sum_{j=1}^{n} v_j \otimes f_{ij} \quad (i = 1, \dots, n)$$

These correspondences give an isomorphism of categories with duality over $\mathbb{C}$ between $\mathsf{com}_{fd}(R(G))$ and the category of continuous representations of G.

Now let V be a continuous representation of G with character χ_V. Regarding V as a right $R(G)$-comodule via the above category isomorphism, we obtain $\nu(V) = \lambda(t_{V(1)} t_{V(2)})$ by Corollary 4.8. To compute this value, we fix a basis $\{v_i\}_{i=1,\dots,n}$ of V and define f_{ij} by (4.7). Then $t_V = f_{11} + \cdots + f_{nn}$. Hence, by (4.2), we compute

$$\nu(V) = \lambda(t_{V(1)} t_{V(2)}) = \int_G \left(\sum_{i,j=1}^{n} f_{ij}(x) f_{ji}(x) \right) d\mu(x) \tag{4.8}$$

Since the action of $x \in G$ is represented by $\rho(x) = (f_{ij}(x))_{i,j=1,\dots,n}$, we have

$$\chi_V(x^2) = \mathrm{Tr}\left(\rho(x)^2\right) = \sum_{i,j=1}^{n} f_{ij}(x) f_{ji}(x)$$

Substituting this to (4.8), we obtain (1.1).

5. Quantum SL_2

5.1. *The Hopf Algebra $\mathcal{O}_q(SL_2)$*

In this section, we give some applications of our results to the quantum coordinate algebra $\mathcal{O}_q(SL_2)$ and the quantized universal enveloping algebra $U_q(\mathfrak{sl}_2)$. For details on these Hopf algebras, we refer the reader to [42] and [43].

Throughout, the base field k is assumed to be an algebraically closed field of characteristic zero. $q \in k$ denotes a fixed non-zero parameter that is not a root of unity. We use the following standard notations:

$$[n]_q = \frac{q^n - q^{-n}}{q - q^{-1}}, \quad [n]_q! = [n]_q \cdot [n-1]_q! \quad (n \geq 1), \quad [0]_q! = 1$$

for $n \in \mathbb{N}_0 = \{0, 1, 2, \dots\}$.

The quantum coordinate algebra $\mathcal{O}_q(SL_2)$ is a Hopf algebra defined as follows: As an algebra, it is generated by a, b, c and d with relations

$$ab = qba, \quad ac = qca, \quad bd = qdb, \quad cd = qdc, \quad bc = cb$$
$$ad - qbc = 1 = da - q^{-1}bc$$

The comultiplication Δ and the counit ε are defined by

$$\Delta(a) = a \otimes a + b \otimes c, \quad \Delta(b) = a \otimes b + b \otimes d, \quad \varepsilon(a) = 1, \quad \varepsilon(b) = 0$$
$$\Delta(c) = c \otimes a + d \otimes c, \quad \Delta(d) = c \otimes b + d \otimes d, \quad \varepsilon(c) = 0, \quad \varepsilon(d) = 1$$

and the antipode S is given by

$$S(a) = d, \quad S(b) = -q^{-1}b, \quad S(c) = -qc, \quad S(d) = a$$

We define an algebra map $\gamma : \mathcal{O}_q(SL_2) \to k$ by

$$\gamma(a) = q^{-1}, \quad \gamma(b) = \gamma(c) = 0, \quad \gamma(d) = q$$

One can check that $\mathcal{O}_q(SL_2)$ is a copivotal Hopf algebra with γ. In what follows, we determine the FS indicator of simple $\mathcal{O}_q(SL_2)$-comodules.

For each $\ell \in \frac{1}{2}\mathbb{N}_0$, we put $I_\ell = \{-\ell, -\ell+1, \dots, \ell-1, \ell\}$ and

$$X_\ell = \mathrm{span}_k\{a^{\ell-i}b^{\ell+i} \mid i \in I_\ell\} \subset \mathcal{O}_q(SL_2)$$

X_ℓ is a right coideal and hence it is a right $\mathcal{O}_q(SL_2)$-comodule. It is known that each X_ℓ is simple and $\{X_\ell \mid \ell \in \frac{1}{2}\mathbb{N}_0\}$ is a complete set of representatives of the isomorphism classes of simple right $\mathcal{O}_q(SL_2)$-comodules. This implies, in particular, that X_ℓ is self-dual.

In this section, we first prove the following result:

Theorem 5.1. $\nu(X_\ell) = (-1)^{2\ell}$.

By Theorem 4.3, this result reads as follows: For each $\ell \in \frac{1}{2}\mathbb{N}_0$, there exists a non-degenerate bilinear form β on X_ℓ satisfying $\beta(x_{(0)}, y)x_{(1)} = \beta(x, y_{(0)})S(y_{(1)})$ and $b(y, \gamma \rightharpoonup x) = (-1)^{2\ell} \cdot b(x, y)$ for all $x, y \in X_\ell$.

To prove Theorem 5.1, we need a matrix corepresentation of X_ℓ. For each $i \in I_\ell$, we fix a square root $\mu_i \in k$ of $[\ell+i]_{q^{-2}}!$ and take

$$x_i^{(\ell)} = \begin{bmatrix} 2\ell \\ \ell+i \end{bmatrix}_{q^{-2}}^{1/2} a^{\ell-i}b^{\ell+i} \quad (i \in I_\ell), \quad \text{where} \quad \begin{bmatrix} 2\ell \\ \ell+i \end{bmatrix}_{q^{-2}}^{1/2} := \frac{\mu_\ell}{\mu_i \cdot \mu_{-i}}$$

as a basis of X_ℓ. Define $c_{ij}^{(\ell)}$ by $\Delta(x_j^{(\ell)}) = \sum x_j^{(\ell)} \otimes c_{ij}^{(\ell)}$. The matrix $(c_{ij}^{(\ell)})$ has been explicitly determined and well-studied in relation to unitary representations of a real form of $\mathcal{O}_q(SL_2)$; see, e.g., [43, §4]. Following *loc. cit.*, we have

$$c_{ij}^{(\ell)} = \begin{cases} N_{ij\ell}^{+} \cdot a^{-i-j} c^{i-j} \cdot p_{\ell+j}(\zeta; q^{-2(i-j)}, q^{2(i+j)} | q^{-2}) & (i+j \le 0, i \ge j) \\ N_{ji\ell}^{+} \cdot a^{-i-j} b^{j-i} \cdot p_{\ell+i}(\zeta; q^{-2(j-i)}, q^{2(i+j)} | q^{-2}) & (i+j \le 0, i \le j) \\ N_{ji\ell}^{-} \cdot p_{\ell-i}(\zeta; q^{-2(i-j)}, q^{2(i+j)} | q^{-2}) \cdot c^{i-j} d^{i+j} & (i+j \ge 0, i \ge j) \\ N_{ij\ell}^{-} \cdot p_{\ell-j}(\zeta; q^{-2(i-j)}, q^{2(i+j)} | q^{-2}) \cdot b^{i-j} d^{i+j} & (i+j \ge 0, i \le j) \end{cases}$$

where $\zeta = -qbc$,

$$N_{ij\ell}^{+} = \frac{q^{-(\ell+j)(j-i)}}{[i-j]_{q^{-2}}!} \cdot \frac{\mu_{+i}\,\mu_{-j}}{\mu_{+j}\,\mu_{-i}}, \quad N_{ij\ell}^{-} = \frac{q^{(\ell-j)(j-i)}}{[j-i]_{q^{-2}}!} \cdot \frac{\mu_{-i}\,\mu_{+j}}{\mu_{-j}\,\mu_{+i}} \quad (= N_{-i,-j,\ell}^{+})$$

and p_m is the *little q-Jacobi polynomial* [43, §2]. We omit the definition of p_m; in what follows, we need only the fact that $p_m(\zeta; q_1, q_2 | q_3)$ ($q_i \in k$) is a polynomial of ζ. Note that, since $S(\zeta) = \zeta$, we have

$$S(p_m(\zeta; q_1, q_2 | q_3)) = p_m(S(\zeta); q_1, q_2 | q_3) = p_m(\zeta; q_1, q_2 | q_3) \tag{5.1}$$

Since $f \mapsto (\gamma \rightharpoonup f)$ ($f \in \mathcal{O}_q(SL_2)$) is an algebra map, we also have

$$\gamma \rightharpoonup p_m(\zeta; q_1, q_2 | q_3) = p_m(\gamma \rightharpoonup \zeta; q_1, q_2 | q_3) = p_m(\zeta; q_1, q_2 | q_3) \tag{5.2}$$

Proof of Theorem 5.1. Let C_ℓ be the coefficient subcoalgebra of X_ℓ. Define

$$Q_\ell : C_\ell \to C_\ell \quad Q_\ell(f) = S(\gamma \rightharpoonup f) \quad (f \in C_\ell)$$

Q_ℓ is well-defined since X_ℓ is self-dual. By Theorem 4.5, we have

$$\nu(X_\ell) = \frac{\mathrm{Tr}(Q_\ell)}{\dim_k(X_\ell)} = \frac{\mathrm{Tr}(Q_\ell)}{2\ell + 1}$$

By (5.1), (5.2) and the above description of $c_{ij}^{(\ell)}$, we have

$$Q_\ell(c_{ij}^{(\ell)}) = S(\gamma \rightharpoonup c_{ij}^{(\ell)}) = (\text{constant}) \times S(c_{ij}^{(\ell)}) = (\text{constant}) \times c_{-j,-i}^{(\ell)}$$

for all $i, j \in I_\ell$. Recall that $\{c_{ij}^{(\ell)}\}$ is a basis of C_ℓ since X_ℓ is simple. The above computation means that Q_ℓ is represented by a generalized permutation matrix with respect to this basis.

Note that $(i,j) = (-j,-i)$ if and only if $j = -i$. If $i \ge 0$, then

$$\begin{aligned} Q_\ell(c_{i,-i}^{(\ell)}) &= q^{-2i} \cdot S\Big(N_{i,-i,\ell}^{+} \cdot c^{2i} \cdot p_{\ell-i}(\zeta; q^{-4i}, 1 | q^{-2})\Big) \\ &= q^{-2i} \cdot N_{i,-i,\ell}^{+} \cdot p_{\ell-i}(\zeta; q^{-4i}, 1 | q^{-2}) \cdot (-q)^{2i} = (-1)^{2i} \cdot c_{i,-i}^{(\ell)} \end{aligned}$$

Since $\ell - i \in \mathbb{Z}$, we have $(-1)^{2i} = (-1)^{2\ell}$. In a similar way, we also have $Q_\ell(c_{i,-i}^{(\ell)}) = (-1)^{2\ell}$ for $i < 0$. Hence we obtain $\mathrm{Tr}(Q_\ell) = (-1)^{2\ell} \cdot (2\ell + 1)$. □

$\mathcal{O}_q(SL_2)$ has a Hopf algebra automorphism τ given by

$$\tau(a) = a, \quad \tau(b) = -b, \quad \tau(c) = -c, \quad \tau(d) = d$$

τ is an involution such that $\gamma \circ \tau = \gamma$ and hence the τ-twisted FS indicator $\nu^{\tau}(X)$ is defined for each $X \in \mathsf{com}_{fd}(\mathcal{O}_q(SL_2))$. Replacing S in the proof of Theorem 5.1 with $S \circ \tau$, we have the following theorem:

Theorem 5.2. $\nu^{\tau}(X_\ell) = +1$.

By Theorem 4.3, this result reads as follows: For each $\ell \in \frac{1}{2}\mathbb{N}_0$, there exists a non-degenerate bilinear form β on X_ℓ satisfying $\beta(x_{(0)}, y)\tau(x_{(1)}) = \beta(x, y_{(0)})S(y_{(1)})$ and $\beta(w, \gamma \rightharpoonup v) = \beta(v, w)$ for all $v, w \in X_\ell$.

Remark 5.3. The character t_ℓ of X_ℓ is given by

$$t_\ell = \sum_{i \in I_\ell} c_{ii}^{(\ell)} = \sum_{i \in I_\ell, i \geq 0} a^{2i} p_{\ell+i}(\zeta; 1, q^{-4i} | q^{-2}) + \sum_{i \in I_\ell, i > 0} p_{\ell+i}(\zeta; 1, q^{4i} | q^{-2}) d^{2i}$$

One can prove Theorems 5.1 and 5.2 by Theorem 4.7 and its corollary (see [43, §4] for a description of the Haar functional on $\mathcal{O}_q(SL_2)$). However, the computation will become more difficult than the above proof.

5.2. The Hopf Algebra $U_q(\mathfrak{sl}_2)$

The quantized enveloping algebra $U_q(\mathfrak{sl}_2)$ is a Hopf algebra defined as follows: As an algebra, it is generated by E, F, K and K^{-1} with relations $KK^{-1} = 1 = K^{-1}K$,

$$KEK^{-1} = q^2 E, \quad KFK^{-1} = q^{-2} F \quad \text{and} \quad EF - FE = \frac{K - K^{-1}}{q - q^{-1}}$$

The comultiplication Δ, the counit ε and the antipode S are given by

$$\Delta(K) = K \otimes K, \quad \Delta(E) = E \otimes K + 1 \otimes E, \quad \Delta(F) = F \otimes 1 + K^{-1} \otimes F$$
$$S(K) = K^{-1},\ S(E) = -EK^{-1},\ S(F) = -KF,\ \varepsilon(K) = 1,\ \varepsilon(E) = \varepsilon(F) = 0$$

We have $S^2(u) = KuK^{-1}$ for all $u \in U_q(\mathfrak{sl}_2)$. Hence the Hopf algebra $U_q(\mathfrak{sl}_2)$ is pivotal with pivotal grouplike element K.

For each $\ell \in \frac{1}{2}\mathbb{N}_0$, we define a left $U_q(\mathfrak{sl}_2)$-module V_ℓ as follows: As a vector space, it has a basis $\{v_i\}_{i \in I_\ell}$. The action of $U_q(\mathfrak{sl}_2)$ on V_ℓ is defined by

$$K \cdot v_i = q^{2i} v_i,\ E \cdot v_i = [\ell - i + 1]_q v_{i-1},\ F \cdot v_i = [\ell + i + 1]_q v_{i+1}\ (i \in I_\ell)$$

where $v_{\ell+1} = v_{-(\ell+1)} = 0$.

There is a unique Hopf pairing $\langle -, - \rangle : U_q(\mathfrak{sl}_2) \times \mathcal{O}_q(SL_2) \to k$ such that

$$\langle K, a \rangle = q^{-1}, \quad \langle K, d \rangle = q, \quad \langle E, c \rangle = 1, \quad \langle F, b \rangle = 1$$
$$\langle K, b \rangle = \langle K, c \rangle = \langle E, a \rangle = \langle E, b \rangle = \langle E, d \rangle = \langle F, a \rangle = \langle F, c \rangle = \langle F, d \rangle = 0$$

see [43, §4] and [42, V.7]. This pairing induces an algebra map $\varphi : U_q(\mathfrak{sl}_2) \to \mathcal{O}_q(SL_2)^{\vee}$. Since $\varphi(K) = \gamma$, φ is in fact a morphism of pivotal algebras. Hence we obtain a k-linear duality preserving functor

$$\Phi : \mathsf{com}_{fd}\Big(\mathcal{O}_q(SL_2)\Big) \xrightarrow[(4.1)]{} \mathsf{mod}_{fd}\Big(\mathcal{O}_q(SL_2)^{\vee}\Big) \xrightarrow{\varphi^{\natural}} \mathsf{mod}_{fd}\Big(U_q(\mathfrak{sl}_2)\Big)$$

One has $\Phi(X_\ell) \cong V_\ell$. In particular, Φ maps simple objects to simple objects. Since $\mathcal{O}_q(SL_2)$ is cosemisimple, the functor Φ is fully faithful and therefore Φ preserves the FS indicator. Hence, by Theorem 5.1, we have:

Theorem 5.4. $\nu(V_\ell) = (-1)^{2\ell}$.

By Theorem 3.8, this result reads as follows: For each $\ell \in \frac{1}{2}\mathbb{N}_0$, there exists a non-degenerate bilinear form β on V_ℓ satisfying $\beta(uv, w) = \beta(v, S(u)w)$ and $\beta(w, Kv) = (-1)^{2\ell} \cdot \beta(v, w)$ for all $u \in U_q(\mathfrak{sl}_2)$ and $v, w \in V_\ell$.

$U_q(\mathfrak{sl}_2)$ has a Hopf algebra automorphism τ defined by $\tau(E) = -E$, $\tau(F) = -F$, $\tau(K) = K$. It is obvious that τ is an involution of the pivotal algebra $U_q(\mathfrak{sl}_2)$. Since $\langle \tau(u), f \rangle = \langle u, \tau(f) \rangle$ for all $u \in U_q(\mathfrak{sl}_2)$ and $f \in \mathcal{O}_q(SL_2)$, Φ also preserves the τ-twisted FS indicator. Therefore we have:

Theorem 5.5. $\nu^\tau(V_\ell) = +1$.

This result reads as follows: For each $\ell \in \frac{1}{2}\mathbb{N}_0$, there exists a non-degenerate bilinear form β on V_ℓ satisfying $\beta(\tau(u)v, w) = b(v, S(u)w)$ and $\beta(w, Kv) = b(v, w)$ for all $u \in U_q(\mathfrak{sl}_2)$ and $v, w \in V_\ell$.

6. Conclusions

As we have briefly reviewed in Section 1, the celebrated theorem of Frobenius and Schur has several generalizations. To give a category-theoretical understanding of these generalizations, in Section 2 we have introduced the FS indicator for categories with duality over a field k; if $\mathcal{C}$ is a category with duality over k, then a linear map $\mathrm{T}_{X,Y} : \mathrm{Hom}_\mathcal{C}(X, Y^\vee) \to \mathrm{Hom}_\mathcal{C}(Y, X^\vee)$, $f \mapsto f^\vee \circ j$ is defined for each $X, Y \in \mathcal{C}$. We call $\mathrm{T}_{X,Y}$ the transposition map. The FS indicator $\nu(X)$ of $X \in \mathcal{C}$ is defined to be the trace of $\mathrm{T}_{X,X} : \mathrm{Hom}_\mathcal{C}(X, X^\vee) \to \mathrm{Hom}_\mathcal{C}(X, X^\vee)$. We have also introduced a general method to twist the given duality by an adjunction, which is a category-theoretical counterpart of several twisted versions of the Frobenius–Schur theorem.

In Section 3, we have introduced the notion of a pivotal algebra. The representation category of a pivotal algebra has duality and therefore the FS indicator is defined for each of its representation. We have given a representation-theoretic interpretation of the FS indicator and a formula of the FS indicator for separable pivotal algebras. These results yield the Frobenius–Schur-type theorems for Hopf algebras, quasi-Hopf algebras, weak Hopf C^*-algebras and Doi's group-like algebras. The notion of pivotal algebras is useful to deal with the twisted FS indicator; as a demonstration, we have constructed the twisted Frobenius–Schur theory for quasi-Hopf algebras.

In Section 4, we have introduced the notion of a copivotal coalgebra as the dual notion of a pivotal algebra and gave results for copivotal coalgebras analogous to pivotal algebras. In particular, we have given a representation-theoretic interpretation of the FS indicator and a formula of the FS indicator for coseparable copivotal coalgebras.

In Section 5, we have applied our results to the quantum coordinate ring $\mathcal{O}_q(SL_2)$ and the quantum enveloping algebra $U_q(\mathfrak{sl}_2)$. For each $\ell \in \frac{1}{2}\mathbb{N}_0$, $\mathcal{O}_q(SL_2)$ has a unique simple right comodule X_ℓ of dimension 2ℓ. We have proved $\nu(X_\ell) = (-1)^{2\ell}$ and analogous results for the twisted case and $U_q(\mathfrak{sl}_2)$ case. As we have remarked, the Haar functional on $\mathcal{O}_q(SL_2)$ is not used in our proof. We expect that the FS indicator for general $\mathcal{O}_q(G)$ will be determined by using the Haar functional.

Acknowledgments

The author would like to thank the referees for careful reading the manuscript. The author is supported by Grant-in-Aid for JSPS Fellows (24·3606).

References

1. Linchenko, V.; Montgomery, S. A Frobenius-Schur theorem for Hopf algebras. *Algebr. Represent. Theory* **2000**, *3*, 347–355.
2. Mason, G.; Ng, S.H. Central invariants and Frobenius-Schur indicators for semisimple quasi-Hopf algebras. *Adv. Math.* **2005**, *190*, 161–195.
3. Schauenburg, P. On the Frobenius-Schur indicators for quasi-Hopf algebras. *J. Algebra* **2004**, *282*, 129–139.

4. Bantay, P. The Frobenius-Schur indicator in conformal field theory. *Phys. Lett. B* **1997**, *394*, 87–88.
5. Fuchs, J.; Ganchev, A.C.; Szlachányi, K.; Vecsernyés, P. S_4 symmetry of $6j$ symbols and Frobenius-Schur indicators in rigid monoidal C^* categories. *J. Math. Phys.* **1999**, *40*, 408–426.
6. Fuchs, J.; Schweigert, C. Category theory for conformal boundary conditions. In *Vertex Operator Algebras in Mathematics and Physics (Toronto, ON, 2000)*; American Mathematical Society: Providence, RI, USA, 2003; Volume 39, pp. 25–70.
7. Ng, S.H.; Schauenburg, P. Frobenius-Schur indicators and exponents of spherical categories. *Adv. Math.* **2007**, *211*, 34–71.
8. Ng, S.H.; Schauenburg, P. Higher Frobenius-Schur indicators for pivotal categories. In *Hopf Algebras and Generalizations*; American Mathematical Society: Providence, RI, USA, 2007; Volume 441, pp. 63–90.
9. Ng, S.H.; Schauenburg, P. Central invariants and higher indicators for semisimple quasi-Hopf algebras. *Trans. Amer. Math. Soc.* **2008**, *360*, 1839–1860.
10. Ng, S.H.; Schauenburg, P. Congruence subgroups and generalized Frobenius-Schur indicators. *Comm. Math. Phys.* **2010**, *300*, 1–46.
11. Kashina, Y.; Sommerhäuser, Y.; Zhu, Y. Self-Dual modules of semisimple Hopf algebras. *J. Algebra* **2002**, *257*, 88–96.
12. Kashina, Y.; Mason, G.; Montgomery, S. Computing the Frobenius-Schur indicator for abelian extensions of Hopf algebras. *J. Algebra* **2002**, *251*, 888–913.
13. Kashina, Y.; Sommerhäuser, Y.; Zhu, Y. On higher Frobenius-Schur indicators. *Mem. Amer. Math. Soc.* **2006**, *181*, 65.
14. Kashina, Y.; Montgomery, S.; Ng, S.H. On the trace of the antipode and higher indicators. *Israel J. Math.* **2012**, *188*, 57–89.
15. Natale, S. Frobenius-Schur indicators for a class of fusion categories. *Pacific J. Math.* **2005**, *221*, 353–377.
16. Shimizu, K. Frobenius-Schur indicators in Tambara-Yamagami categories. *J. Algebra* **2011**, *332*, 543–564.
17. Shimizu, K. Some computations of Frobenius-Schur indicators of the regular representations of Hopf algebras. *Algebr. Represent. Theory* **2012**, *15*, 325–357.
18. Junya, T. Frobenius-Schur Theorem for Association Schemes (in Japanese). *RIMS Kokyuroku* **2006**, *1476*, 21–27.
19. Doi, Y. Group-Like algebras and their representations. *Comm. Algebra* **2010**, *38*, 2635–2655.
20. Geck, M. Kazhdan–Lusztig cells and the Frobenius–Schur indicator. 2011, arXiv:1110.5672v2. Available online: http://arxiv.org/abs/1110.5672 (accessed on 10 October 2012).
21. Kawanaka, N.; Matsuyama, H. A twisted version of the Frobenius-Schur indicator and multiplicity-free permutation representations. *Hokkaido Math. J.* **1990**, *19*, 495–508.
22. Bump, D.; Ginzburg, D. Generalized Frobenius-Schur numbers. *J. Algebra* **2004**, *278*, 294–313.
23. Sage, D.S.; Vega, M.D. Twisted Frobenius-Schur indicators for Hopf algebras. *J. Algebra* **2012**, *354*, 136–147.
24. Balmer, P. Witt groups. In *Handbook of K-Theory*; Springer: Berlin, Germany, 2005; pp. 539–576.
25. Calmès, B.; Hornbostel, J. Tensor-Triangulated categories and dualities. *Theory Appl. Categ.* **2009**, *22*, 136–200.
26. Mac Lane, S. *Categories for the Working Mathematician*, 2nd ed.; Springer-Verlag: New York, NY, USA, 1998; p. 314.
27. Caenepeel, S.; Militaru, G.; Zhu, S. *Frobenius and Separable Functors for Generalized Module Categories and Nonlinear Equations*; *Lecture Notes in Mathematics*; Springer-Verlag: Berlin, Germany, 2002; Volume 1787.
28. Larson, R.G.; Radford, D.E. Finite-Dimensional cosemisimple Hopf algebras in characteristic 0 are semisimple. *J. Algebra* **1988**, *117*, 267–289.
29. Mizukawa, H. Wreath product generalizations of the triple (S_{2n}, H_n, ϕ) and their spherical functions. *J. Algebra* **2011**, *334*, 31–53.
30. Tambara, D. Invariants and semi-direct products for finite group actions on tensor categories. *J. Math. Soc. Japan* **2001**, *53*, 429–456.
31. Nikshych, D. Non-Group-Theoretical semisimple Hopf algebras from group actions on fusion categories. *Selecta Math. (N.S.)* **2008**, *14*, 145–161.
32. Jedwab, A. A trace-like invariant for representations of Hopf algebras. *Comm. Algebra* **2010**, *38*, 3456–3468.
33. Jedwab, A.; Krop, L. Trace-Like invariant for representations of nilpotent liftings of quantum planes. *Comm. Algebra* **2011**, *39*, 1465–1475.

34. Doi, Y. Bi-Frobenius algebras and group-like algebras. In *Hopf Algebras*; Dekker: New York, NY, USA, 2004; Volume 237, pp. 143–155.
35. Böhm, G.; Nill, F.; Szlachányi, K. Weak Hopf algebras. I. Integral theory and C^*-structure. *J. Algebra* **1999**, *221*, 385–438.
36. Böhm, G.; Szlachányi, K. Weak Hopf algebras. II. Representation theory, dimensions, and the Markov trace. *J. Algebra* **2000**, *233*, 156–212.
37. Etingof, P.; Nikshych, D.; Ostrik, V. On fusion categories. *Ann. Math.* **2005**, *162*, 581–642.
38. Drinfeld, V.G. Quasi-Hopf algebras. *Algebra i Analiz* **1989**, *1*, 114–148.
39. Hausser, F.; Nill, F. Integral theory for quasi-Hopf algebras. 1999, arXiv:math/9904164. Available online: http://arxiv.org/abs/math/9904164 (accessed on 10 October 2012).
40. Panaite, F. A Maschke-type theorem for quasi-Hopf algebras. In *Rings, Hopf algebras, and Brauer groups (Antwerp/Brussels, 1996)*; Dekker: New York, NY, USA, 1998; Volume 197, pp. 201–207.
41. Dăscălescu, S.; Năstăsescu, C.; Raianu, Ş. *Hopf Algebras; Monographs and Textbooks in Pure and Applied Mathematics*; Marcel Dekker Inc.: New York, NY, USA, 2001; Volume 235, p. 401.
42. Kassel, C. *Quantum Groups; Graduate Texts in Mathematics*; Springer-Verlag: New York, NY, USA, 1995; Volume 155.
43. Klimyk, A.; Schmüdgen, K. *Quantum Groups and Their Representations; Texts and Monographs in Physics*; Springer-Verlag: Berlin, Germany, 1997.

Hasse-Schmidt Derivations and the Hopf Algebra of Non-Commutative Symmetric Functions

Michiel Hazewinkel

Burg. 's Jacob Laan 18, NL-1401BR BUSSUM, The Netherlands; michhaz@xs4all.nl

Abstract: Let **NSymm** be the Hopf algebra of non-commutative symmetric functions (in an infinity of indeterminates): NSymm=Z. It is shown that an associative algebra A with a Hasse-Schmidt derivation $d = (id, d_1, d_2, ...)$ on it is exactly the same as an **NSymm** module algebra. The primitives of **NSymm** act as ordinary derivations. There are many formulas for the generators Z_i in terms of the primitives (and vice-versa). This leads to formulas for the higher derivations in a Hasse-Schmidt derivation in terms of ordinary derivations, such as the known formulas of Heerema and Mirzavaziri (and also formulas for ordinary derivations in terms of the elements of a Hasse-Schmidt derivation). These formulas are over the rationals; no such formulas are possible over the integers. Many more formulas are derivable.

Keywords: non-commutative symmetric functions; Hasse-Schmidt derivation; higher derivation; Heerema formula; iMirzavaziri formula; non-commutative Newton formulas

MSC: 05E05; 16W25

1. Introduction

Let A be an associative algebra (or any other kind of algebra for that matter). A derivation on A is an endomorphism ∂ of the underlying Abelian group of A such that

$$\partial(ab) = a(\partial b) + (\partial a)b \text{ for all } a,b \in A \tag{1.1}$$

A Hasse-Schmidt derivation is a sequence($d_0 = id, d_1, d_2, ..., d_n, ...$) of endomorphisms of the underlying Abelian group such that for all $n \geq 1$.

$$d_n(ab) = \sum_{i=0}^{n} (d_i a)(d_{n-i} b) \tag{1.2}$$

Note that d_1 is a derivation as defined by Equation 1.1. The individual d_n that occurs in a Hasse-Schmidt derivation is also sometimes called a higher derivation.

A question of some importance is whether Hasse-Schmidt derivations can be written down in terms of polynomials in ordinary derivations. For instance, in connection with automatic continuity for Hasse-Schmidt derivations on Banach algebras.

Such formulas have been written down by, for instance, Heerema and Mirzavaziri in [1,2]. They also will be explicitly given below.

It is the purpose of this short note to show that such formulas follow directly from some easy results about the Hopf algebra **NSymm** of non-commutative symmetric functions. In fact this Hopf algebra constitutes a universal example concerning the matter.

2. Hopf Algebras and Hopf Module Algebras

Everything will take place over a commutative associative unital base ring k; unadorned tensor products will be tensor products over k. In this note k will be the ring of integers **Z**, or the field of rational numbers **Q**.

Recall that a Hopf algebra over k is a k-module H together with five k-module morphisms $m: H \otimes H \longrightarrow H$, $e: k \longrightarrow H$, $\mu: H \longrightarrow H \otimes H$, $\varepsilon: H \longrightarrow k$, $\iota: H \longrightarrow H$, such that (H,m,e) is an associative k-algebra with unit, (H,μ,ε) is a co-associative co-algebra with co-unit, μ and ε are algebra morphisms (or, equivalently, that m and e are co-algebra morphisms), and such that ι satisfies $m(\iota \otimes \mathrm{id})\mu = \varepsilon e$, $m(\mathrm{id} \otimes \iota)\mu = \varepsilon e$. The antipode ι will play no role in what follows. If there is no antipode (specified) one speaks of a bi-algebra. For a brief introduction to Hopf algebras (and co-algebras) with plenty of examples see Chapters 2 and 3 of [3].

Recall also that an element $p \in H$ is called primitive if $\mu(p) = p \otimes 1 + 1 \otimes p$. These form a sub-$k$-module of H and form a Lie algebra under the commutator difference product $(p,p') \mapsto pp' - p'p$. I shall use $Prim(H)$ to denote this k-Lie-algebra.

Given a Hopf algebra over k, a Hopf module algebra is a k-algebra A together with an action of the underlying algebra of H on (the underlying module of) A such that:

$$h(ab) = \sum_{(h)} (h_{(1)}a)(h_{(2)}b) \text{ for all } a,b \in A \text{, and } h(1) = \varepsilon(h)1 \text{ where } \mu(h) = \sum_{(h)} h_{(1)} \otimes h_{(2)} \tag{2.1}$$

and where I have used Sweedler-Heynemann notation for the co-product.

Note that this means that the primitive elements of H act as derivations.

3. The Hopf Algebra NSymm of Non-Commutative Symmetric Functions

As an algebra over the integers **NSymm** is simply the free associative algebra in countably many (non-commuting) indeterminates, $\mathbf{NSymm} = \mathbf{Z}\langle Z\rangle = \mathbf{Z}\langle Z_1, Z_2, \ldots\rangle$. The comultiplication and counit are given by

$$\mu(Z_n) = \sum_{i+j=n} Z_i \otimes Z_j \text{, where } Z_0 = 1 \text{, } \varepsilon(1) = 1, \ \varepsilon(Z_n) = 0 \text{ for } n \geq 1 \tag{3.1}$$

As **NSymm** is free as an associative algebra, it is no trouble to verify that this defines a bi-algebra. The seminal paper [4] started the whole business of non-commutative symmetric functions, and is now a full-fledged research area in its own right.

Now consider an **NSymm** Hopf module, algebra A. Then, by Equations 2.1 and 3.1 the module endomorphims defined by the actions of the Z_n, $n \geq 1$, $d_n(a) = Z_n a$, define a Hasse-Schmidt derivation. Conversely, if A is a k-algebra together with a Hasse-Schmidt derivation one defines a **NSymm** Hopf module algebra structure on A by setting $Z_n a = d_n(a)$. This works because **NSymm** is free as an algebra.

Thus an **NSymm** Hopf module algebra A is precisely the same thing as a k-algebra A together with a Hasse-Schmidt derivation on it and the matter of writing the elements of the sequence of morphisms that make up the Hasse-Schmidt derivation in terms of ordinary derivations comes down to the matter of finding enough primitives of **NSymm** so that the generators, Z_n, can be written as polynomials in these primitives.

4. The Newton Primitives of NSymm

Define the non-commutative polynomials P_n and P'_n by the recursion formulas

$$P_n = nZ_n - (Z_{n-1}P_1 + Z_{n-2}P_2 + \ldots + Z_1P_{n-1})$$
$$P_n' = nZ_n - (P_1'Z_{n-1} + P_2'Z_{n-2} + \ldots + P'_{n-1}Z_1) \tag{4.1}$$

These are non-commutative analogues of the well known Newton formulas for the power sums in terms of the complete symmetric functions in the usual commutative theory of symmetric functions. It is not difficult to write down an explicit expression for these polynomials:

$$P_n(Z) = \sum_{\substack{i_1+\ldots+i_m=n \\ i_j \in \mathbf{N}}} (-1)^{m+1} i_m Z_{i_1} Z_{i_2} \ldots Z_{i_m} \tag{4.2}$$

Nor is it difficult to write down a formula for the Z_n in terms of the P's or P''s. However, to do that one definitely needs to use rational numbers and not just integers [5]. For instance

$$Z_2 = \frac{P_1^2 + P_2}{2}$$

The key observation is now:

4.1. Proposition

The elements P_n and P_n' are primitive elements of the Hopf algebra **NSymm**.

The proof is a straightforward uncomplicated induction argument using the recursion Formulas 4.1. See e.g., [3], page 147.

Using the P_n' an immediate corollary is the following main theorem from [2].

4.2. Theorem

Let A be an associative algebra over the rational numbers $\mathbf{Q}$ and let $(id, d_1, d_2, \ldots, d_n, \ldots)$ be a Hasse-Schmidt derivation on it. Then the δ_n defined recursively by

$$\delta_n = nd_n - \delta_1 d_{n-1} - \ldots - \delta_{n-1} d_1 \tag{4.5}$$

are ordinary derivations and

$$d_n = \sum_{\substack{r_1+r_2+\ldots+r_m=n \\ r_j \in \mathbf{N}}} c_{r_1,r_2,\ldots,r_m} \delta_{r_1} \delta_{r_2} \ldots \delta_{r_m} \tag{4.6}$$

Where

$$c_{r_1,r_2,\ldots,r_m} = \frac{1}{r_1 + r_2 + \ldots + r_m} \frac{1}{r_2 + \ldots + r_m} \cdots \frac{1}{r_{m-1} + r_m} \frac{1}{r_m} \tag{4.7}$$

4.3. Comment

Because

$$P_n' \equiv nZ_n \bmod(Z_1, Z_2, \dots, Z_{n-1})$$

the formulas expressing the Z_n in terms of the P_n' are unique and so denominators are really needed.

4.4. Comment and Example

There are many more primitive elements in **NSymm** than just the P_n' and P_n. One could hope that by using all of them, integral formulas for the Z_n in terms of primitives would become possible. This is not the case. The full Lie algebra of primitives of **NSymm** was calculated in [6]. It readily follows from the description there that $\mathbf{Z}\langle Prim(\mathbf{NSymm})\rangle$, the sub-algebra of **NSymm** generated by all primitive elements is strictly smaller than **NSymm**. In fact much smaller in a sense that is specified in locus citandi. Thus the theorem does not hold over the integers.

A concrete example of a Hasse-Schmidt derivation of which the constituting endomorphisms cannot be written as integral polynomials in derivations can be given in terms of **NSymm** itself, as follows: The Hopf algebra **NSymm** is graded by giving Z_n degree n. Note that each graded piece is a free **Z**-module of finite rank. Let **QSymm**, often called the Hopf algebra of quasi-symmetric functions, be the graded dual Hopf algebra. Then each Z_n defines a functional $\alpha_n : \mathbf{QSymm} \longrightarrow \mathbf{Z}$. Now define an endomorphism d_n of **QSymm** as the composed morphism

$$\mathbf{QSymm} \xrightarrow{\mu_{\mathbf{QSymm}}} \mathbf{QSymm} \otimes_{\mathbf{Z}} \mathbf{QSymm} \xrightarrow{\mathrm{id} \otimes \alpha_n} \mathbf{QSymm}$$

Then the d_n form a Hasse-Schmidt derivation of which the components cannot be written as integer polynomials in ordinary derivations.

5. The Hopf Algebra LieHopf

In [1] a formula for manufacturing Hasse-Schmidt derivations from a collection of ordinary derivations is shown that is more pleasing—at least to me—than 4.6. This result from locus citandi can be strengthened to give a theorem similar to Theorem 4.4 but with more symmetric formulae. This involves another Hopf algebra over the integers which I like to call **LieHopf**.

As an algebra **LieHopf** is again the free associative algebra in countably many indeterminates $\mathbf{Z}\langle U\rangle = \mathbf{Z}\langle U_1, U_2, \dots\rangle$. However, this time the co-multiplication and co-unit are defined by

$$\mu(U_n) = U_n \otimes 1 + 1 \otimes U_n, \ \varepsilon(U_n) = 0 \tag{5.1}$$

so that all the U_n are primitive. Also, in fact the Lie algebra of primitives of this Hopf algebra is the free Lie algebra on countably many generators.

Over the integers **LieHopf** and **NSymm** are very different but over the rationals they become isomorphic. There are very many isomorphisms. A particularly nice one is given in considering the power series identity

$$1 + Z_1 t + Z_2 t^2 + Z_3 t^3 + \dots = \exp(U_1 t + U_2 t^2 + U_3 t^3 + \dots) \tag{5.2}$$

which gives the following formulae for the U's in terms of the Z's and vice versa.

$$Z_n(U) = \sum_{r_1+\ldots+r_m=n} \frac{U_{r_1}U_{r_2}\ldots U_{r_m}}{m!} \tag{5.3}$$

$$U_n(Z) = \sum_{r_1+\ldots+r_m=n} (-1)^{m+1} \frac{Z_{r_1}Z_{r_2}\ldots Z_{r_m}}{m} \tag{5.4}$$

For two detailed proofs that these formulas do indeed give an isomorphism of Hopf algebras see [7]; or see Chapter 6 of [3]. In terms of derivations, reasoning as above in Section 4, this gives the following theorem.

5.1. Theorem

Let A be an algebra over the rationals and let $(id, d_1, d_2, \ldots)$ be a Hasse-Schmidt derivation on it. Then the ∂_n defined by

$$\partial_n = \sum_{r_1+\ldots+r_m=n} (-1)^{m+1} \frac{d_{r_1}d_{r_2}\ldots d_{r_m}}{m} \tag{5.6}$$

are (ordinary) derivations and

$$d_n = \sum_{r_1+\ldots+r_m=n} \frac{\partial_{r_1}\partial_{r_2}\ldots\partial_{r_m}}{m!} \tag{5.7}$$

5.2. Comment

Perhaps I should add that for any given collection of ordinary derivations, Formula 5.7 yields a Hasse-Schmidt derivation. That is the theorem from [1] with which I started this section.

6. Conclusions

Hasse-Schmidt derivations on an associative algebra A are exactly the same as Hopf module algebra structures on A for the Hopf algebra **NSymm**. This leads to formulas connecting ordinary derivations to higher derivations.

It remains to explore this phenomenon for other kinds of algebras.

The dual of **NSymm** is **QSymm**, the Hopf algebra of quasi-symmetric functions. It remains to be clarified what a coalgebra comodule over **QSymm** means in terms of coderivations. There are also other (mixed) variants to be further explored.

References and Notes

1. Heerema, N. Higher derivations and automorphisms. *Bull. American Math. Soc.* **1970**, 1212–1225. [CrossRef]
2. Mirzavaziri, M. Characterization of higher derivations on algebras. *Comm. Algebra* **2010**, *38*, 981–987. [CrossRef]
3. Hazewinkel, M.; Nadiya, G.; Vladimir, V.K. *Algebras, Rings, and Modules:Lie Algebras and Hopf Algebras.*; American Mathematicial Society: Providence, RI, USA, 2010.
4. Gel'fand, I.M.; Daniel, K.; Alain, L.; Bernard, L.; Vladimir, S.R.; Jean-Yves, T. Noncommutative symmetric functions. *Adv. Math.* **1995**, *112*, 218–348.

5. This is an instance where the noncommutative formulas are more elegant and also easier to prove than their commutative analogues. In the commutative case there are all kinds of multiplicities that mess things up.
6. Hazewinkel, M. The primitives of the Hopf algebra of noncommutative symmetric functions. *São Paulo J. Math. Sci.* **2007**, *1*, 175–203.
7. Hazewinkel, M. *The Leibniz Hopf Algebra and Lyndon Words*; CWI: Amsterdam, The Netherlands, 1996.

8

Hopf Algebra Symmetries of an Integrable Hamiltonian for Anyonic Pairing

Jon Links

Centre for Mathematical Physics, School of Mathematics and Physics, The University of Queensland, Brisbane 4072, Australia; E-Mail: jrl@maths.uq.edu.au

Abstract: Since the advent of Drinfel'd's double construction, Hopf algebraic structures have been a centrepiece for many developments in the theory and analysis of integrable quantum systems. An integrable anyonic pairing Hamiltonian will be shown to admit Hopf algebra symmetries for particular values of its coupling parameters. While the integrable structure of the model relates to the well-known six-vertex solution of the Yang–Baxter equation, the Hopf algebra symmetries are not in terms of the quantum algebra $U_q(sl(2))$. Rather, they are associated with the Drinfel'd doubles of dihedral group algebras $D(D_n)$.

Keywords: Hopf algebra; Drinfel'd double construction; quantum integrability; Yang–Baxter equation

1. Introduction

Integrable quantum systems which admit exact solutions are central in advancing understanding of many-body systems. Classic examples are provided by the Heisenberg spin chain [1], the Bose [2] and Fermi [3] gases with delta-function interactions, the Bardeen–Cooper–Schrieffer pairing Hamiltonian with uniform scattering interactions [4], and the Hubbard model in one dimension [5]. With the development of the Quantum Inverse Scattering Method [6] as a systematic prescription for constructing integrable quantum systems through the Yang–Baxter equation [3,7,8], and solving them through the algebraic Bethe ansatz, it subsequently emerged that Hopf algebraic structures are fundamental in quantum integrability. The works of Jimbo [9] and Drinfel'd [10] were instrumental in formulating the notion of quantum algebras $U_q(g)$, deformations of the universal enveloping algebras of a Lie algebra g, which have the structure of a *quasi-triangular* Hopf algebra. The significance of the quasi-triangular structure is that it affords an algebraic solution of the Yang–Baxter equation. Matrix solutions of the Yang–Baxter equation are then generated through representations of these algebras. The simplest example of the two-dimensional loop representation of the untwisted affine quantum algebra $U_q(sl(2)^{(1)})$ leads to the six-vertex model solution of the Yang–Baxter equation, which establishes integrability of the anisotropic (XXZ) Heisenberg chain. The precise form of six-vertex solution obtained depends on the choice of *gradation* for $U_q(sl(2)^{(1)})$. The principal gradation leads to the symmetric solution, while the homogeneous gradation leads to an asymmetric solution [11]. Only in the latter case is the solution invariant with respect to the action of the non-affine subalgebra $U_q(sl(2))$.

The work of Drinfel'd [10] also provides a means to construct a quasi-triangular Hopf algebra from any Hopf algebra and the dual algebra, through a procedure known as the double construction. The double construction applied to finite group algebras [12] yields a framework in which to develop anyonic models that lead to notions of topological quantum computation [13]. In a series of works [14–16], solutions of the Yang–Baxter associated with Drinfel'd doubles of dihedral group algebras, denoted $D(D_n)$, have been studied. In particular, it was found that two-dimensional representations of these algebras belong to the aforementioned six-vertex model solution in the symmetric case. The symmetric solution was employed in [17] to construct an integrable anyonic pairing Hamiltonian, which generalises the pairing Hamiltonian with uniform scattering interactions solved by Richardson [4]. Below, this integrable anyonic pairing Hamiltonian will be shown to admit Hopf algebra symmetries given by $D(D_n)$ for particular values of the coupling parameters.

2. The Integrable Hamiltonian for Anyonic Pairing

Consider a general anyonic pairing Hamiltonian of the reduced Bardeen–Cooper–Schrieffer form, which acts on a Hilbert space $\mathcal{H}$ of dimension 4^L, given by

$$H = \frac{1}{2}\sum_{j=1}^{L} \varepsilon_j \left(a_{j+}^\dagger a_{j+} + a_{j-}^\dagger a_{j-}\right) - \sum_{k>l}^{L} \left(G_{kl} a_{l+}^\dagger a_{l-}^\dagger a_{k-} a_{k+} + \text{h.c.}\right) \tag{1}$$

Above, $\{\varepsilon_j : j = 1, ..., L\}$ represent single-particle energy levels (two-fold denegerate labelled by $\pm$) and G_{kl} are the pairing interaction coupling parameters of the model. For $q = \exp(i\beta)$, $\beta \in \mathbb{R}$ the operators $\{a_{j\pm}, a_{j\pm}^\dagger : j = 1, ..., L\}$ satisfy the relations

$$\begin{aligned}
\{a_{j\sigma}, a_{j\rho}\} &= \{a_{j+}, a_{k-}\} = 0, \\
\{a_{j\sigma}, a_{j\rho}^\dagger\} &= \delta_{\sigma\rho} I, \\
a_{j\sigma} a_{k\sigma} &= -q a_{k\sigma} a_{j\sigma} \qquad j > k \\
a_{j\sigma}^\dagger a_{k\sigma} &= -q^{-1} a_{k\sigma} a_{j\sigma}^\dagger \qquad j > k
\end{aligned}$$

and those relations obtained by taking Hermitian conjugates. Throughout, I is used to denote an identity operator. These types of anyonic operators are considered as q-deformations of fermionic operators, with the usual fermionic commutation relations recovered in the limit $q \to 1$. The anyonic creation and annihilation operators may be realised in terms of the canonical fermionic operators $\{c_{j\pm}, c_{j\pm}^\dagger : j = 1, ..., L\}$ through a generalised Jordan–Wigner transformation

$$a_{j\sigma} = c_{j\sigma} \prod_{k=j+1}^{L} q^{2n_{k\sigma} - I}$$

$$a_{j\sigma}^\dagger = c_{j\sigma}^\dagger \prod_{k=j+1}^{L} q^{I - 2n_{k\sigma}}$$

where $n_{j\sigma} = c_{j\sigma}^\dagger c_{j\sigma}$.

As with the more familiar fermionic pairing Hamiltonians, one of the notable features of Equation (1) is the *blocking effect*. For any unpaired anyon at level j, the action of the pairing interaction is zero since only paired anyons interact. This means that the Hilbert space can be decoupled into a product of paired and unpaired anyonic states in which the action of the Hamiltonian on the space for the unpaired anyons is automatically diagonal in the natural basis. In view of this property, the pair number operator

$$N = \sum_{j=1}^{L} a_{j+}^\dagger a_{j+} a_{j-}^\dagger a_{j-}$$

commutes with Equation (1) and thus provides a good quantum number. Below, M will be used to denote the eigenvalues of the pair number operator.

In [17] it was shown that, for a suitable restriction on the coupling parameters, the Hamiltonian is integrable in the sense of the Quantum Inverse Scattering Method and admits an exact solution derived through the algebraic Bethe ansatz. To characterise the integrable manifold of the coupling parameter space, the set of parameters $\{\alpha\} \cup \{z_j : j = 1, ..., L\}$ are introduced with the following constraints imposed:

$$\varepsilon_j = z_j^2, \tag{2}$$

$$G_{kl} = \frac{z_k z_l \sin(2\beta) \exp(-i\alpha)}{\sin(\alpha - 2\beta)} \qquad k > l. \tag{3}$$

The conserved operators for this integrable model are obtained via the Quantum Inverse Scattering Method in a standard manner. Here, the key steps are noted. A transfer matrix $t(u) \in \mathrm{End}(\mathcal{H})$ is constructed as

$$t(x) = \mathrm{tr}_a\left(T(x)\right) \tag{4}$$

where $T(x)$ is the monodromy matrix and tr_a is the partial trace over an auxiliary space labelled by a. The monodromy matrix is required to satisfy the relation

$$R_{ab}(x/y)T_a(x)T_b(y) = T_b(y)T_a(x)R_{ab}(x/y) \tag{5}$$

which is an operator equation on $V \otimes V \otimes \mathcal{H}$, with the two auxiliary spaces labelled by a and b. Above,

$$R(x) = \left(\begin{array}{cc|cc} q^2x - q^{-2}x^{-1} & 0 & 0 & 0 \\ 0 & x - x^{-1} & q^2 - q^{-2} & 0 \\ \hline 0 & q^2 - q^{-2} & x - x^{-1} & 0 \\ 0 & 0 & 0 & q^2x - q^{-2}x^{-1} \end{array}\right) \tag{6}$$

is the six-vertex solution (it is convenient for our purposes to express the deformation parameter as q^2 rather than the more familiar q) of the Yang–Baxter equation [3,7,8]

$$R_{12}(x/y)R_{13}(x)R_{23}(y) = R_{23}(y)R_{13}(x)R_{12}(x/y) \tag{7}$$

which acts on the three-fold space $V \otimes V \otimes V$. The subscripts above refer to the spaces on which the operators act, e.g.,

$$R_{12}(x) = R(x) \otimes I$$

Two important properties of $R(x)$, which will be called upon later, are

$$R_{21}(x) = R_{12}(x), \tag{8}$$

$$R(x)^{t_2} = (x - x^{-1})(q^{-4}x^{-1} - q^4x)\left[R(q^{-4}x^{-1})^{t_2}\right]^{-1} \tag{9}$$

where t_2 denotes partial transposition in the second space of the tensor product.

The monodromy matrix is

$$T_a(x) = L_{a1}(xz_1^{-1})L_{a2}(xz_2^{-1})....L_{aL}(xz_L^{-1})U_a \tag{10}$$

where

$$U = \begin{pmatrix} \exp(i(\beta(L - 2M + 2) - \alpha)) & 0 \\ 0 & \exp(i(\alpha - \beta(L - 2M + 2))) \end{pmatrix} \tag{11}$$

and $L(x) = R(q^{-1}x)$. Bearing in mind the earlier comments regarding the blocking effect, we may write

$$L_{aj}(x) = x\begin{pmatrix} q^{n_j-I} & 0 \\ 0 & q^{I-n_j}\end{pmatrix} + (q^2 - q^{-2})\begin{pmatrix} 0 & b_j \\ b_j^\dagger & 0\end{pmatrix} - x^{-1}\begin{pmatrix} q^{I-n_j} & 0 \\ 0 & q^{n_j-I}\end{pmatrix}$$

where

$$n_j = c_{j+}^\dagger c_{j+} + c_{j-}^\dagger c_{j-}$$
$$b_j = c_{j-}c_{j+}$$

Note that U is defined in a sector-dependent manner in terms of the eigenvalues M of N, which is legitimate since N is conserved.

A consequence of Equation (10), and the diagonal form of U, is that the transfer matrices form a commutative family; *i.e.*,

$$[t(x),\, t(y)] = 0 \qquad \forall\, x,\, y \in \mathbb{C} \tag{12}$$

The transfer matrices can be expanded in a Laurent series

$$t(x) = \sum_{j=-L}^{L} t^{(j)}x^j$$

such that, because of Equation (12), the co-efficients commute

$$\left[t^{(j)},\, t^{(k)}\right] = 0, \qquad -L \le j,\, k \le L$$

Finally it can be verified that the Hamiltonian Equation (1), subject to the constraints of Equations (2,3), is expressible as (the corresponding expression in [17] contains typographical errors, which are corrected here)

$$H = \frac{1}{4\sin(2\beta)\sin(\alpha - 2\beta)}\left[t^{(L-2)}\prod_{j=1}^{L} z_j + 2\cos(\alpha - 4\beta)\sum_{j=1}^{L} z_j^2\right] \tag{13}$$

establishing that $\{t^{(j)} : j = 1, ..., L\}$ provides a set of Abelian conserved operators for the system. In this sense the system is said to be integrable.

In the remainder of this work it will be shown that for certain further restrictions on the coupling parameters there are additional Hopf algebraic symmetries of the system. These non-Abelian symmetries are not related to a quantum algebra $U_q(sl(2))$ structure, but are realised through the Drinfel'd doubles of dihedral group algebras.

3. Drinfel'd Doubles of Dihedral Group Algebras

The dihedral group D_n has two generators $\sigma\tau$ satisfying:

$$\sigma^n = e,\ \tau^2 = e,\ \tau\sigma = \sigma^{n-1}\tau$$

where e denotes the group identity. Considering D_n as a group algebra, the Drinfel'd double [10] of D_n, denoted $D(D_n)$, has basis

$$\{gh^* | g, h \in D_n\}$$

where g are the group elements and g^* are their dual elements. This gives an algebra of dimension $4n^2$. Multiplication of dual elements is defined by

$$g^*h^* = \delta(g,h)g^* \tag{14}$$

where δ is the Kronecker delta function. The products h^*g are computed using

$$h^*g = g(g^{-1}hg)^* \tag{15}$$

The algebra $D(D_n)$ becomes a Hopf algebra by imposing the following coproduct, antipode and counit respectively:

$$\begin{aligned}\Delta(gh^*) &= \sum_{k\in D_n} g(k^{-1}h)^* \otimes gk^* = \sum_{k\in D_n} gk^* \otimes g(hk^{-1})^* \\ S(gh^*) &= (h^{-1})^*g^{-1} = g^{-1}(gh^{-1}g^{-1})^* \\ \varepsilon(gh^*) &= \delta(h,e)\end{aligned} \tag{16}$$

An important property of $D(D_n)$ which will be called upon later is

$$S^2(a) = a \qquad \forall\, a \in D(D_n) \tag{17}$$

Defining $\bar{e} = \sum_{g\in D_n} g^*$ the universal R-matrix is given by

$$\mathcal{R} = \sum_{g\in D_n} g\bar{e} \otimes eg^* \tag{18}$$

This can be shown to satisfy the relations for a quasi-triangular Hopf algebra as defined in [10]:

$$\begin{aligned}\mathcal{R}\Delta(a) &= \Delta^T(a)\mathcal{R}, \quad \forall\, a \in D(D_n) \\ (\Delta \otimes \mathrm{id})\mathcal{R} &= \mathcal{R}_{13}\mathcal{R}_{23} \\ (\mathrm{id} \otimes \Delta)\mathcal{R} &= \mathcal{R}_{13}\mathcal{R}_{12}\end{aligned} \tag{19}$$

where Δ^T is the opposite coproduct

$$\Delta^T(gh^*) = \sum_{k\in D_n} gk^* \otimes g(k^{-1}h)^* = \sum_{k\in D_n} g(hk^{-1})^* \otimes gk^*$$

When n is even, $D(D_n)$ admits eight one-dimensional irreducible representations, $(n^2-4)/2$ two-dimensional irreducible representations, and eight $n/2$-dimensional irreducible representations. When n is odd, $D(D_n)$ admits two one-dimensional irreducible representations, $(n^2-1)/2$ two-dimensional irreducible representations, and two n-dimensional irreducible representations. The explicit irreducible representations are given in [14]. Our interest will be in the two-dimensional irreducible representations. To describe them, let $\omega = e^{2\pi i/n}$. Then these representations have the form

$$\pi_{1(j)}(\sigma\bar{e}) = \begin{pmatrix} \omega^j & 0 \\ 0 & \omega^{-j} \end{pmatrix}, \quad \pi_{1(j)}(\tau\bar{e}) = \begin{pmatrix} 0 & 1 \\ 1 & 0 \end{pmatrix}, \quad \pi_{1(j)}(eg^*) = \delta(g,e)\begin{pmatrix} 1 & 0 \\ 0 & 1 \end{pmatrix}$$

for $j = 1, \ldots, (n-2)/2$ if n is even and $j = 1, \ldots, (n-1)/2$ if n is odd,

$$\pi_{2(j)}(\sigma\bar{e}) = \begin{pmatrix} \omega^j & 0 \\ 0 & \omega^{-j} \end{pmatrix}, \quad \pi_{2(j)}(\tau\bar{e}) = \begin{pmatrix} 0 & 1 \\ 1 & 0 \end{pmatrix}, \quad \pi_{2(j)}(eg^*) = \delta(g,\sigma^{n/2})\begin{pmatrix} 1 & 0 \\ 0 & 1 \end{pmatrix}$$

for $j = 1, ..., (n-2)/2$ if n is even, and

$$\pi_{(j,k)}(\sigma\bar{e}) = \begin{pmatrix} \omega^j & 0 \\ 0 & \omega^{-j} \end{pmatrix}, \quad \pi_{(j,k)}(\tau\bar{e}) = \begin{pmatrix} 0 & 1 \\ 1 & 0 \end{pmatrix}, \quad \pi_{(j,k)}(eg^*) = \begin{pmatrix} \delta(g,\sigma^k) & 0 \\ 0 & \delta(g,\sigma^{-k}) \end{pmatrix}$$

for $j = 1, ..., n$ and where $k = 1, ..., (n-2)/2$ if n is even, and $k = 1, ..., (n-1)/2$ if n is odd.

For any of the above two-dimensional representations π_μ, $\mu = 1(j), 2(j), (j,k)$ the tensor product representation applied to the universal R-matrix Equation (18) yields the general form

$$(\pi_\mu \otimes \pi_\mu)R = \left(\begin{array}{cc|cc} \omega^l & 0 & 0 & 0 \\ 0 & \omega^{-l} & 0 & 0 \\ \hline 0 & 0 & \omega^{-l} & 0 \\ 0 & 0 & 0 & \omega^l \end{array}\right) \tag{20}$$

for some $l = 1, ..., n$. Choosing $q = \omega^l$ in Equation (6) we then find

$$\begin{aligned} R(x) &= \left(\begin{array}{cc|cc} \omega^{2l}x - \omega^{-2l}x^{-1} & 0 & 0 & 0 \\ 0 & x - x^{-1} & \omega^{2l} - \omega^{-2l} & 0 \\ \hline 0 & \omega^{2l} - \omega^{-2l} & x - x^{-1} & 0 \\ 0 & 0 & 0 & \omega^{2l}x - \omega^{-2l}x^{-1} \end{array}\right) \\ &= (x-1)\omega^l(\pi_\alpha \otimes \pi_\alpha)\mathcal{R} - (x^{-1}+1)\omega^{-l}(\pi_\alpha \otimes \pi_\alpha)\mathcal{R}^{-1} + (\omega^{2l} - \omega^{-2l})P \end{aligned} \tag{21}$$

where

$$P = \left(\begin{array}{cc|cc} 1 & 0 & 0 & 0 \\ 0 & 0 & 1 & 0 \\ \hline 0 & 1 & 0 & 0 \\ 0 & 0 & 0 & 1 \end{array}\right)$$

is the permutation operator on the tensor product space. This shows that the Baxterisation of the $D(D_n)$ R-matrix in two-dimensional representations leads to the symmetric six-vertex model at q a root of unity, which was previously reported in [14]. Baxterisation of the $D(D_n)$ R-matrix in higher-dimensional representations lead to the Fateev–Zamolodchikov solution of the Yang–Baxter equation, as discussed in [15,16].

Having identified the relationship Equation (21) between the solution Equation (6) of the Yang–Baxter equation and representations of the universal R-matrix Equation (18) for $D(D_n)$, we can now proceed to determine when $D(D_n)$ is a symmetry algebra of the transfer matrix associated to the Hamiltonian Equation (1) subject to the constraints Equations 2 and 3.

4. Symmetries of the Transfer Matrix and Hamiltonian

First we define

$$R^{+} = q^{-1}\lim_{x\to\infty}\frac{1}{x}R(x) = \left(\begin{array}{cc|cc} q & 0 & 0 & 0 \\ 0 & q^{-1} & 0 & 0 \\ \hline 0 & 0 & q^{-1} & 0 \\ 0 & 0 & 0 & q \end{array}\right)$$

$$R^{-} = -q\lim_{x\to 0}xR(x) = \left(\begin{array}{cc|cc} q^{-1} & 0 & 0 & 0 \\ 0 & q & 0 & 0 \\ \hline 0 & 0 & q & 0 \\ 0 & 0 & 0 & q^{-1} \end{array}\right) = (R^{+})^{-1}$$

It follows from Equation (7) that

$$R_{12}(u)R^{+}_{13}R^{+}_{23} = R^{+}_{23}R^{+}_{13}R_{12}(u) \tag{22}$$

$$R^{+}_{12}R_{13}(u)R^{-}_{23} = R^{-}_{23}R_{13}(u)R^{+}_{12} \tag{23}$$

We then define a modified monodromy matrix

$$T_a(x) = L_{a1}(xz_1^{-1})L_{a2}(xz_2^{-1})....L_{aL}(xz_L^{-1})R^{+}_{aL}....R^{+}_{a2}R^{+}_{a1} \tag{24}$$

Through use of Equations (7,22,23) it can be shown that this monodromy matrix satisfies a generalised version of Equation (5):

$$R_{ab}(x/y)T_a(x)R^{+}_{ab}T_b(y) = T_b(y)R^{+}_{ab}T_a(x)R_{ab}(x/y)$$

The transfer matrix is again defined by Equation (4). From the results of [18] it is known that Equation (12) still holds by use of Equation (9).

The action of $D(D_n)$ on an L-fold tensor product space is given through iterated use of the co-product action Equation (16):

$$\Delta^{(L)} = (\Delta\otimes \mathrm{id})\Delta^{(L-1)} = (\mathrm{id}\otimes\Delta)\Delta^{(L-1)}, \qquad \Delta^{(2)} = \Delta$$

Below, for ease of notation, we will omit the representation symbols π_μ when dealing with tensor product representations obtained through this action. Whenever we have

$$\beta = \frac{2\pi l}{n}, \qquad l = 1, ..., n \tag{25}$$

the monodromy matrix Equation (24) commutes with the action of $D(D_n)$ as a consequence of Equations (8–21). From the results of [19], the transfer matrix obtained from Equation (24) also commutes with the action of $D(D_n)$ due to Equation (17).

Observing that we may write

$$R^{+}_{aj} = \begin{pmatrix} q^{n_j - I} & 0 \\ 0 & q^{I-n_j} \end{pmatrix}$$

we may simplify Equation (24) as

$$T_a(x) = L_{a1}(xz_1^{-1})L_{a2}(xz_2^{-1})....L_{aL}(xz_L^{-1})\tilde{U}_a$$

where

$$\tilde{U} = \begin{pmatrix} \exp(i\beta(2M-L)) & 0 \\ 0 & \exp(i\beta(L-2M)) \end{pmatrix} \tag{26}$$

Comparing Equations 11 and 26 and taking note of Equation (25), these matrices are made equal by choosing

$$\alpha = \frac{4\pi l(L-2M+1)}{n} \tag{27}$$

meaning that the transfer matrices obtained from the monodromy matrices Equations 10 and 24 are equal. Thus we have established that the transfer matrix associated to the integrable Hamiltonian Equation (1) subject to the constraints of Equations 2 and 3 commutes with action of the quasi-triangular Hopf algebra $D(D_n)$ whenever Equations 25 and 27 hold.

A crucial point to bear in mind is that the transfer matrices were defined in a sector-dependent manner, where each sector is associated with a fixed number of Cooper pairs. However the $D(D_n)$ action does not preserve sectors, and specifically $\tau\bar{e}$ acts as a particle-hole transformation:

$$\Delta^{(L)}(\tau\bar{e})N = (L-N)\Delta^{(L)}(\tau\bar{e}) \tag{28}$$

whereas

$$\Delta^{(L)}(\sigma\bar{e})N = N\Delta^{(L)}(\sigma\bar{e}) \tag{29}$$
$$\Delta^{(L)}(eg^{*})N = N\Delta^{(L)}(eg^{*}) \tag{30}$$

These relations follow from the above two-dimensional matrix representations for which it is seen that representations of $\sigma\bar{e}$ and eg^{*} are always diagonal in the basis in which the action of N is diagonal. In the same basis, representations of $\tau\bar{e}$ are orthogonal matrices with non-zero off-diagonal entries.

Recall that the Hamiltonian is defined through the transfer matrix by Equation (13). Consequently, while $D(D_n)$ is a symmetry of the transfer matrix obtained from Equation (24) in the conventional sense, the interpretation of $D(D_n)$ as a symmetry of the Hamiltonian is more subtle as the choice Equation (27) is sector-dependent and thus α needs to be treated as an operator-valued quantity. From Equation (28) we have for α given by Equation (27) that for each sector where N has eigenvalue M

$$\begin{aligned} \Delta^{(L)}(\tau\bar{e})\alpha &= \Delta^{(L)}(\tau\bar{e})2\beta(L-2M+1) \\ &= 2\beta(L-2(L-M)+1)\Delta^{(L)}(\tau\bar{e}) \\ &= (4\beta-\alpha)\Delta^{(L)}(\tau\bar{e}) \end{aligned}$$

Using Equation (13) we obtain

$$\Delta^{(L)}(\tau\bar{e})t^{(L-2)} = t^{(L-2)}\Delta^{(L)}(\tau\bar{e})$$

$$\Delta^{(L)}(\tau\bar{e})\left(2\sin(2\beta)\sin(\alpha-2\beta)H - \cos(\alpha-4\beta)\sum_{j=1}^{L} z_j^2\right)$$
$$= \left(2\sin(2\beta)\sin(\alpha-2\beta)H - \cos(\alpha-4\beta)\sum_{j=1}^{L} z_j^2\right)\Delta^{(L)}(\tau\bar{e})$$

$$2\sin(2\beta)\sin(2\beta-\alpha)\Delta^{(L)}(\tau\bar{e})H - \cos(\alpha)\sum_{j=1}^{L} z_j^2\Delta^{(L)}(\tau\bar{e})$$
$$= \left(2\sin(2\beta)\sin(\alpha-2\beta)H - \cos(\alpha-4\beta)\sum_{j=1}^{L} z_j^2\right)\Delta^{(L)}(\tau\bar{e})$$

From the trigonometric identity

$$\cos(\alpha) - \cos(\alpha-4\beta) = 2\sin(2\beta)\sin(\alpha-2\beta)$$

this then leads to the following *anti-symmetry* relation for the integrable Hamiltonian Equations (1–3) whenever Equations 25 and 27 hold

$$\Delta^{(L)}(\tau\bar{e})\left(H - \frac{1}{2}\sum_{j=1}^{L} z_j^2 I\right) = -\left(H - \frac{1}{2}\sum_{j=1}^{L} z_j^2 I\right)\Delta^{(L)}(\tau\bar{e})$$

This relation shows how the spectrum of the Hamiltonian maps under a particle-hole transformation $M \mapsto L - M$ induced by Equation (28). On the other hand,

$$\Delta^{(L)}(\sigma\bar{e})\left(H - \frac{1}{2}\sum_{j=1}^{L} z_j^2 I\right) = \left(H - \frac{1}{2}\sum_{j=1}^{L} z_j^2 I\right)\Delta^{(L)}(\sigma\bar{e}),$$
$$\Delta^{(L)}(eg^*)\left(H - \frac{1}{2}\sum_{j=1}^{L} z_j^2 I\right) = \left(H - \frac{1}{2}\sum_{j=1}^{L} z_j^2 I\right)\Delta^{(L)}(eg^*), \qquad \forall g \in D_n$$

Thus as a result of Equations 29 and 30, the action of $\sigma\bar{e}$ and eg^* leaves the spectrum of the Hamiltonian invariant in each sector with fixed M.

Finally, if the above procedure is followed using the asymmetric R-matrix

$$R(x) = \begin{pmatrix} (q^2x - q^{-2}x^{-1}) & 0 & | & 0 & 0 \\ 0 & (x - x^{-1}) & | & x^{-1}(q^2 - q^{-2}) & 0 \\ - & - & & - & - \\ 0 & x(q^2 - q^{-2}) & | & (x - x^{-1}) & 0 \\ 0 & 0 & | & 0 & (q^2x - q^{-2}x^{-1}) \end{pmatrix}$$

a transfer matrix is obtained which commutes with the co-product action of $U_q(sl(2))$ [20,21]. However in this setting the corresponding conserved operator $t^{(L-2)}$ contains additional interaction terms. As a result, an expression analogous to Equation (13) does not yield an operator in the form of Equation (1).

5. Conclusions

An analysis of an integrable Hamiltonian for anyonic pairing, as given by Equation (1) subject to Equations 2 and 3, was undertaken. Values of the coupling parameters were identified for which the model admits Hopf algebraic symmetries. In Section 2 the construction of the integrable model was outlined in terms of the Quantum Inverse Scattering Method. This was achieved through the symmetric, six-vertex solution of the Yang–Baxter equation. The Hamiltonian was identified through a conserved operator associated to the corresponding transfer matrix. In Section 3 a description of the quasi-triangular Hopf algebra $D(D_n)$ was presented, including explicit expressions for all irreducible, two-dimensional representations. Through these representations it was established that the symmetric, six-vertex solution of the Yang–Baxter equation is related to representations of the universal R-matrix for $D(D_n)$. These results were utilised in Section 4 to construct a transfer matrix which preserved the $D(D_n)$ symmetry. From this transfer matrix, values of the coupling parameters were identified for which the Hamiltonian Equation (1) subject to Equations 2 and 3 has $D(D_n)$ as a symmetry algebra. However the interpretation of $D(D_n)$ as a symmetry algebra for the Hamiltonian is somewhat unconventional in that both commuting and anti-commuting actions for the generators were found. The anti-commuting action is associated with a particular $D(D_n)$ generator that induces a particle-hole transformation.

Acknowledgments

This work was supported by the Australian Research Council through Discovery Project *Topological properties of exactly solvable, two-dimensional quantum systems* (DP110101414).

References

1. Bethe, H. Zur Theorie der Metalle: I. Eigenwerte und Eigenfunktionen der linearen Atomkette. *Zeitschrift für. Physik* **1931**, *71*, 205–226.
2. Lieb, E.H.; Liniger, W. Exact analysis of an interacting Bose gas. I. The general solution and the ground state. *Phys. Rev.* **1963**, *130*, 1605–1616.
3. Yang, C.N. Some exact results for the many-body problem in one dimension with repulsive delta-function interaction. *Phys. Rev. Lett.* **1967**, *19*, 1312–1315.
4. Richardson, R.W. A restricted class of exact eigenstates of the pairing-force Hamiltonian. *Phys. Lett.* **1963**, *3*, 277–279.
5. Lieb, E.H.; Wu, F.Y. Absence of Mott transition in an exact solution of the short-range, one-band model in one dimension. *Phys. Rev. Lett.* **1968**, *20*, 1445–1448.
6. Takhtadzhan, L.A.; Faddeev, L.D. The quantum method of the inverse problem and the Heisenberg XYZ model. *Russ. Math. Surveys* **1979**, *34*, 11–68.
7. McGuire, J.B. Study of exactly soluble one-dimensional N-body problems. *J. Math. Phys.* **1964**, *5*, 622–636.
8. Baxter, R.J. Partition function of the eight-vertex lattice model. *Ann. Phys.* **1972**, *70*, 193–228.
9. Jimbo, M. A q-difference analog of $U(g)$ and the Yang-Baxter equation. *Lett. Math. Phys.* **1985**, *10*, 63–69.
10. Drinfel'd, V.G. Quantum Groups. In *Proceedings of the International Congress of Mathematicians*; Gleason, A.M., Ed.; American Mathematical Society: Providence, Rhode Island, 1986; pp. 798–820.
11. Bracken, A.J.; Delius, G.W.; Gould, M.D.; Zhang, Y.-Z. Infinite families of gauge equivalent R-matrices and gradations of quantized affine algebras. *Int. J. Mod. Phys. B* **1994**, *8*, 3679–3691.
12. Gould, M.D. Quantum double finite group algebras and their representations. *Bull. Aust. Math. Soc.* **1993**, *48*, 275–301.
13. Kitaev, A.Y. Fault-Tolerant quantum computation by anyons. *Ann. Phys.* **2003**, *303*, 2–30.
14. Dancer, K.A.; Isaac, P.; Links, J. Representations of the quantum doubles of finite group algebras and spectral parameter dependent solutions of the Yang-Baxter equation. *J. Math. Phys.* **2006**, *47*, 1–18.
15. Finch, P.E.; Dancer, K.A.; Isaac, P.; Links, J. Solutions of the Yang–Baxter equation: Descendents of the six-vertex model from the Drinfeld doubles of dihedral group algebras. *Nucl. Phys. B* **2011**, *847*, 387–412.
16. Finch, P.E. Integrable Hamiltonians with $D(D_n)$ symmetry from the Fateev-Zamolodchikov model. *J. Stat. Mech. Theory Exp.* **2011**, doi:10.1088/1742-5468/2011/04/P04012.
17. Dunning, C.; Ibañez, M.; Links, J.; Sierra, G.; Zhao, S.-Y. Exact solution of the $p + ip$ pairing Hamiltonian and a hierarchy of integrable models. *J. Stat. Mech. Theory Exp.* **2010**, doi:10.1088/1742-5468/2010/08/P08025.

18. Links, J.; Foerster, A. On the construction of integrable closed chains with quantum supersymmetry. *J. Phys. A Math. Gen.* **1997**, *30*, 2483–2487.
19. Links, J.R.; Gould, M.D. Casimir invariants for Hopf algebras. *Rep. Math. Phys.* **1992**, *31*, 91–111.
20. Grosse, H.; Pallua, S.; Prester, P.; Raschhofer, E. On a quantum group invariant spin chain with non-local boundary conditions. *J. Phys. A Math. Gen.* **1994**, *27*, 4761–4771.
21. Karowski, M.; Zapletal, A. Quantum group invariant integrable n-state vertex models with periodic boundary conditions. *Nucl. Phys. B* **1994**, *419*, 567–588.

9

Quasitriangular Structure of Myhill–Nerode Bialgebras

Robert G. Underwood

Department of Mathematics/Informatics Institute, Auburn University Montgomery, P.O. Box 244023, Montgomery, AL 36124, USA; E-Mail: runderwo@aum.edu

Abstract: In computer science the Myhill–Nerode Theorem states that a set L of words in a finite alphabet is accepted by a finite automaton if and only if the equivalence relation $\sim_L$, defined as $x \sim_L y$ if and only if $xz \in L$ exactly when $yz \in L, \forall z$, has finite index. The Myhill–Nerode Theorem can be generalized to an algebraic setting giving rise to a collection of bialgebras which we call Myhill–Nerode bialgebras. In this paper we investigate the quasitriangular structure of Myhill–Nerode bialgebras.

Keywords: algebra; coalgebra; bialgebra; Myhill–Nerode theorem; Myhill–Nerode bialgebra; quasitriangular structure

1. Introduction

Let Σ_0 be a finite alphabet and let $\hat{\Sigma}_0$ denote the set of words formed from the letters in Σ_0. Let $L \subseteq \hat{\Sigma}_0$ be a language, and let $\sim_L$ be the equivalence relation defined as $x \sim_L y$ if and only if $xz \in L$ exactly when $yz \in L, \forall z \in \hat{\Sigma}_0$. The Myhill–Nerode Theorem of computer science states that L is accepted by a finite automaton if and only if $\sim_L$ has finite index (cf. [1, 1, Chapter III, §9, Proposition 9.2], [2, §3.4, Theorem 3.9]). In [3, Theorem 5.4] the authors generalize the Myhill–Nerode theorem to an algebraic setting in which a finiteness condition involving the action of a semigroup on a certain function plays the role of the finiteness of the index of $\sim_L$, while a bialgebra plays the role of the finite automaton which accepts the language. We call these bialgebras *Myhill–Nerode bialgebras*.

The purpose of this paper is to investigate the quasitriangular structure of Myhill–Nerode bialgebras.

By construction, a Myhill–Nerode bialgebra B is cocommutative and finite dimensional over its base field. Thus B admits (at least) the trivial quasitriangular structure $(B, 1 \otimes 1)$. We ask: does B (or its linear dual B^*) have any non-trivial quasitriangular structures?

Towards a solution to this problem, we construct a class of commutative Myhill–Nerode bialgebras and give a complete account of the quasitriangular structure of one of them. We begin with some background information regarding algebras, coalgebras, and bialgebras.

2. Algebras, Coalgebras and Bialgebras

Let K be an arbitrary field of characteristic 0 and let A be a vector space over K with scalar product ra for all $r \in K$, $a \in A$. Scalar product defines two maps $s_1 : K \otimes A \to A$ with $r \otimes a \mapsto ra$ and $s_2 : A \otimes K \to A$ with $a \otimes r \mapsto ra$, for $a \in A$, $r \in K$. Let $I_A : A \to A$ denote the identity map. A *K-algebra* is a triple (A, m_A, η_A) where $m_A : A \otimes A \to A$ is a K-linear map which satisfies

$$m_A(I_A \otimes m_A)(a \otimes b \otimes c) = m_A(m_A \otimes I_A)(a \otimes b \otimes c) \tag{1}$$

and $\eta_A : K \to A$ is a K-linear map for which

$$m_A(I_A \otimes \eta_A)(a \otimes r) = ra = m_A(\eta_A \otimes I_A)(r \otimes a) \tag{2}$$

for all $r \in K$, $a, b, c \in A$. The map m_A is the *multiplication map* of A and η_A is the *unit map* of A. Condition (1) is the *associative property* and Condition (2) is the *unit property*.

We write $m_A(a \otimes b)$ as ab. The element $1_A = \eta_A(1_K)$ is the unique element of A for which $a1_A = a = 1_Aa$ for all $a \in A$. Let A, B be algebras. An *algebra homomorphism* from A to B is a K-linear map $\phi : A \to B$ such that $\phi(m_A(a_1 \otimes a_2)) = m_B(\phi(a_1) \otimes \phi(a_2))$ for all a_1, $a_2 \in A$, and $\phi(1_A) = 1_B$. In particular, for A to be a subalgebra of B we require $1_A = 1_B$.

For any two vector spaces V, W let $\tau : V \otimes W \to W \otimes V$ denote the *twist map* defined as $\tau(a \otimes b) = b \otimes a$, for $a \in V$, $b \in W$. For K-algebras A, B, we have that $A \otimes B$ is a K-algebra with multiplication

$$m_{A\otimes B} : (A \otimes B) \otimes (A \otimes B) \to A \otimes B$$

defined by

$$\begin{aligned} m_{A\otimes B}((a \otimes b) \otimes (c \otimes d)) &= (m_A \otimes m_B)(I_A \otimes \tau \otimes I_B)(a \otimes (b \otimes c) \otimes d) \\ &= (m_A \otimes m_B)((a \otimes c) \otimes (b \otimes d)) = ac \otimes bd \end{aligned}$$

for $a, c \in A$, $b, d \in B$. The unit map $\eta_{A\otimes B} : K \to A \otimes B$ given as

$$\eta_{A\otimes B}(r) = \eta_A(r) \otimes 1_B$$

for $r \in K$.

Let C be a K-vector space. A *K-coalgebra* is a triple $(C, \Delta_C, \epsilon_C)$ in which $\Delta_C : C \to C \otimes C$ is K-linear and satisfies

$$(I_C \otimes \Delta_C)\Delta_C(c) = (\Delta_C \otimes I_C)\Delta_C(c) \tag{3}$$

and $\epsilon_C : C \to K$ is K-linear with

$$s_1(\epsilon_C \otimes I_C)\Delta_C(c) = c = s_2(I_C \otimes \epsilon_C)\Delta_C(c) \tag{4}$$

for all $c \in C$. The maps Δ_C and ϵ_C are the *comultiplication* and *counit* maps, respectively, of the coalgebra C. Condition (3) is the *coassociative property* and Condition (4) is the *counit property*.

We use the notation of M. Sweedler [4, §1.2] to write

$$\Delta_C(c) = \sum_{(c)} c_{(1)} \otimes c_{(2)}$$

Note that Condition (4) implies that

$$\sum_{(c)} \epsilon_C(c_{(1)})c_{(2)} = c = \sum_{(c)} \epsilon_C(c_{(2)})c_{(1)} \tag{5}$$

Let C be a K-coalgebra. A nonzero element c of C for which $\Delta_C(c) = c \otimes c$ is a *grouplike element* of C. If c is grouplike, then

$$\begin{aligned} c &= s_1(\epsilon_C \otimes I_C)\Delta_C(c) \\ &= s_1(\epsilon_C \otimes I_C)(c \otimes c) = \epsilon_C(c)c \end{aligned}$$

and so, $\epsilon_C(c) = 1$. The grouplike elements of C are linearly independent [4, Proposition 3.2.1].

Let C, D be coalgebras. A K-linear map $\phi : C \to D$ is a *coalgebra homomorphism* if $(\phi \otimes \phi)\Delta_C(c) = \Delta_D(\phi(c))$ and $\epsilon_C(c) = \epsilon_D(\phi(c))$ for all $c \in C$. The tensor product $C \otimes D$ of two coalgebras is again a coalgebra with comultiplication map

$$\Delta_{C\otimes D} : C \otimes D \to (C \otimes D) \otimes (C \otimes D)$$

defined by

$$\begin{aligned}
\Delta_{C\otimes D}(c \otimes d) &= (I_C \otimes \tau \otimes I_D)(\Delta_C \otimes \Delta_D)(c \otimes d) \\
&= (I_C \otimes \tau \otimes I_D)(\Delta_C(c) \otimes \Delta_D(d)) \\
&= (I_C \otimes \tau \otimes I_D)(\sum_{(c),(d)} c_{(1)} \otimes c_{(2)} \otimes d_{(1)} \otimes d_{(2)}) \\
&= \sum_{(c),(d)} c_{(1)} \otimes d_{(1)} \otimes c_{(2)} \otimes d_{(2)}
\end{aligned}$$

for $c \in C, d \in D$. The counit map $\epsilon_{C\otimes D} : C \otimes D \to K$ is defined as

$$\epsilon_{C\otimes D}(c \otimes d) = \epsilon_C(c)\epsilon_D(d)$$

for $c \in C, d \in D$.

A *K-bialgebra* is a K-vector space B together with maps $m_B, \eta_B, \Delta_B, \epsilon_B$ for which (B, m_B, η_B) is a K-algebra and $(B, \Delta_B, \epsilon_B)$ is a K-coalgebra and for which Δ_B and ϵ_B are algebra homomorphisms. Let B, B' be bialgebras. A K-linear map $\phi : B \to B'$ is a *bialgebra homomorphism* if ϕ is both an algebra and coalgebra homomorphism.

A *K-Hopf algebra* is a bialgebra H together with an additional K-linear map $\sigma_H : H \to H$ that satisfies

$$m_H(I_H \otimes \sigma_H)\Delta_H(h) = \epsilon_H(h)1_H = m_H(\upsilon_H \otimes I_H)\Delta_H(h) \tag{6}$$

for all $h \in H$. The map σ_H is the *coinverse* (or *antipode*) map and property Condition (6) is the *coinverse* (or *antipode*) *property*. Though we will not consider Hopf algebras here, more details on the subject can be found in [5–8].

An important example of a K-bialgebra is given as follows. Let G be a semigroup with unity, 1. Let KG denote the semigroup algebra. Then KG is a bialgebra with comultiplication map

$$\Delta_{KG} : KG \to KG \otimes KG$$

defined by $x \mapsto x \otimes x$, for all $x \in G$, and counit map $\epsilon_{KG} : KG \to K$ given by $x \mapsto 1$, for all $x \in G$. The bialgebra KG is the *semigroup bialgebra on G*.

Let B be a bialgebra, and let A be an algebra which is a left B-module with action denoted by "$\cdot$". Suppose that

$$b \cdot (aa') = \sum_{(b)} (b_{(1)} \cdot a)(b_{(2)} \cdot a')$$

and

$$b \cdot 1_A = \epsilon_B(b)1_A$$

for all $a, a' \in A, b \in B$. Then A is a *left B-module algebra*. A K-linear map $\phi : A \to A'$ is a *left B-module algebra homomorphism* if ϕ is both an algebra and a left B-module homomorphism.

Let C be a coalgebra and a right B-module with action denoted by "$\cdot$". Suppose that for all $c \in C$, $b \in B$,

$$\Delta_C(c \cdot b) = \sum_{(c),(b)} c_{(1)} \cdot b_{(1)} \otimes c_{(2)} \cdot b_{(2)}$$

and

$$\epsilon_C(c \cdot b) = \epsilon_C(c)\epsilon_B(b)$$

Then C is a *right B-module coalgebra*. A K-linear map $\phi : C \to C'$ is a *right B-module coalgebra homomorphism* if ϕ is both a coalgebra and a right B-module homomorphism.

Let C be a coalgebra and let $C^* = \mathrm{Hom}_K(C, K)$ denote the linear dual of C. Then the coalgebra structure of C induces an algebra structure on C^*.

Proposition 1. *If C is a coalgebra, then C^* is an algebra.*

Proof. Recall that C is a triple $(C, \Delta_C, \epsilon_C)$ where $\Delta_C : C \to C \otimes C$ is K-linear and satisfies the coassociativity property, and $\epsilon_C : C \to K$ is K-linear and satisfies the counit property. The dual map of Δ_C is a K-linear map

$$\Delta_C^* : (C \otimes C)^* \to C^*$$

Since $C^* \otimes C^* \subseteq (C \otimes C)^*$, we define the multiplication map of C^*, denoted as m_{C^*}, to be the restriction of Δ_C^* to $C^* \otimes C^*$. For $f, g \in C^*, c \in C$,

$$(fg)(c) = m_{C^*}(f \otimes g)(c) = \Delta_C^*(f \otimes g)(c) = (f \otimes g)(\Delta_C(c)) = \sum_{(c)} f(c_{(1)})g(c_{(2)})$$

The coassociatively property of Δ_C yields the associative property of m_{C^*}. Indeed, for $f, g, h \in C^*$, $c \in C$,

$$\begin{aligned}
m_{C^*}(I_{C^*} \otimes m_{C^*})(f \otimes g \otimes h)(c) &= \Delta_C^*(I_{C^*} \otimes \Delta_C^*)(f \otimes g \otimes h)(c) \\
&= \Delta_C^*(f \otimes \Delta_C^*(g \otimes h))(c) \\
&= (f \otimes \Delta_C^*(g \otimes h))\Delta_C(c) \\
&= \sum_{(c)} f(c_{(1)})\Delta_C^*(g \otimes h)(c_{(2)}) \\
&= \sum_{(c)} f(c_{(1)})(g \otimes h)\Delta_C(c_{(2)}) \\
&= (f \otimes g \otimes h)(\sum_{(c)} c_{(1)} \otimes \Delta_C(c_{(2)})) \\
&= (f \otimes g \otimes h)(\sum_{(c)} \Delta_C(c_{(1)}) \otimes c_{(2)}) \quad \text{by Condition (3)} \\
&= \sum_{(c)} (f \otimes g)\Delta_C(c_{(1)}) \otimes h(c_{(2)}) \\
&= \sum_{(c)} \Delta_C^*(f \otimes g)(c_{(1)}) \otimes h(c_{(2)}) \\
&= (\Delta_C^*(f \otimes g) \otimes h)\Delta_C(c) \\
&= \Delta_C^*(\Delta_C^*(f \otimes g) \otimes h)(c) \\
&= \Delta_C^*(\Delta_C^* \otimes I_{C^*})(f \otimes g \otimes h)(c) \\
&= m_{C^*}(m_{C^*} \otimes I_{C^*})(f \otimes g \otimes h)(c)
\end{aligned}$$

In addition, the counit map of C dualizes to yield

$$\epsilon_C^* : K := K^* \to C^*$$

defined as $\epsilon_C^*(k)(c) = k(\epsilon(c)) = k\epsilon(c)$. Thus we define the unit map η_{C^*} to be ϵ_C^*. One can show that the counit property of ϵ_C implies the unit property for η_{C^*}. To this end, for $f \in C^*, r \in K, c \in C$,

$$\begin{aligned} m_{C^*}(I_{C^*} \otimes \eta_{C^*})(f \otimes r)(c) &= \Delta_C^*(I_{C^*} \otimes \epsilon_C^*)(f \otimes r)(c) \\ &= \Delta_C^*(f \otimes \epsilon_C^*(r))(c) \\ &= (f \otimes \epsilon_C^*(r))(\Delta_C(c)) \\ &= \sum_{(c)} f(c_{(1)})\epsilon_C^*(r)(c_{(2)}) \\ &= \sum_{(c)} f(c_{(1)})r(\epsilon_C(c_{(2)})) \\ &= r\sum_{(c)} f(c_{(1)})\epsilon_C(c_{(2)}) \\ &= r\sum_{(c)} \epsilon_C(c_{(2)})f(c_{(1)}) \\ &= r\sum_{(c)} f(\epsilon_C(c_{(2)})c_{(1)}) \\ &= rf(\sum_{(c)} \epsilon_C(c_{(2)})c_{(1)}) \\ &= rf(c) \quad \text{by Condition (5)} \end{aligned}$$

In a similar manner, one obtains

$$m_{C^*}(\eta_{C^*} \otimes I_{C^*})(r \otimes f) = rf$$

Thus $(C^*, m_{C^*}, \eta_{C^*})$ is an algebra. Note that $\eta_{C^*}(1_K)(c) = \epsilon_C(c), \forall c$, and so, ϵ_C is the unique element of C^* for which $\epsilon_C f = f = f\epsilon_C$ for all $f \in C^*$. ◇

Let (A, m_A, η_A) be a K-algebra. Then one may wonder if A^* is a K-coalgebra. The multiplication map $m_A : A \otimes A \to A$ dualizes to yield $m_A^* : A^* \to (A \otimes A)^*$. Unfortunately, if A is infinite dimensional over K, then $A^* \otimes A^*$ is a proper subset of $(A \otimes A)^*$, and hence m_A^* may not induce the required comultiplication map $A^* \to A^* \otimes A^*$.

There is still however a K-coalgebra arising via duality from the algebra A. An ideal I of A is *cofinite* if $\dim(A/I) < \infty$. The *finite dual* A° of A is defined as

$$A^\circ = \{f \in A^* : f(I) = 0 \text{ for some cofinite ideal } I \text{ of } A\}$$

Note that A° is the largest subspace W of A^* for which $m_A^*(W) \subseteq W \otimes W$.

Proposition 2. *If A is an algebra, then A° is a coalgebra.*

Proof. The proof is similar to the method used in Proposition 2.1. We restrict the map m_A^* to A° to yield the K-linear map $m_A^* : A^\circ \to (A \otimes A)^*$. Now by [4, Proposition 6.0.3], $m_A^*(A^\circ) \subseteq A^\circ \otimes A^\circ$. Let Δ_{A° denote the restriction of m_A^* to A°. We show that Δ_{A° satisfies the coassociative condition. For $f \in A^\circ, a, b, c \in A$, we have

$$\begin{aligned}(I \otimes \Delta_{A^\circ})\Delta_{A^\circ}(f)(a \otimes b \otimes c) &= (I \otimes m_A^*)m_A^*(f)(a \otimes b \otimes c) \\ &= m_A^*(f)((I \otimes m_A)(a \otimes b \otimes c)) \\ &= m_A^*(f)(a \otimes bc) \\ &= f(m_A(a \otimes bc)) \\ &= f(a(bc)) \\ &= f((ab)c) \\ &= f(m_A(ab \otimes c)) \\ &= m_A^*(f)(ab \otimes c) \\ &= m_A^*(f)((m_A \otimes I)(a \otimes b \otimes c)) \\ &= (m_A^* \otimes I)m_A^*(f)(a \otimes b \otimes c) \\ &= (\Delta_{A^\circ} \otimes I)\Delta_{A^\circ}(f)(a \otimes b \otimes c)\end{aligned}$$

For the counit map of A°, we consider the dual map $\eta_A^* : A^* \to K^* := K$. Now η_A^* restricts to a map $\eta_A^* : A^\circ \to K$. We let ϵ_{A° denote the restriction of η_A^* to A°. For $f \in A^\circ, r \in K$,

$$\epsilon_{A^\circ}(f)(r) = f(\eta_A(r)) = f(r1_A) = rf(1_A) = f(1_A)(r)$$

and so, $\epsilon_{A^\circ}(f) = f(1_A)$. We show that ϵ_{A° satisfies the counit property. First let $s_1 : K \otimes A^\circ \to A^\circ$ be defined by the scalar multiplication of A°. For $f \in A^\circ, r \in K, a \in A$,

$$\begin{aligned}s_1((\epsilon_{A^\circ} \otimes I)\Delta_{A^\circ}(f))(a) &= s_1((\eta_A^* \otimes I)m_A^*(f))(a) \\ &= (\eta_A^* \otimes I)m_A^*(f)(s_1^*(a)) \\ &= (\eta_A^* \otimes I)m_A^*(f)(1 \otimes a) \\ &= m_A^*(f)((\eta_A \otimes I)(1 \otimes a)) \\ &= f(m_A(\eta_A \otimes I)(1 \otimes a)) \\ &= f(a)\end{aligned}$$

In a similar manner, one obtains

$$s_2((I \otimes \epsilon_{A^\circ})\Delta_{A^\circ}(f))(a) = f(a)$$

where $s_2 : A^\circ \otimes K \to A^\circ$ is given by scalar multiplication. Thus A° is a coalgebra.

◇

Proposition 3. *If B is a bialgebra, then B° is a bialgebra.*

Proof. As a coalgebra, B is a triple $(B, \Delta_B, \epsilon_B)$. By Proposition 2.1, B^* is an algebra with maps $m_{B^*} = \Delta_B^*$ and $\eta_{B^*} = \epsilon_B^*$. Let m_{B° denote the restriction of m_{B^*} to $B^\circ \otimes B^\circ$, and let η_{B° denote the restriction of η_{B^*} to B°. Then the triple $(B^\circ, m_{B^\circ}, \eta_{B^\circ})$ is a K-algebra.

As an algebra, B is a triple (B, m_B, η_B). By Proposition 2.2, B° is a coalgebra with maps Δ_{B° and ϵ_{B°. It remains to show that Δ_{B° and ϵ_{B° are algebra homomorphisms. First observe that for $f, g \in B^\circ$, $a, b \in B$ one has

$$(fg)(a) = m_{B^\circ}(f \otimes g)(a) = \Delta_B^*(f \otimes g)(a) = (f \otimes g)\Delta_B(a)$$

and

$$\Delta_{B^\circ}(f)(a \otimes b) = m_B^*(f)(a \otimes b) = f(m_B(a \otimes b)) = f(ab)$$

We have

$$\begin{aligned}
\Delta_{B^\circ}(fg)(a \otimes b) &= (fg)(ab) \\
&= (f \otimes g)(\Delta_B(ab)) \\
&= (f \otimes g)(\Delta_B(a)\Delta_B(b)) \\
&= (f \otimes g)(m_{B\otimes B}(\Delta_B(a) \otimes \Delta_B(b)) \\
&= m^*_{B\otimes B}(f \otimes g)(\Delta_B(a) \otimes \Delta_B(b)) \\
&= (I \otimes \tau \otimes I)(\Delta_{B^\circ} \otimes \Delta_{B^\circ})(f \otimes g)(\Delta_B(a) \otimes \Delta_B(b)) \\
&= (\Delta_{B^\circ}(f) \otimes \Delta_{B^\circ}(g))(I \otimes \tau \otimes I)(\Delta_B \otimes \Delta_B)(a \otimes b) \\
&= (\Delta_{B^\circ}(f) \otimes \Delta_{B^\circ}(g))(\Delta_{B\otimes B}(a \otimes b)) \\
&= \Delta^*_{B\otimes B}(\Delta_{B^\circ}(f) \otimes \Delta_{B^\circ}(g))(a \otimes b) \\
&= m_{B^\circ \otimes B^\circ}(\Delta_{B^\circ}(f) \otimes \Delta_{B^\circ}(g))(a \otimes b) \\
&= (\Delta_{B^\circ}(f)\Delta_{B^\circ}(g))(a \otimes b)
\end{aligned}$$

and so Δ_{B° is an algebra map. We next show that ϵ_{B° is an algebra map. For $f, g \in B^\circ$,

$$\epsilon_{B^\circ}(f) = \epsilon_{B^\circ}(f)(1) = f(\eta_B(1)) = f(1_B)$$

Thus

$$\begin{aligned}
\epsilon_{B^\circ}(fg) &= (fg)(1_B) \\
&= f(1_B)g(1_B) \\
&= \epsilon_{B^\circ}(f)\epsilon_{B^\circ}(g)
\end{aligned}$$

and so, ϵ_{B° is an algebra map.

◇

Proposition 4. *Suppose that B is a bialgebra that is finite dimensional over K. Then B^* is a bialgebra.*

Proof. If $\dim(B) < \infty$, then $B^\circ = B^*$. The result then follows from Proposition 2.3.

◇

Let $G = \{x_1, x_2, \ldots, x_n\}$ be a finite semigroup with unity element $1_{KG} = x_1$, and let KG denote the semigroup bialgebra. By Proposition 2.4 KG^* is a bialgebra of dimension n over K. Let $\{e_1, e_2, \ldots, e_n\}$ be the dual basis for KG^* defined as $e_i(x_j) = \delta_{i,j}$.

Proposition 5. *The comultiplication map $\Delta_{KG^*} : KG^* \to KG^* \otimes KG^*$ is given as*

$$\Delta_{KG^*}(e_i) = \sum_{x_i = x_j x_k} e_j \otimes e_k$$

and the counit map $\epsilon_{KG^*} : KG^* \to K$ is defined as $\epsilon_{KG^*}(e_i) = e_i(x_1) = \delta_{i,1}$.

Proof, See [7, (1.3.7)]. ◇

Let B be a K-bialgebra. Then B is *cocommutative* if

$$\tau(\Delta_B(b)) = \Delta_B(b)$$

for all $b \in B$.

Proposition 6. *If B is cocommutative, then B° is a commutative algebra. If B is a commutative algebra, then B° is cocommutative.*

Proof. See [7, Lemma 1.2.2, Proposition 1.2.4].

◇

3. Quasitriangular Bialgebras

Let B be a bialgebra and let $B \otimes B$ be the tensor product algebra. Let $U(B \otimes B)$ denote the group of units in $B \otimes B$ and let $R \in U(B \otimes B)$. The pair (B, R) is *almost cocommutative* if the element R satisfies

$$\tau(\Delta_B(b)) = R\Delta_B(b)R^{-1} \tag{7}$$

for all $b \in B$.

If the bialgebra B is cocommutative, then the pair $(B, 1 \otimes 1)$ is almost cocommutative since $R = 1 \otimes 1$ satisfies Condition (7). However, if B is commutative and non-cocommutative, then (B, R) cannot be almost cocommutative for any $R \in U(B \otimes B)$ since Condition (7) in this case reduces to the condition for cocommutativity.

Write $R = \sum_{i=1}^{n} a_i \otimes b_i \in U(B \otimes B)$. Let

$$R^{12} = \sum_{i=1}^{n} a_i \otimes b_i \otimes 1 \in B \otimes B \otimes B$$

$$R^{13} = \sum_{i=1}^{n} a_i \otimes 1 \otimes b_i \in B \otimes B \otimes B$$

$$R^{23} = \sum_{i=1}^{n} 1 \otimes a_i \otimes b_i \in B \otimes B \otimes B$$

The pair (B, R) is *quasitriangular* if (B, R) is almost cocommutative and the following conditions hold

$$(\Delta_B \otimes I)R = R^{13}R^{23} \tag{8}$$

$$(I \otimes \Delta_B)R = R^{13}R^{12} \tag{9}$$

Clearly, if B is cocommutative then $(B, 1 \otimes 1)$ is quasitriangular.

Let B be a bialgebra. A *quasitriangular structure* is an element $R \in U(B \otimes B)$ so that (B, R) is quasitriangular. Let (B, R) and (B', R') be quasitriangular bialgebras. Then (B, R), (B', R') are *isomorphic as quasitriangular bialgebras* if there exists a bialgebra isomorphism $\phi : B \to B'$ for which $R' = (\phi \otimes \phi)(R)$. Two quasitriangular structures R, R' on a bialgebra B are *equivalent quasitriangular structures* if $(B, R) \cong (B, R')$ as quasitriangular bialgebras.

The following proposition shows that every bialgebra isomorphism $\phi : B \to B'$ with B quasitriangular extends to an isomorphism of quasitriangular bialgebras.

Proposition 7. *Suppose (B, R) is quasitriangular and suppose that $\phi : B \to B'$ is an isomorphism of K-bialgebras. Let $R' = (\phi \otimes \phi)(R)$. Then (B', R') is quasitriangular.*

Proof. Note that $(\phi \otimes \phi)(R^{-1}) = ((\phi \otimes \phi)(R))^{-1}$. Let $b' \in B'$. Then there exists $b \in B$ for which $\phi(b) = b'$. Now

$$\begin{aligned}
\tau\Delta_{B'}(b') &= \tau\Delta_{B'}(\phi(b)) \\
&= \tau(\phi\otimes\phi)\Delta_B(b) \\
&= (\phi\otimes\phi)\tau\Delta_B(b) \\
&= (\phi\otimes\phi)(R\Delta_B(b)R^{-1}) \\
&= (\phi\otimes\phi)(R)(\phi\otimes\phi)\Delta_B(b)(\phi\otimes\phi)(R^{-1}) \\
&= (\phi\otimes\phi)(R)\Delta_{B'}(\phi(b))((\phi\otimes\phi)(R))^{-1} \\
&= (\phi\otimes\phi)(R)\Delta_{B'}(b')((\phi\otimes\phi)(R))^{-1} \\
&= R'\Delta_{B'}(b')(R')^{-1}
\end{aligned}$$

and so, (B, R') is almost cocommutative. Moreover,

$$\begin{aligned}
(\Delta_{B'}\otimes I)(R') &= (\Delta_{B'}\otimes I)(\phi\otimes\phi)(R) \\
&= (\Delta_{B'}\otimes I)(\sum_{i=1}^{n}\phi(a_i)\otimes\phi(b_i)) \\
&= \sum_{i=1}^{n}\Delta_{B'}\phi(a_i)\otimes\phi(b_i)) \\
&= \sum_{i=1}^{n}(\phi\otimes\phi)\Delta_B(a_i)\otimes\phi(b_i)) \\
&= (\phi\otimes\phi\otimes\phi)(\sum_{i=1}^{n}\Delta_B(a_i)\otimes b_i) \\
&= (\phi\otimes\phi\otimes\phi)(\Delta_B\otimes I)(R) \\
&= (\phi\otimes\phi\otimes\phi)(R^{13}R^{23}) \\
&= (\phi\otimes\phi\otimes\phi)((\sum_{i=1}^{n}a_i\otimes 1\otimes b_i)(\sum_{i=1}^{n}1\otimes a_i\otimes b_i)) \\
&= (\sum_{i=1}^{n}\phi(a_i)\otimes 1\otimes\phi(b_i))(\sum_{i=1}^{n}1\otimes\phi(a_i)\otimes\phi(b_i)) \\
&= ((\phi\otimes\phi)(R))^{13}((\phi\otimes\phi)(R))^{23} \\
&= (R')^{13}(R')^{23}
\end{aligned}$$

In a similar manner one shows that

$$(I\otimes\Delta_{B'})(R') = (R')^{13}(R')^{12}$$

Thus (B', R') is quasitriangular. ◇

Quasitriangular bialgebras are important since they give rise to solutions of the equation

$$R^{12}R^{13}R^{23} = R^{23}R^{13}R^{12} \tag{10}$$

which is known as the *quantum Yang–Baxter equation (QYBE)*. The QYBE was first introduced in statistical mechanics, see [9]. An element $R \in B\otimes B$ which satisfies (10) is a *solution to the QYBE*.

Certainly, the QYBE admits the trivial solution $R = 1\otimes 1$, and of course, if B is commutative, then any $R \in B\otimes B$ is a solution to the QYBE. For B non-commutative, it is of great interest to find non-trivial solutions $R \in B\otimes B$ to the QYBE. We have the following result due to V. G. Drinfeld [10].

Proposition 8. *(Drinfeld) Suppose (B, R) is quasitriangular. Then R is a solution to the QYBE.*

Proof. One has

$$\begin{aligned}
R^{12}R^{13}R^{23} &= R^{12}(\Delta \otimes I)(R) \quad \text{by (8)} \\
&= (R \otimes 1)(\sum_{i=1}^{n} \Delta(a_i) \otimes b_i) \\
&= \sum_{i=1}^{n} R\Delta(a_i) \otimes b_i \\
&= \sum_{i=1}^{n} \tau\Delta(a_i)R \otimes b_i \quad \text{by (7)} \\
&= (\sum_{i=1}^{n} \tau\Delta(a_i) \otimes b_i)(R \otimes 1) \\
&= (\tau\Delta \otimes I)(R)R^{12} \\
&= (\tau \otimes I)(\Delta \otimes I)(R)R^{12} \\
&= (\tau \otimes I)(R^{13}R^{23})R^{12} \quad \text{by (8)} \\
&= R^{23}R^{13}R^{12}
\end{aligned}$$

◇

The following proposition provides necessary conditions on $R \in U(B \otimes B)$ in order for (B, R) to be quasitriangular.

Proposition 9. *Suppose (B, R) is quasitriangular. Then*

(i) $s_1(\epsilon \otimes I)(R) = 1,$

(ii) $s_2(I \otimes \epsilon)(R) = 1.$

Proof. For (i) one has

$$\begin{aligned}
(s_1 \otimes I)(\epsilon \otimes I \otimes I)(\Delta \otimes I)(R) &= (s_1 \otimes I)(\epsilon \otimes I \otimes I)(\sum_{i=1}^{n} \Delta(a_i) \otimes b_i) \\
&= (s_1 \otimes I)(\sum_{i=1}^{n} (\epsilon \otimes I)\Delta(a_i) \otimes b_i) \\
&= \sum_{i=1}^{n} s_1(\epsilon \otimes I)\Delta(a_i) \otimes b_i \\
&= \sum_{i=1} a_i \otimes b_i \\
&= R
\end{aligned}$$

In view of Condition (8)

$$\begin{aligned}
R &= (s_1 \otimes I)(\epsilon \otimes I \otimes I)(R^{13}R^{23}) \\
&= (s_1 \otimes I)(\epsilon \otimes I \otimes I)(R^{13})(s_1 \otimes I)(\epsilon \otimes I \otimes I)(R^{23}) \\
&= (s_1 \otimes I)(\epsilon \otimes I \otimes I)(\sum_{i=1}^{n} a_i \otimes 1 \otimes b_i)(s_1 \otimes I)(\epsilon \otimes I \otimes I)(\sum_{i=1}^{n} 1 \otimes a_i \otimes b_i) \\
&= (\sum_{i=1}^{n} \epsilon(a_i)1 \otimes b_i)(\sum_{i=1}^{n} a_i \otimes b_i) \\
&= (\sum_{i=1}^{n} 1 \otimes \epsilon(a_i)b_i)R
\end{aligned}$$

Thus

$$1 \otimes \sum_{i=1} \epsilon(a_i)b_i = 1 \otimes 1$$

and consequently,

$$1 = s_1(\sum_{i=1}^{n} \epsilon(a_i) \otimes b_i) = s_1(\epsilon \otimes I)(R)$$

A similar argument is used to prove (ii).

◇

4. Myhill–Nerode Bialgebras

In this section we review the main result of [3] in which the authors give a bialgebra version of the Myhill–Nerode Therorem. Let G be a semigroup with unity, 1 and let $H = KG$ be the semigroup bialgebra. There is a right H-module structure on H^* defined as

$$(p \leftharpoonup x)(y) = p(xy)$$

for all $x, y \in H$, $p \in H^*$. For $x \in H$, $p \in H^*$, the element $p \leftharpoonup x$ is the *right translate of p by x*.

Proposition 10. *([3, Proposition 5.4].) Let G be a semigroup with 1, let $H = KG$ denote the semigroup bialgebra. Let $p \in H^*$. Then the following are equivalent.*

(i) The set $\{p \leftharpoonup x : x \in G\}$ of right translates is finite.

(ii) There exists a finite dimensional bialgebra B, a bialgebra homomorphism $\Psi : H \to B$, and an element $f \in B^*$ so that $p(h) = f(\Psi(h))$ for all $h \in H$.

(Note: The bialgebras of (ii) are defined to be *Myhill–Nerode bialgebras.*)

Proof. *(i)* $\Longrightarrow$ *(ii)*. Let $Q = \{p \leftharpoonup x : x \in G\}$ be the finite set of right translates. For each $u \in G$, we define a right operator $r_u : Q \to Q$ by the rule

$$(p \leftharpoonup x)r_u = (p \leftharpoonup x) \leftharpoonup u = p \leftharpoonup xu$$

Observe that the set $\{r_u : u \in G\}$ is finite with $|\{r_u : u \in G\}| \leq |Q|^{|Q|}$. The set $\{r_u : u \in G\}$ is a semigroup with unity, $1 = r_1$ under composition of operators. Indeed,

$$(p \leftharpoonup x)(r_u r_v) = (p \leftharpoonup xu)r_v = p \leftharpoonup xuv = (p \leftharpoonup x)r_{uv}$$

Thus $r_u r_v = r_{uv}$, for all $u, v \in G$. Let B denote the semigroup bialgebra on $\{r_u : u \in G\}$. Let $\Psi : H \to B$ be the K-linear map defined by $\Psi(u) = r_u$. Then

$$\Psi(uv) = r_{uv} = r_u r_v = \Psi(u)\Psi(v)$$

and

$$\begin{aligned} \Delta_B(\Psi(u)) &= \Delta_B(r_u) \\ &= r_u \otimes r_u \\ &= \Psi(u) \otimes \Psi(v) \\ &= (\Psi \otimes \Psi)(u \otimes u) \\ &= (\Psi \otimes \Psi)\Delta_H(u) \end{aligned}$$

and so, Ψ is a homomorphism of bialgebras.

Let $f \in B^*$ be defined by

$$\begin{aligned} f(r_u) &= ((p \leftharpoonup 1)r_u)(1) \\ &= (p \leftharpoonup u)(1) \\ &= p(u) \end{aligned}$$

Then $p(h) = f(\Psi(h))$, for all $h \in H$, as required.

$(ii) \Longrightarrow (i)$. Suppose there exists a finite dimensional bialgebra B, a bialgebra homomorphism $\Psi : H \to B$, and an element $f \in B^*$ so that $p(h) = f(\Psi(h))$ for all $h \in H$. Define a right H-module action $\cdot$ on B as

$$b \cdot h = b\Psi(h)$$

for all $b \in B, h \in H$. Then for $b \in B$, $x \in G$,

$$\begin{aligned}
\Delta_B(b \cdot x) &= \Delta_B(b\Psi(x)) \\
&= \Delta_B(b)\Delta_B(\Psi(x)) \\
&= (\sum_{(b)} b_{(1)} \otimes b_{(2)})(\Psi \otimes \Psi)\Delta_H(x) \\
&= (\sum_{(b)} b_{(1)} \otimes b_{(2)})(\Psi(x) \otimes \Psi(x)) \\
&= \sum_{(b)} b_{(1)}\Psi(x) \otimes b_{(2)}\Psi(x) \\
&= \sum_{(b)} b_{(1)} \cdot x \otimes b_{(2)} \cdot x
\end{aligned}$$

and

$$\epsilon_B(b \cdot x) = \epsilon_B(b\Psi(x)) = \epsilon_B(b)\epsilon_B(\Psi(x)) = \epsilon_B(b)\epsilon_H(x)$$

Thus B is a right H-module coalgebra.

Now, let Q be the collection of grouplike elements of B. Since Q is a linearly independent subset of B and B is finite dimensional, Q is finite. Since B is a right H-module coalgebra with action "$\cdot$",

$$\Delta_B(q \cdot x) = q \cdot x \otimes q \cdot x$$

for $q \in Q$, $x \in G$. Thus $\cdot$ restricts to give an action (also denoted by "$\cdot$") of G on Q. Now for $x, y \in G$,

$$\begin{aligned}
(p \leftharpoonup x)(y) &= p(xy) \\
&= f(\Psi(xy)) \\
&= f(\Psi(x)\Psi(y)) \\
&= f((1_B\Psi(x))\Psi(y)) \\
&= f((1_B \cdot x) \cdot y) \qquad (11)
\end{aligned}$$

Let

$$S = \{q \in Q : \; q = 1_B \cdot x \text{ for some } x \in G\}$$

In view of Condition (11) there exists a function

$$\varrho : S \to \{p \leftharpoonup x : \; x \in G\}$$

defined as

$$\varrho(1_B \cdot x)(y) = f((1_B \cdot x) \cdot y) = (p \leftharpoonup x)(y)$$

Since ϱ is surjective and S is finite, $\{p \leftharpoonup x : \; x \in G\}$ is finite.

◇

We illustrate the connection between Proposition 4.1 and the usual Myhill–Nerode Theorem. Let $\hat{\Sigma}_0$ denote the set of words in a finite alphabet Σ_0. Let $L \subseteq \hat{\Sigma}_0$ be a language. Suppose that the equivalence relation $\sim_L$ (as in the Introduction) has finite index. Then the usual Myhill–Nerode

Theorem says that there exists a finite automaton which accepts L. We show how to construct this finite automaton using Proposition 4.1.

Consider $G = \hat{\Sigma}_0$ as a semigroup with unity where the semigroup operation is concatenation and the unity element is the empty word. Let $H = KG$ denote the semigroup bialgebra. Then the characteristic function of L extends to an element $p \in H^*$. Since $\sim_L$ has finite index, the set of right translates $\{p \leftharpoonup x : x \in G\}$ is finite [3, Proposition 2.3]. Now Proposition 4.1 (i)$\Longrightarrow$ (ii) applies to show that there exists a finite dimensional bialgebra B, a bialgebra homomorphism $\Psi : H \to B$ and an element $f \in B^*$ so that $p(h) = f(\Psi(h))$, for all $h \in H$.

This bialgebra determines a finite automaton $\langle Q, \Sigma, \delta, q_0, F\rangle$, where Q is the finite set of states, Σ is the input alphabet, δ is the transition function, q_0 is the initial state, and F is the set of final states (see [2, Chapter 2] for details on finite automata.)

For the states of the automata, we let Q be the (finite) set of grouplike elements of B. For the input alphabet, we choose $\Sigma = \Sigma_0$. As we have seen, the right H-module structure of B restricts to an action "$\cdot$" of G on Q, and so we define the transition function $\delta : Q \times \Sigma_0 \to Q$ by the rule $\delta(q, x) = q \cdot x$, for $q \in Q$, $x \in \Sigma_0$. The initial state is $q_0 = 1_B$, and the set of final states F is the subset of Q of the form $1_B \cdot x$, $x \in G$ for which

$$p(x) = f(\Psi(x)) = f1_B\Psi(x)) = f(1_B \cdot x) = 1$$

By construction, the finite automaton $\langle Q, \Sigma_0, \delta, 1_B, F\rangle$ accepts L.

5. Quasitriangular Structure of Myhill–Nerode Bialgebras

In this section we use Proposition 4.1 to construct a collection of Myhill–Nerode bialgebras. We then compute the quasitriangular structure of one of these bialgebras.

Let $\Sigma_0 = \{a\}$ be the alphabet on a single letter a. Let $\hat{\Sigma}_0 = \{1, a, aa, aaa, \dots\}$ denote the collection of all words of finite length formed from Σ_0. Here 1 denotes the empty word of length 0. For convenience, we shall write

$$a^i = \underbrace{aaa\cdots a}_{i \text{ times}},$$

for $i \geq 0$.

Fix an integer $i \geq 0$ and let $L_i = \{a^i\} \subseteq \hat{\Sigma}_0$. Then the language L_i is accepted by the finite automaton given in Figure 1.

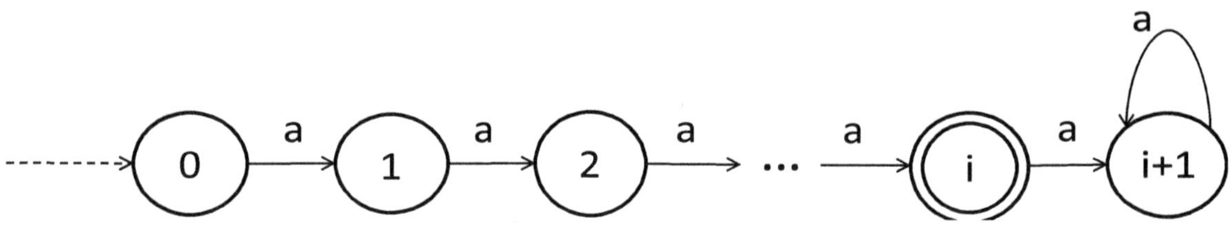

Figure 1. Finite automaton accepting $L_i = \{a^i\}$, accepting state is i.

By the usual Myhill–Nerode Theorem, the equivalence relation $\sim_{L_i}$, defined as $x \sim_{L_i} y$ if and only if $xz \in L_i$ exactly when $yz \in L_i, \forall z$, has finite index. If $p_i : \hat{\Sigma}_0 \to \{0,1\} \subseteq K$ is the characteristic function of L_i, then $\sim_{L_i}$ is equivalent to the relation $\sim_{p_i}$ defined as: $x \sim_{p_i} y$ if and only if $p_i(xz) = p_i(yz), \forall z \in \hat{\Sigma}_0$. Let $[x]_{p_i}$ denote the equivalence class of x under $\sim_{p_i}$. The Myhill–Nerode theorem now says that the set $\{[x]_{p_i} : x \in \hat{\Sigma}_0\}$ is finite.

Now we consider $G = \hat{\Sigma}_0$ as a semigroup with unity 1 with concatenation as the binary operation. Let $H = KG$ be the semigroup bialgebra. The characteristic function p_i of L_i extends to an element of H^*. By [3, Proposition 2.3], the set of right translates $\{p_i \leftharpoonup x : x \in G\}$ is finite. Thus by Proposition 4.1, there exists a finite dimensional bialgebra B_i, a bialgebra homomorphism $\Psi : H \to B_i$, and an element $f_i \in B_i^*$ so that $p_i(h) = f_i(\Psi(h))$ for all $h \in H$.

In what follows, we give the bialgebra structure of the collection $\{B_i : i \geq 0\}$ and compute the quasitriangular structure of the bialgebra B_0.

For $i \geq 0$, the finite set of right translates of $p_i \in H^*$ is

$$Q_i = \{p_i \leftharpoonup 1, p_i \leftharpoonup a, p_i \leftharpoonup a^2, \ldots, p_i \leftharpoonup a^i, p_i \leftharpoonup a^{i+1}\}$$

One finds that the set of right operators on Q_i is $\{r_1, r_a, r_{a^2}, \ldots, r_{a^i}, r_{a^{i+1}}\}$. Under composition, the set of right operators is a semigroup with unity r_1. We have, for $0 \leq m, n \leq i+1$,

$$r_{a^m} r_{a^n} = \begin{cases} r_{a^{m+n}} & \text{if } 0 \leq m+n \leq i+1 \\ r_{a^{i+1}} & \text{if } m+n > i+1 \end{cases}$$

By construction, B_i is the semigroup bialgebra on $\{r_1, r_a, r_{a^2}, \ldots, r_{a^i}, r_{a^{i+1}}\}$.

5.1. Quasitriangular Structure of B_0

In the case $i = 0$, B_0 is the semigroup bialgebra on $\{r_1, r_a\}$ with algebra structure defined by $r_1 r_1 = r_1, r_1 r_a = r_a, r_a r_1 = r_a, r_a r_a = r_a$. Let $\{e_0, e_1\}$ be the dual basis defined as $e_0(r_1) = 1, e_0(r_a) = 0$, $e_1(r_1) = 0, e_1(r_a) = 1$. Then $\{e_0, e_1\}$ is the set of minimal idempotents for B_0^*. Comultiplication on B_0^* is given as

$$\Delta_{B_0^*}(e_0) = e_0 \otimes e_0$$

$$\Delta_{B_0^*}(e_1) = e_0 \otimes e_1 + e_1 \otimes e_0 + e_1 \otimes e_1$$

and the counit map is defined by

$$\epsilon_{B_0^*}(e_0) = 1, \quad \epsilon_{B_0^*}(e_1) = 0$$

Proposition 11. *Let B_0 be the K-bialgebra as above. Then there is exactly one quasitriangular structure on B_0, namely, $R = 1_{B_0} \otimes 1_{B_0}$.*

Proof. Certainly, $1 \otimes 1 = 1_{B_0} \otimes 1_{B_0}$ is a quasitriangular structure for B_0. We claim that $1 \otimes 1$ is the only quasitriangular structure. Observe that there is bialgebra isomorphism $\phi : B_0 \to B_0^*$ defined as $\phi(r_1) = e_0 + e_1$, $\phi(r_a) = e_0$. Thus if (B_0, R) is quasitriangular, then (B_0^*, R'), $R' = (\phi \otimes \phi)(R)$, is quasitriangular by Proposition 3.1. So, we first compute all of the quasitriangular structures of B_0^*. To this end, suppose that (B_0^*, R') is quasitriangular for some element $R' \in B_0^* \otimes B_0^*$. Since

$$B_0^* \otimes B_0^* = K(e_0 \otimes e_0) \oplus K(e_0 \otimes e_1) \oplus K(e_1 \otimes e_0) \oplus K(e_1 \otimes e_1)$$

$$R' = w(e_0 \otimes e_0) + x(e_0 \otimes e_1) + y(e_1 \otimes e_0) + z(e_1 \otimes e_1)$$

for $w, x, y, z \in K$. By Proposition 3.3(i),

$$\begin{aligned} 1_{B_0^*} &= e_0 + e_1 \\ &= s_1(\epsilon \otimes I)(w(e_0 \otimes e_0) + x(e_0 \otimes e_1) + y(e_1 \otimes e_0) + z(e_1 \otimes e_1)) \\ &= we_0 + xe_1 \end{aligned}$$

and so, $w = x = 1$. From Proposition 3.3(ii), one also has $y = 1$. Thus

$$R' = e_0 \otimes e_0 + e_0 \otimes e_1 + e_1 \otimes e_0 + z(e_1 \otimes e_1)$$

for $z \in K$. Now,

$$\begin{aligned}(\Delta \otimes I)(R') &= (\Delta \otimes I)(e_0 \otimes e_0 + e_0 \otimes e_1 + e_1 \otimes e_0 + z(e_1 \otimes e_1)) \\ &= (e_0 \otimes e_0) \otimes e_0 + (e_0 \otimes e_0) \otimes e_1 + (e_0 \otimes e_1 + e_1 \otimes e_0 + e_1 \otimes e_1) \otimes e_0 \\ &+ z((e_0 \otimes e_1 + e_1 \otimes e_0 + e_1 \otimes e_1) \otimes e_1) \\ &= e_0 \otimes e_0 \otimes e_0 + e_0 \otimes e_0 \otimes e_1 + e_0 \otimes e_1 \otimes e_0 + e_1 \otimes e_0 \otimes e_0 + e_1 \otimes e_1 \otimes e_0 \\ &+ z(e_0 \otimes e_1 \otimes e_1) + z(e_1 \otimes e_0 \otimes e_1) + z(e_1 \otimes e_1 \otimes e_1) \end{aligned} \tag{12}$$

Moreover,

$$\begin{aligned}(R')^{13}(R')^{23} &= (e_0 \otimes (e_0 + e_1) \otimes e_0 + e_0 \otimes (e_0 + e_1) \otimes e_1 + e_1 \otimes (e_0 + e_1) \otimes e_0 \\ &+ z(e_1 \otimes (e_0 + e_1) \otimes e_1)) \cdot ((e_0 + e_1) \otimes e_0 \otimes e_0 + (e_0 + e_1) \otimes e_0 \otimes e_1 \\ &+ (e_0 + e_1) \otimes e_1 \otimes e_0 + z((e_0 + e_1) \otimes e_1 \otimes e_1)) \\ &= (e_0 \otimes e_0 \otimes e_0 + e_0 \otimes e_1 \otimes e_0 + e_0 \otimes e_0 \otimes e_1 + e_0 \otimes e_1 \otimes e_1 \\ &+ e_1 \otimes e_0 \otimes e_0 + e_1 \otimes e_1 \otimes e_0 + z(e_1 \otimes e_0 \otimes e_1) \\ &+ z(e_1 \otimes \otimes e_1 \otimes e_1)) \cdot (e_0 \otimes e_0 \otimes e_0 + e_1 \otimes e_0 \otimes e_0 + e_0 \otimes e_0 \otimes e_1 + e_1 \otimes e_1 \otimes e_0 \\ &+ e_0 \otimes e_1 \otimes e_0 + e_1 \otimes e_1 \otimes e_0 + z(e_0 \otimes e_1 \otimes e_1) + z(e_1 \otimes \otimes e_1 \otimes e_1)) \\ &= e_0 \otimes e_0 \otimes e_0 + e_0 \otimes e_1 \otimes e_0 + e_0 \otimes e_0 \otimes e_1 + z(e_0 \otimes e_1 \otimes e_1) \\ &+ e_1 \otimes e_0 \otimes e_0 + e_1 \otimes e_1 \otimes e_0 + z(e_1 \otimes e_0 \otimes e_1) + z^2(e_1 \otimes e_1 \otimes e_1) \end{aligned} \tag{13}$$

Equations 12 and 13 yield the relation $z^2 = z$. Thus either $z = 0$ or $z = 1$. If $z = 0$, then R' is not a unit in $B_0^* \otimes B_0^*$. Thus

$$R' = e_0 \otimes e_0 + e_0 \otimes e_1 + e_1 \otimes e_0 + e_1 \otimes e_1 = 1 \otimes 1$$

is the only quasitriangular structure for B_0^*.

Consequently, if (B_0, R) is quasitriangular, then $(\phi \otimes \phi)(R) = 1_{B_0^*} \otimes 1_{B_0^*}$. It follows that $R = 1_{B_0} \otimes 1_{B_0}$.

◇

5.2. Questions for Future Research

Though the Myhill–Nerode bialgebra B_0 has only the trivial quasitriangular structure, it remains to compute the quasitriangular structure of B_i for $i \geq 1$. Moreover, the linear dual B_i^* is a commutative, cocommutative K-bialgebra and it would be of interest to find its quasitriangular structure. Unlike the $i = 0$ case, we may have $B_i \not\cong B_i^*$ (for instance, $B_1 \not\cong B_1^*$) and so this is indeed a separate problem.

Suppose that L is a language of words built from the alphabet $\Sigma_0 = \{a, b\}$. If L is accepted by a finite automaton, then by Proposition 4.1, L gives rise to a Myhill–Nerode bialgebra B (see for example, [3, §6].) By construction, B is a cocommutative K-bialgebra and hence B has at least the trivial quasitriangular structure. Are there any other structures? Note that B^* is a commutative K-algebra. For which R (if any) is (B^*, R) quasitriangular?

References

1. Eilenberg, S. *Automata, Languages, and Machines, Vol. A*; Academic Press: New York, NY, USA, 1974.
2. Hopcroft, J.E.; Ullman, J.D. *Introduction to Automata Theory, Languages, and Computation*; Addison-Wesley: Upper Saddle River, NJ, USA, 1979.
3. Nichols, W.D.; Underwood, R.G. Algebraic Myhill-Nerode theorems. *Theor. Comp. Sci.* **2011**, *412*, 448–457.
4. Sweedler, M. *Hopf Algebras*; W. A. Benjamin: New York, NY, USA, 1969.
5. Abe, E. Hopf Algebras. In *Cambridge Tracts in Mathematics, 74*; Cambridge University Press: Cambridge, UK; New York, NY, USA, 1980.

6. Childs, L.N. *Taming Wild Extensions: Hopf Algebras and Local Galois Module Theory (Surveys and Monographs) 80*; American Mathematical Society: Providence, RI, USA, 2000.
7. Montgomery, S. *Hopf Algebras and Their Actions on Rings. CBMS 82*; American Mathematical Society: Providence, RI, USA, 1993.
8. Underwood, R.G. *An Introduction to Hopf Algebras*; Springer: New York, NY, USA, 2011.
9. Nichita, F. Introduction to the yang-baxter equation with open problems. *Axioms* **2012**, *1*, 33–37.
10. Drinfeld, V.G. On almost cocommutative Hopf algebras. *Leningr. Math. J.* **1990**, *1*, 321–342.

10

The Duality between Corings and Ring Extensions

Florin F. Nichita [1,*] and Bartosz Zieliński [2]

[1] Institute of Mathematics of the Romanian Academy, 21 Calea Grivitei Street, 010702 Bucharest, Romania
[2] Department of Theoretical Physics and Computer Science, University of Łódź, Pomorska 149/153, 90-236, Łódź, Poland; E-Mail: bzielinski@uni.lodz.pl
* E-Mail: florin.nichita@imar.ro

Abstract: We study the duality between corings and ring extensions. We construct a new category with a self-dual functor acting on it, which extends that duality. This construction can be seen as the non-commutative case of another duality extension: the duality between finite dimensional algebras and coalgebra. Both these duality extensions have some similarities with the Pontryagin-van Kampen duality theorem.

Keywords: corings; ring extension; duality; Yang–Baxter equation

MSC: 16T25; 16T15

---

1. Introduction

Non-commutative geometry is a branch of mathematics concerned with geometric approach to non-commutative algebras, and with constructions of spaces which are locally presented by non-commutative algebras of functions. Its main motivation is to extend the commutative duality between spaces and functions to the non-commutative setting.

More specifically, in topology, compact Hausdorff topological spaces can be reconstructed from the Banach algebra of functions on the space. The Pontryagin duality theorem refers to the duality between the category of compact Hausdorff Abelian groups and the category of discrete Abelian groups. The Pontryagin–van Kampen duality theorem extends this duality to all locally compact Hausdorff Abelian topological groups by including the categories of compact Hausdorff Abelian groups and discrete Abelian groups into the category of locally compact Hausdorff Abelian topological groups (see [1]). This can be illustrated by the following diagram:

$$\begin{array}{ccc}
\mathbf{LCA} & \underset{(-,T)}{\overset{(-,T)}{\leftrightarrows}} & \mathbf{LCA} \\
\uparrow & & \uparrow \\
\mathbf{Cpct} & \underset{(-,T)}{\overset{(-,T)}{\leftrightarrows}} & \mathbf{Disc.}
\end{array}$$

Taking the Pontryagin–van Kampen duality theorem as a model, an extension for the duality between finite dimensional algebras and coalgebras to the category of finite dimensional Yang–Baxter structures was constructed in [2]. The resulting duality theorem can be illustrated by the following diagram:

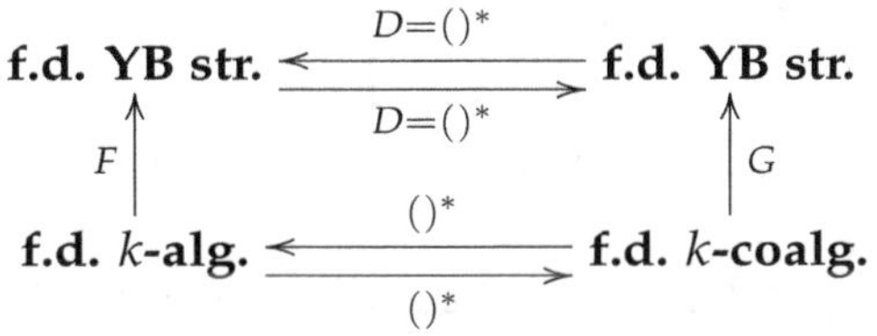

Our motivation in this paper is to extend the above duality to the non-commutative setting.

In Section 2, we present in a new fashion the duality between right finitely generated projective corings and ring extensions (compare with [3]).

In Section 3, we define the category of (right finitely generated projective) generalized Yang–Baxter structures. We construct full and faithful embeddings from the categories of ring extensions and corings to the category of generalized Yang–Baxter structures. We show that taking the right dual is a duality functor in the category of right finitely generated projective generalized Yang–Baxter structures. Then we conclude that the duality between right finitely generated projective corings and ring extensions can be lifted up to the category of right finitely generated projective generalized Yang–Baxter structures.

There are some more comments to be made.

(i) We propose as a research project the investigation of other connections between the duality of (co)algebras and the Pontryagin duality. (For example, one might try to endow the (co)algebra structures with some topological structures.)
(ii) At the epistemologic level, the extension of the duality of (co)algebra structures seems to be a model for the relation between interdisciplinarity, pluridisciplinarity and transdisciplinarity (see [4]).
(iii) This paper explains that taking the dual of some objects can be seen a "continuous" process. Let us visualize this statement by considering an example from geometry. We take a triangular prism: We can see it as two parallel triangles joint by 3 segments. In total it has 5 planar geometric figures, 9 edges and 6 vertices. The geometric dual of the triangular prism has 6 planar geometric figures, 9 edges and 5 vertices. Now, one can start with a triangular prism, "shave" its corners, and then continuously deform that figure in order to obtain the geometric dual of the triangular prism.

2. Notations and Preliminaries

Throughout this paper $\mathbb{K}$ is a commutative ring, and all $\mathbb{K}$-modules M are such that for all $m \in M$, $2m = 0$ implies $m = 0$.

Let A, B, C, *etc.* be algebras over ground commutative ring $\mathbb{K}$. Unadorned tensor product will denote the tensor product over $\mathbb{K}$. For modules M in ${}_A\mathcal{M}_B$, symbols M^*, ${}^*M$, ${}^*M^*$ denote right dual, left dual and bidual of M, and ${}_A\mathcal{M}_B(M,N)$ denotes the $\mathbb{K}$-module of (A,B)-bimodule maps $M \to N$. In what follows we shall concentrate on right dual of M but similar observations can be made for the left dual as well.

For all $\phi \in {}_A\mathcal{M}_B(M,N)$, let $\phi^* : N^* \to M^*$ denote the right adjoint of ϕ *i.e.*, $\phi^*(g)(m) := g \circ \phi(m)$.

We denote by $(\cdot)^{\mathrm{op}} : A \to A^{\mathrm{op}}$ the canonical anti-algebra isomorphism from the algebra A into its opposite A^{op} (which is the identity on the underlying $\mathbb{K}$-modules), *i.e.*, $a = a^{\mathrm{op}}$ as module elements and $(aa')^{\mathrm{op}} = a'^{\mathrm{op}}a^{\mathrm{op}}$ for all $a, a' \in A$.

The following facts are well known, but we recall them to set up the notation:

(i) If $M \in {}_A\mathcal{M}_B$ then $M^* \in {}_{A^{\mathrm{op}}}\mathcal{M}_{B^{\mathrm{op}}}$ with $(a^{\mathrm{op}} f b^{\mathrm{op}})(m) = bf(am)$.

Assume that $M \in {}_A\mathcal{M}_B$ is also finitely generated projective as a right B-module, *i.e.*, there exists a dual basis $\hat{f}_i \in M^*$, $\hat{m}_i \in M$, $i \in I$ such that for any $m \in M$, $m = \sum_i \hat{m}_i \hat{f}_i(m)$. Then

(ii) The mapping $\kappa_M : M \to M^{**}$, $\kappa_M(m)(f) = f(m)^{\mathrm{op}}$ is an isomorphism in ${}_A\mathcal{M}_B$, with the inverse $\kappa_M^{-1}(\tilde{m}) = \sum_i \hat{m}_i \tilde{m}(\hat{f}_i)^{\mathrm{op}}$. In fact κ is a natural morphism between identity functor in ${}_A\mathcal{M}_B$ and the functor $()^{**} : {}_A\mathcal{M}_B \to {}_A\mathcal{M}_B$.

(iii) If $N \in {}_B\mathcal{M}_C$ then $\kappa_{M,N} : M^* \otimes_{B^{\mathrm{op}}} N^* \to (M \otimes_B N)^*$, given by $\kappa_{M,N}(f \otimes_{B^{\mathrm{op}}} g)(m \otimes n) = g(f(m)n)$, is an isomorphism in ${}_{A^{\mathrm{op}}}\mathcal{M}_{C^{\mathrm{op}}}$ with the inverse

$$\kappa_{M,N}^{-1}(\alpha) = \sum_i \hat{f}_i \otimes_{B^{\mathrm{op}}} \alpha(\hat{m}_i \otimes_B \cdot) \tag{1}$$

(iv) Let $M \in {}_A\mathcal{M}_B$, $N \in {}_B\mathcal{M}_C$, $P \in {}_C\mathcal{M}_D$, where A, B, C, D are algebras. Then the following diagram is commutative:

$$\begin{array}{ccc} M^* \otimes_{B^{\mathrm{op}}} N^* \otimes_{C^{\mathrm{op}}} P^* & \xrightarrow{\kappa_{M,N} \otimes_{C^{\mathrm{op}}} P^*} & (M \otimes_B N)^* \otimes_{C^{\mathrm{op}}} P^* \\ \Big\downarrow {\scriptstyle M^* \otimes_{B^{\mathrm{op}}} \kappa_{N,P}} & & \Big\downarrow {\scriptstyle \kappa_{M \otimes_B N, P}} \\ M^* \otimes_{B^{\mathrm{op}}} (N \otimes_C P)^* & \xrightarrow{\kappa_{M, N \otimes_C P}} & (M \otimes_B N \otimes_C P)^* \end{array} \tag{2}$$

(v) Let $M \in {}_A\mathcal{M}_B$ be finitely generated projective as B-module, with dual basis $\hat{m}_i \in M$, $\hat{f}_i \in M^*$, $i \in I$, and let $N \in {}_B\mathcal{M}_C$ be finitely generated projective as a C-module with dual basis $\hat{n}_i \in N$, $\hat{g}_i \in N^*$, $i \in J$. Then $M \otimes_B N \in {}_A\mathcal{M}_C$ is finitely generated projective as a C-module with a dual basis

$$m_i \otimes_B n_j \in M \otimes_B N,\ \kappa_{M,N}(\hat{f}_i \otimes_{B^{\mathrm{op}}} \hat{g}_j) \in (M \otimes_B N)^*,\ i \in I,\ j \in J \tag{3}$$

The following terminology and theorems concerning corings and ring extensions are needed in this paper. For a review on coalgebras see: [5–7]. For a review on corings see [3].

Definition 1. $\mathcal{C} \in {}_B\mathcal{M}_B$ is called a *B-coring* if there exist morphisms $\Delta^{\mathcal{C}}, \varepsilon^{\mathcal{C}} \in {}_B\mathcal{M}_B$, $\Delta^{\mathcal{C}} : \mathcal{C} \to \mathcal{C} \otimes_B \mathcal{C}$, $\varepsilon^{\mathcal{C}} : \mathcal{C} \to B$ such that

$$(\Delta^{\mathcal{C}} \otimes_B \mathcal{C}) \circ \Delta^{\mathcal{C}} = (\mathcal{C} \otimes_B \Delta^{\mathcal{C}}) \circ \Delta^{\mathcal{C}} \tag{4}$$

$$(\varepsilon^{\mathcal{C}} \otimes_B \mathcal{C}) \circ \Delta^{\mathcal{C}} = \mathcal{C} = (\mathcal{C} \otimes_B \varepsilon^{\mathcal{C}}) \circ \Delta^{\mathcal{C}} \tag{5}$$

In the sequel we shall use Sweedler's notation $\Delta^{\mathcal{C}}(c) = c_{(1)} \otimes_B c_{(2)}$. Given B-corings $\mathcal{C}$ and $\mathcal{D}$, a map $\phi \in {}_B\mathcal{M}_B(\mathcal{C}, \mathcal{D})$ is called a *morphism of B-corings* if $(\phi \otimes_B \phi) \circ \Delta^{\mathcal{C}} = \Delta^{\mathcal{D}} \circ \phi$ and $\varepsilon^{\mathcal{D}} \circ \phi = \varepsilon^{\mathcal{C}}$. The category of B-corings is denoted by $\mathbf{Crg}_B$.

Definition 2. Ring $\mathcal{R}$ is called an extension of a ring B if there exists an injective unital ring morphism $\imath_{\mathcal{R}} : B \to \mathcal{R}$. Observe that $\mathcal{R} \in {}_B\mathcal{M}_B$ by $\imath_{\mathcal{R}}$. Given ring extensions $\imath_{\mathcal{R}} : B \to \mathcal{R}$ and $\imath_{\mathcal{P}} : B \to \mathcal{P}$, a ring morphism $\alpha : \mathcal{R} \to \mathcal{P}$ is called a *morphism of ring extensions* if $\alpha \circ \imath_{\mathcal{R}} = \imath_{\mathcal{P}}$ or, equivalently, if $\alpha \in {}_B\mathcal{M}_B(\mathcal{R}, \mathcal{P})$. The category of ring extensions of B is denoted by $\mathbf{Rge}_B$.

The full subcategory of $\mathbf{Crg}_B$ (resp. $\mathbf{Rge}_B$) consisting of those B-corings (resp. ring extensions of B) that are finitely generated projective as right B-modules is denoted by $\mathbf{r.f.g.pCrg}_B$ (resp. $\mathbf{r.f.g.pRge}_B$).

Lemma 1. *(i) If $\mathcal{C} \in \mathbf{Crg}_B$ then $\mathcal{C}^* \in {}_{B^{\mathrm{op}}}\mathcal{M}_{B^{\mathrm{op}}}$ is a ring extension of B^{op} with multiplication*

$$(rr')(c) := r'(r(c_{(1)})c_{(2)}), \quad \textit{for all } r, r' \in \mathcal{C}^* \tag{6}$$

unit $1_{\mathcal{C}^*} := \varepsilon^{\mathcal{C}}$ and embedding map

$$\imath_{\mathcal{C}^*} : B \to \mathcal{C}^*, \quad b^{op} \mapsto b^{op} 1_{\mathcal{C}^*} \tag{7}$$

(ii) *If $\phi : \mathcal{C} \to \mathcal{D}$ is any coring morphism then $\phi^* : \mathcal{D}^* \to \mathcal{C}^*$ is a ring extension morphism.*

(iii) *If $\mathcal{R} \in$ **r.f.g.pRge**$_B$ then $\mathcal{R}^*$ is a B^{op}-coring with comultiplication and counit*

$$\Delta^{\mathcal{R}^*} : \mathcal{R}^* \to \mathcal{R}^* \otimes_{B^{op}} \mathcal{R}^* \quad c \mapsto \sum_i \hat{f}_i \otimes_{B^{op}} c(\hat{r}_i \cdot) \tag{8}$$

$$\varepsilon^{\mathcal{R}^*} : \mathcal{R}^* \to B^{op}, \quad c \mapsto c(1_{\mathcal{R}})^{op} \tag{9}$$

where $\hat{r}_i \in \mathcal{R}$, $\hat{f}_i \in \mathcal{R}^*$, $i \in I$ is a (finite) dual basis of $\mathcal{R}$.

(iv) *If $\phi : \mathcal{R} \to \mathcal{S}$ is a morphism of right finitely generated projective ring extensions of B, then $\phi : \mathcal{S}^* \to \mathcal{R}^*$ is a morphism of B^{op}-corings.*

(v) *Functor $()^{**}$: **r.f.g.pRge**$_B \to$ **r.f.g.pRge**$_B$ is equivalent to the identity functor on **r.f.g.pRge**$_B$. For all $\mathcal{R} \in$ **r.f.g.pRge**$_B$, $\kappa_{\mathcal{R}} : \mathcal{R} \to \mathcal{R}^{**}$ is a ring extension isomorphism facilitating this equivalence.*

(vi) *Functor $()^{**}$: **r.f.g.pCrg**$_B \to$ **r.f.g.pCrg**$_B$ is equivalent to the identity functor on **r.f.g.pCrg**$_B$. For all $\mathcal{C} \in$ **r.f.g.pCrg**$_B$, $\kappa_{\mathcal{C}} : \mathcal{C} \to \mathcal{C}^{**}$ is a B-coring isomorphism facilitating this equivalence.*

Proof. The statements (i) and (ii) are contained in Proposition 3.2 [8], while (iii) and (v) are rephrasings of Theorem 3.7 [8] (cf. [3], 17.8–17.13)
(iv) Consider any ring extension morphism $\phi : \mathcal{R} \to \mathcal{S}$. Let $\hat{r}_i \in \mathcal{R}$, $\hat{f}_i \in \mathcal{R}^*$, $i \in I$ be any finite dual basis of $\mathcal{R}$, and let $\hat{s}_i \in \mathcal{S}$, $\hat{g}_i \in \mathcal{S}^*$, $i \in J$ be any finite dual basis of $\mathcal{S}$. For all $s \in \mathcal{S}^*$,

$$\begin{aligned}\Delta^{\mathcal{R}^*} \circ \phi^*(s) &= \sum_i \hat{f}_i \otimes_{B^{op}} s(\phi(\hat{r}_i)\phi(\cdot)) = \sum_{ij} \hat{f}_i \otimes_{B^{op}} s(\hat{s}_j\hat{g}_j(\phi(\hat{r}_i))\phi(\cdot)) \\ &= \sum_{ij} \hat{f}_i \otimes_{B^{op}} s(\hat{s}_j\phi(\hat{g}_j(\phi(\hat{r}_i))\cdot)) = \sum_{ij} \hat{f}_i \otimes_{B^{op}} \hat{g}_j(\phi(\hat{r}_i))^{op} s(\hat{s}_j\phi(\cdot)) \\ &= \sum_{ij} \hat{f}_i\hat{g}_j(\phi(\hat{r}_i))^{op} \otimes_{B^{op}} s(\hat{s}_j\phi(\cdot)) = \sum_j \hat{g}_j \circ \phi \otimes_{B^{op}} s(\hat{s}_j\phi(\cdot)) \\ &= (\phi^* \otimes_{B^{op}} \phi^*)(\sum_j \hat{g}_j \otimes_{B^{op}} s(\hat{s}_j\cdot)) = (\phi^* \otimes_{B^{op}} \phi^*) \circ \Delta^{\mathcal{S}}(s)\end{aligned}$$

and

$$\varepsilon^{\mathcal{R}^*} \circ \phi^*(s) = \varepsilon^{\mathcal{R}^*}(s \circ \phi) = (s \circ \phi)(1_{\mathcal{R}}) = s(1_{\mathcal{S}}) = \varepsilon^{\mathcal{S}^*}(s)$$

Hence ϕ^* is a coring map.
(vi) It is enough to prove that $\kappa_{\mathcal{C}}$, is a coring map for any $\mathcal{C} \in$ **r.f.g.pCrg**$_B$. Let $\mathcal{C}$ be a B-coring, and let $\hat{c}_i \in \mathcal{C}$, $\hat{f}_i \in \mathcal{C}^*$, $i \in I$, be any finite dual basis of $\mathcal{C}$. Observe that $\hat{f}_i \in \mathcal{C}^*$, $\kappa_{\mathcal{C}}(\hat{c}) \in \mathcal{C}^{**}$, $i \in I$ is a dual basis of $\mathcal{C}^*$. Indeed, for any $g \in \mathcal{C}^*$,

$$g = \sum_i \hat{f}_i g(\hat{m}_i)^{op} = \sum_i \hat{f}_i \kappa_{\mathcal{C}}(\hat{m}_i)$$

Hence, for all $c \in \mathcal{C}$,

$$\begin{aligned}\Delta^{\mathcal{C}^{**}} \circ \kappa_{\mathcal{C}}(c) &= \sum_i \kappa_{\mathcal{C}}(\hat{c}_i) \otimes_B \kappa_{\mathcal{C}}(c)(\hat{f}_i\cdot) = \sum_i \kappa_{\mathcal{C}}(\hat{c}_i) \otimes_B (\hat{f}_i\cdot)(c)^{op} \\ &= \sum_i \kappa_{\mathcal{C}}(\hat{c}_i) \otimes_B \cdot(\hat{f}_i(c_{(1)})c_{(2)})^{op} = \sum_i \kappa_{\mathcal{C}}(\hat{c}_i) \otimes_B \kappa_{\mathcal{C}}(\hat{f}_i(c_{(1)})c_{(2)}) \\ &= \sum_i \kappa_{\mathcal{C}}(\hat{c}_i\hat{f}_i(c_{(1)}) \otimes_B \kappa_{\mathcal{C}}(c_{(2)}) = (\kappa_{\mathcal{C}} \otimes_B \kappa_{\mathcal{C}}) \circ \Delta^{\mathcal{C}}(c)\end{aligned}$$

and

$$\varepsilon^{\mathcal{C}^{**}} \circ \kappa_{\mathcal{C}}(c) = \kappa_{\mathcal{C}}(c)(1_{\mathcal{C}^*})^{op} = 1_{\mathcal{C}^*}(c)^{opop} = \varepsilon^{\mathcal{C}}(c)$$

□

Corollary 1. *$()^*$ is a duality functor between* $\mathbf{r.f.g.pRge}_B$ *and* $\mathbf{r.f.g.pCrg}_{B^{op}}$*:*

$$\mathbf{r.f.g.pRge}_B \underset{()^*}{\overset{()^*}{\rightleftarrows}} \mathbf{r.f.g.pCrg}_{B^{op}} \tag{10}$$

3. An Extension for the Duality between Corings and Ring Extensions

Our aim in this section is to extend the duality between right finitely generated projective ring extensions and corings to the category of right finitely generated projective generalized Yang–Baxter structures.

We use the following terminology concerning the Yang–Baxter equation. Some references on this topic are: [9–11], *etc.*

Let B be a $\mathbb{K}$-algebra. Given a (B,B)-bimodule V and a (B,B)-bilinear map $R : V \otimes_B V \to V \otimes_B V$ we write $R^{12} = R \otimes_B id$, $R^{23} = id \otimes_B R : V \otimes_B V \otimes_B V \to V \otimes_B V \otimes_B V$ where $id : V \to V$ is the identity map.

Definition 3. An invertible (B,B)-linear map $R : V \otimes_B V \to V \otimes_B V$ is called a *generalized Yang–Baxter operator* (or simply a *generalised YB operator*) if it satisfies the equation

$$R^{12} \circ R^{23} \circ R^{12} = R^{23} \circ R^{12} \circ R^{23} \tag{11}$$

Definition 4. For an algebra B, we define the category $\mathbf{YB\ str}_B$ whose objects are 4-tuples $(V, \varphi, e, \varepsilon)$, where

(i) V is a (B,B)-bimodule;
(ii) $\varphi : V \otimes_B V \to V \otimes_B V$ is a generalized YB operator;
(iii) $e \in V$ such that for all $b \in B$, $eb = be$, and for all $x \in V$, $\varphi(x \otimes e) = e \otimes_B x$, $\varphi(e \otimes_B x) = x \otimes_B e$;
(iv) $\varepsilon : V \to B$ is a (B,B)-bimodule map, such that $(id \otimes_B \varepsilon) \circ \varphi = \varepsilon \otimes_B id$, $(\varepsilon \otimes_B id) \circ \varphi = id \otimes_B \varepsilon$.

A morphism $f : (V, \varphi, e, \varepsilon) \to (V', \varphi', e', \varepsilon')$ in the category $\mathbf{YB\ str}_B$ is a (B,B)-bilinear map $f : V \to V'$ such that:

(v) $(f \otimes_B f) \circ \varphi = \varphi' \circ (f \otimes_B f)$,
(vi) $f(e) = e'$,
(vii) $\varepsilon' \circ f = \varepsilon$.

Composition of morphisms is defined as the standard composition of B-linear maps. A full subcategory of $\mathbf{YB\ str}_B$ consisting of all such $(V, \varphi, e, \varepsilon)$ for which V is finitely generated projective as a right B-module is defined by $\mathbf{r.f.g.pYB\ str}_B$.

Remark 1. Let $R : V \otimes_B V \to V \otimes_B V$ be a generalised YB operator . Then $(V, R, 0, 0)$ is an object in the category $\mathbf{YB\ str}_B$.

Theorem 1. *(i) There exists a functor:*

$$F : \mathbf{Rge}_B \to \mathbf{YB\ str}_B, \quad \mathcal{R} \mapsto (\mathcal{R}, \varphi_{\mathcal{R}}, 1_{\mathcal{R}}, 0 \in {}^*\mathcal{R}^*)$$
$$\text{where} \quad \varphi_{\mathcal{R}}(r \otimes_B r') = rr' \otimes_B 1 + 1 \otimes_B rr' - r \otimes_B r' \tag{12}$$

Any ring extension map f is simply mapped into a (B,B) bimodule map.
(ii) F is a full and faithful embedding.

Proof. i) The proof that $\varphi_{\mathcal{R}}$ is a generalised YB operator is left to the reader (cf. Proposition 2.1 from [12], $\varphi_{\mathcal{R}}^{-1} = \varphi_{\mathcal{R}}$). Furthermore $\varphi_{\mathcal{R}}(r \otimes_B 1) = r \otimes_B 1 + 1 \otimes_B r - r \otimes_B 1 = 1 \otimes_B r$, $\varphi_{\mathcal{R}}(1 \otimes_B r) =$

$1 \otimes_B r + r \otimes_B 1 - 1 \otimes_B r = r \otimes_B 1$, $(id \otimes_B 0) \circ \varphi_{\mathcal{R}} = 0 = (0 \otimes_B id)$, $(0 \otimes_B id) \circ \varphi_{\mathcal{R}} = 0 = (id \otimes_B 0)$. Hence $(\mathcal{R}, \varphi_{\mathcal{R}}, 1_{\mathcal{R}}, 0)$ is an object in the category $\mathbf{YB\ str}_B$.

Let $f : \mathcal{R} \to \mathcal{S}$ be a morphism of ring extensions. Then $f(1_{\mathcal{R}}) = 1_{\mathcal{S}}$ and $0 \circ f = 0$. Moreover

$$(f \otimes_B f) \circ \varphi_{\mathcal{R}}(r \otimes_B r') = f(r)f(r') \otimes_B f(1) + f(1) \otimes_B f(r)f(r') - f(r) \otimes_B f(r') \\ = \varphi_{\mathcal{S}} \circ (f \otimes_B f)(r \otimes_B r').$$

Hence $f : (\mathcal{R}, \varphi_{\mathcal{R}}, 1_{\mathcal{R}}, 0) \to (\mathcal{S}, \varphi_{\mathcal{S}}, 1_{\mathcal{S}}, 0)$ is a morphism in the category $\mathbf{YB\ str}_B$.

(ii) If $F\mathcal{R} = F\mathcal{S}$, for some $\mathcal{R}, \mathcal{S} \in \mathbf{Rge}_B$, then obviously $\mathcal{R} = \mathcal{S}$ as (B, B)-bimodules, $1_{\mathcal{S}} = 1_{\mathcal{R}}$, and the only thing which can differ is the multiplication. Denote by $\cdot$ the multiplication in $\mathcal{R}$, and by $\circ$ the multiplication in $\mathcal{S}$. Then, as $\varphi_{\mathcal{R}} = \varphi_{\mathcal{S}}$, for all $r, r' \in \mathcal{R}$,

$$r \cdot r' \otimes_B 1 + 1 \otimes_B r \cdot r' - r \otimes_B r' = r \circ r' \otimes_B 1 + 1 \otimes_B r \circ r' - r \otimes_B r'$$

hence

$$(r \cdot r' - r \circ r') \otimes_B 1 = -1 \otimes_B (r \cdot r' - r \circ r')$$

Multiplying tensor factors on both sides of this equation (whether using multiplication in $\mathcal{R}$ or $\mathcal{S}$ is irrelevant) yields $2(r \cdot r' - r \circ r') = 0$, hence $r \cdot r' = r \circ r'$, and so $\mathcal{R} = \mathcal{S}$ as algebras. Therefore F is an embedding.

Obviously, distinct ring extension maps are also distinct as (B, B)-bimodule morphisms, hence F is a faithful functor.

Let $f : (\mathcal{R}, \varphi_{\mathcal{R}}, 1_{\mathcal{R}}, 0) \to (\mathcal{S}, \varphi_{\mathcal{S}}, 1_{\mathcal{S}}, 0)$ be a morphism in $\mathbf{YB\ str}_B$, where $\mathcal{R}, \mathcal{S} \in \mathbf{Rge}_B$. Then f is unital, and $(f \otimes_B f) \circ \varphi_{\mathcal{R}} = \varphi_{\mathcal{S}} \circ (f \otimes_B f)$, hence, for all $r, r' \in \mathcal{R}$,

$$f(rr') \otimes_B 1 + 1 \otimes_B f(rr') - f(r) \otimes_B f(r') = f(r)f(r') \otimes_B 1 + 1 \otimes_B f(r)f(r') - f(r) \otimes_B f(r').$$

Multiplying factors in tensor products in both sides of the above equation yields $2(f(rr') - f(r)f(r')) = 0$, hence $f(rr') = f(r)f(r')$ and, as f is a (B, B)-bimodule map, it is a ring extension map. Therefore, F is a full functor. □

Theorem 2. *(i) There exists a functor*

$$G : \mathbf{Crg}_B \to \mathbf{YB\ str}_B, \quad \mathcal{C} \mapsto (\mathcal{C}, \psi_{\mathcal{C}}, 0, \varepsilon^{\mathcal{C}}) \\ \textit{where} \quad \psi_{\mathcal{C}} = \Delta^{\mathcal{C}} \otimes_B \varepsilon^{\mathcal{C}} + \varepsilon^{\mathcal{C}} \otimes_B \Delta^{\mathcal{C}} - id \otimes_B id \tag{13}$$

A coring morphism is mapped into a (B, B)-bimodule morphism.

(ii) G is a full and faithful embbeding.

Proof. i) The proof that $\psi_{\mathcal{C}}$ is a generalised YB operator (cf. Proposition 2.3 from [12]) is left to the reader ($\psi_{\mathcal{C}}^{-1} = \psi_{\mathcal{C}}$). Furthermore, for all $c \in \mathcal{C}$, $\psi_{\mathcal{C}}(c \otimes_B 0) = 0 = 0 \otimes_B c$, $\psi_{\mathcal{C}}(0 \otimes_B c) = 0 = c \otimes_B 0$. Moreover, for all $c, c' \in \mathcal{C}$,

$$(id \otimes_B \varepsilon^{\mathcal{C}}) \circ \psi_{\mathcal{C}}(c \otimes_B c') = c_{(1)}\varepsilon^{\mathcal{C}}(c_{(2)})\varepsilon^{\mathcal{C}}(c') + \varepsilon^{\mathcal{C}}(c)c'_{(1)}\varepsilon^{\mathcal{C}}(c'_{(2)}) - c\varepsilon^{\mathcal{C}}(c') = \varepsilon^{\mathcal{C}}(c)c' \\ = (\varepsilon^{\mathcal{C}} \otimes_B id)(c \otimes_B c')$$

and

$$(\varepsilon^{\mathcal{C}} \otimes_B id) \circ \psi_{\mathcal{C}}(c \otimes_B c') = \varepsilon^{\mathcal{C}}(c_{(1)})c_{(2)}\varepsilon^{\mathcal{C}}(c') + \varepsilon^{\mathcal{C}}(c)\varepsilon^{\mathcal{C}}(c'_{(1)})c'_{(2)} - \varepsilon^{\mathcal{C}}(c)c' = c\varepsilon^{\mathcal{C}}(c') \\ = (id \otimes_B \varepsilon^{\mathcal{C}})(c \otimes_B c')$$

Hence $(\mathcal{C}, \psi_{\mathcal{C}}, 0, \varepsilon^{\mathcal{C}})$ is an object in **YB str**$_B$. Let $f : \mathcal{C} \to \mathcal{D}$ be any morphism of B-corings. Then f is also a (B, B)-bimodule morphism, $f(0) = 0$, $\varepsilon^{\mathcal{D}} \circ f = \varepsilon^{\mathcal{C}}$, and,

$$\begin{aligned}\psi_{\mathcal{D}} \circ (f \otimes_B f) &= \Delta^{\mathcal{D}} \circ f \otimes_B \varepsilon^{\mathcal{D}} \circ f + \varepsilon^{\mathcal{D}} \circ f \otimes_B \Delta^{\mathcal{D}} \circ f - f \otimes_B f \\ &= (f \otimes_B f) \circ \Delta^{\mathcal{C}} \otimes_B \varepsilon^{\mathcal{C}} + \varepsilon^{\mathcal{C}} \otimes_B (f \otimes_B f) \circ \Delta^{\mathcal{C}} + f \otimes_B f = (f \otimes_B f) \circ \psi_{\mathcal{C}}\end{aligned}$$

Therefore $f : (\mathcal{C}, \psi_{\mathcal{C}}, 0, \varepsilon^{\mathcal{C}}) \to (\mathcal{D}, \psi_{\mathcal{D}}, 0, \varepsilon^{\mathcal{D}})$ is a morphism in **YB str**$_B$.
(ii) Suppose that $G\mathcal{C} = G\mathcal{D}$ for some B-corings $\mathcal{C}$, $\mathcal{D}$. This means that $\mathcal{C} = \mathcal{D}$ as (B, B)-bimodules, $\varepsilon^{\mathcal{C}} = \varepsilon^{\mathcal{D}}$, and the only things which can differ are comultiplications. However, as $\psi_{\mathcal{C}} = \psi_{\mathcal{D}}$, we have

$$\Delta^{\mathcal{C}} \otimes_B \varepsilon^{\mathcal{C}} + \varepsilon^{\mathcal{C}} \otimes_B \Delta^{\mathcal{C}} - I \otimes_B I = \Delta^{\mathcal{D}} \otimes_B \varepsilon^{\mathcal{C}} + \varepsilon^{\mathcal{C}} \otimes_B \Delta^{\mathcal{D}} - I \otimes_B I$$

hence

$$(\Delta^{\mathcal{C}} - \Delta^{\mathcal{D}}) \otimes_B \varepsilon^{\mathcal{C}} = -\varepsilon^{\mathcal{C}} \otimes_B (\Delta^{\mathcal{C}} - \Delta^{\mathcal{D}})$$

Composing both sides of the above equation with $\Delta^{\mathcal{C}}$ yields $2(\Delta^{\mathcal{C}} - \Delta^{\mathcal{D}}) = 0$ hence $\Delta^{\mathcal{C}} = \Delta^{\mathcal{D}}$ and $\mathcal{C} = \mathcal{D}$ as (B, B)-corings. Hence G is an embedding.

Obviously distinct B-coring morphisms are also distinct as (B, B)-bimodule morphisms, hence G is a faithful functor.

Let $f : (\mathcal{C}, \psi_{\mathcal{C}}, 0, \varepsilon^{\mathcal{C}}) \to (\mathcal{D}, \psi_{\mathcal{D}}, 0, \varepsilon^{\mathcal{D}})$, where $\mathcal{C}, \mathcal{D}$ are corings, be a morphism in **YB str**$_B$. Then (B, B)-bimodule morphism $f : \mathcal{C} \to \mathcal{D}$ is counital, *i.e.*, $\varepsilon^{\mathcal{D}} \circ f = \varepsilon^{\mathcal{C}}$. Furthermore, $(f \otimes_B f) \circ \psi_{\mathcal{C}} = \psi_{\mathcal{D}} \circ (f \otimes_B f)$, and hence $(f \otimes_B f) \circ \psi_{\mathcal{C}} \circ \Delta^{\mathcal{C}} = \psi_{\mathcal{D}} \circ (f \otimes_B f) \circ \Delta^{\mathcal{C}}$. Observe that $\psi_{\mathcal{C}} \circ \Delta^{\mathcal{C}} = \Delta^{\mathcal{C}}$. Therefore

$$\begin{aligned}(f \otimes_B f) \circ \Delta^{\mathcal{C}} &= (f \otimes_B f) \circ \psi_{\mathcal{C}} \circ \Delta^{\mathcal{C}} = \psi_{\mathcal{D}} \circ (f \otimes_B f) \circ \Delta^{\mathcal{C}} \\ &= (\Delta^{\mathcal{D}} \circ f \otimes_B \varepsilon^{\mathcal{D}} \circ f + \varepsilon^{\mathcal{D}} \circ f \otimes_B \Delta^{\mathcal{D}} \circ f - f \otimes_B f) \circ \Delta^{\mathcal{C}} \\ &= (\Delta^{\mathcal{D}} \circ f \otimes_B \varepsilon^{\mathcal{C}} + \varepsilon^{\mathcal{C}} \otimes_B \Delta^{\mathcal{D}} \circ f - f \otimes_B f) \circ \Delta^{\mathcal{C}} = 2\Delta^{\mathcal{D}} \circ f - (f \otimes_B f) \circ \Delta^{\mathcal{C}}\end{aligned}$$

i.e., $2(f \otimes_B f) \circ \Delta^{\mathcal{C}} = 2\Delta^{\mathcal{D}} \circ f$, hence $(f \otimes_B f) \circ \Delta^{\mathcal{C}} = \Delta^{\mathcal{D}} \circ f$, and f is a B-coring map. Therefore G is full. □

Problem 1. *Let* $(V, R, e, \varepsilon) \in$ **r.f.g.pYB str**$_B$. *Then*

$$(V, R, e, \varepsilon)^* := (V^*, R^\dagger, \varepsilon, e^\dagger) \in \textbf{r.f.g.pYB str}_{B^{op}} \tag{14}$$

where $e^\dagger(f) = f(e)$, *and*

$$R^\dagger = \kappa_{V,V}^{-1} \circ R^* \circ \kappa_{V,V} \tag{15}$$

Moreover,

$$\kappa : () \to ()^{**}, \quad \kappa_V : (V, R, e, \varepsilon) \to (V^{**}, R^{\dagger\dagger}, e^\dagger, \varepsilon^\dagger) \tag{16}$$

is a natural isomorphism in **r.f.g.pYB str**$_B$.

Proof. R is invertible, hence $R^{\dagger -1} = \kappa_{V,V}^{-1} \circ (R^{-1})^* \circ \kappa_{V,V}$. We shall prove that $R^\dagger$ satisfies the Yang–Baxter equation. Observe that

$$\kappa_{V \otimes_B V, V}^{-1} \circ (R \otimes_B I)^* \circ \kappa_{V \otimes_B V, V} = R^* \otimes_{B^{op}} I \tag{17}$$

$$\kappa_{V, V \otimes_B V}^{-1} \circ (I \otimes_B R)^* \circ \kappa_{V, V \otimes_B V} = I \otimes_{B^{op}} R^* \tag{18}$$

Indeed, let $\Gamma \in (V \otimes_B V)^*$, $f \in V^*$, and let $\hat{v}_i \in V$, $\hat{f}_i \in V^*$, $i \in I$, be a dual basis of V.

$$\begin{aligned}
&\kappa^{-1}_{V\otimes_B V,V} \circ (R \otimes_B I)^* \circ \kappa_{V\otimes_B V,V}(\Gamma \otimes_{B^{op}} f) \\
&\quad = \kappa^{-1}_{V\otimes_B V,V} \circ (R \otimes_B I)^*(v \otimes_B v' \otimes_B v'' \mapsto f(\Gamma(v \otimes_B v')v'')) \\
&\quad = \kappa^{-1}_{V\otimes_B V,V}(v \otimes_B v' \otimes_B v'' \mapsto f(\Gamma(R(v \otimes_B v'))v'')) \\
&\quad = \sum_{i,j\in I} \kappa_{V,V}(\hat{f}_i \otimes_{B^{op}} \hat{f}_j) \otimes_{B^{op}} f(\Gamma(R(\hat{v}_i \otimes_B \hat{v}_j))\cdot) \\
&\quad = \sum_{i,j\in I} \kappa_{V,V}(\hat{f}_i \otimes_{B^{op}} \hat{f}_j)\Gamma(R(\hat{v}_i \otimes_B \hat{v}_j))^{op} \otimes_{B^{op}} f \\
&\quad = (v \otimes_B v' \mapsto \sum_{i,j\in I} \Gamma(R(\hat{v}_i \otimes_B \hat{v}_j))\hat{f}_j(\hat{f}_i(v)v')) \otimes_{B^{op}} f \\
&\quad = \Gamma \circ R \otimes_{B^{op}} f = (R^* \otimes_{B^{op}} I)(\Gamma \otimes_{B^{op}} f)
\end{aligned}$$

Similarly we can prove the other equality. By virtue of (17,18), we can write

$$R^\dagger \otimes_{B^{op}} I = (\kappa^{-1}_{V,V} \otimes_{B^{op}} I) \circ \kappa^{-1}_{V\otimes_B V,V} \circ (R \otimes_B I)^* \circ \kappa_{V\otimes_B V,V} \circ (\kappa_{V,V} \otimes_{B^{op}} I) \tag{19}$$

$$I \otimes_{B^{op}} R^\dagger = (I \otimes_{B^{op}} \kappa^{-1}_{V,V}) \circ \kappa^{-1}_{V,V\otimes_B V} \circ (I \otimes_B R)^* \circ \kappa_{V,V\otimes_B V} \circ (I \otimes_{B^{op}} \kappa_{V,V}) \tag{20}$$

By (2),

$$\kappa_{V\otimes_B V,V} \circ (\kappa_{V,V} \otimes_{B^{op}} I) \circ (I \otimes_{B^{op}} \kappa^{-1}_{V,V}) = \kappa_{V,V\otimes_B V} \tag{21}$$

$$\kappa_{V,V\otimes_B V} \circ (I \otimes_{B^{op}} \kappa_{V,V})(\kappa^{-1}_{V,V} \otimes_{B^{op}} I) = \kappa_{V\otimes_B V,V} \tag{22}$$

and therefore

$$\begin{aligned}
R^{\dagger 12}R^{\dagger 23}R^{\dagger 12} &= (\kappa^{-1}_{V,V} \otimes_{B^{op}} I) \circ \kappa^{-1}_{V\otimes_B V,V} \circ (R^{12} \circ R^{23} \circ R^{12})^* \circ \kappa_{V\otimes_B V,V} \circ (\kappa_{V,V} \otimes_{B^{op}} I) \\
&= (I \otimes_{B^{op}} \kappa^{-1}_{V,V}) \circ \kappa^{-1}_{V,V\otimes_B V} \circ (R^{23} \circ R^{12} \circ R^{23})^* \circ \kappa_{V,V\otimes_B V} \circ (I \otimes_{B^{op}} \kappa_{V,V}) = R^{\dagger 23}R^{\dagger 12}R^{\dagger 23}
\end{aligned}$$

Hence $R^\dagger$ is a generalised YB operator .

Proofs of bilinearity of e^* and centrality of ε are the same as proofs of analogues properties of duals of units and counits in Lemma 1. Moreover, for all $f \in V^*$,

$$\begin{aligned}
R^\dagger(\varepsilon \otimes_{B^{op}} f) &= \kappa^{-1}_{V,V} \circ R^* \circ \kappa_{V,V}(\varepsilon \otimes_{B^{op}} f) = \kappa^{-1}_{V,V}(f \circ (\varepsilon \otimes_{B^{op}} I) \circ R) \\
&= \kappa^{-1}_{V,V}(f \circ (I \otimes_{B^{op}} \varepsilon)) = \sum_i \hat{f}_i \otimes_{B^{op}} f(\hat{v}_i\varepsilon(\cdot)) = \sum_i \hat{f}_i \otimes_{B^{op}} \varepsilon(f(\hat{v}_i)\cdot) \\
&= \sum_i \hat{f}_i \otimes_{B^{op}} f(\hat{v}_i)^{op}\varepsilon = f \otimes_{B^{op}} \varepsilon
\end{aligned}$$

and

$$\begin{aligned}
R^\dagger(f \otimes_{B^{op}} \varepsilon) &= \kappa^{-1}_{V,V} \circ R^* \circ \kappa_{V,V}(f \otimes_{B^{op}} \varepsilon) = \kappa^{-1}_{V,V}(\varepsilon \circ (f \otimes_{B^{op}} I) \circ R) \\
&= \kappa^{-1}_{V,V}(f \circ (I \otimes_{B^{op}} \varepsilon) \circ R) = \kappa^{-1}_{V,V}(f \circ (\varepsilon \otimes_{B^{op}} I)) = \sum_i \hat{f}_i \otimes_{B^{op}} f(\varepsilon(\hat{v}_i)\cdot) \\
&= \sum_i \hat{f}_i \otimes_{B^{op}} \varepsilon(\hat{v}_i)^{op} f = \varepsilon \otimes_{B^{op}} f
\end{aligned}$$

Furthermore, for all $x = f \otimes_{B^{op}} g \in V^* \otimes_{B^{op}} V^*$,

$$\begin{aligned}(e^\dagger \otimes_{B^{op}} I) \circ R^\dagger(x) &= (e^\dagger \otimes_{B^{op}} I) \circ \kappa_{V,V}^{-1} \circ R^* \circ \kappa_{V,V}(x) = (e^\dagger \otimes_{B^{op}} I) \circ \kappa_{V,V}^{-1}(\kappa_{V,V}(x) \circ R) \\ &= (e^\dagger \otimes_{B^{op}} I)(\sum_i \hat{f}_i \otimes_{B^{op}} \kappa_{V,V}(x) \circ R(\hat{v}_i \otimes_B \cdot) = \sum_i \hat{f}_i(e)^{op} \kappa_{V,V}(x) \circ R(\hat{v}_i \otimes_B \cdot) \\ &= \sum_i \kappa_{V,V}(x) \circ R(\hat{v}_i \otimes_B \hat{f}_i \cdot) = \kappa_{V,V}(x) \circ R(e \otimes_B \cdot) = \kappa_{V,V}(\cdot \otimes_B e) = g(f(\cdot)e) = g(e)f(\cdot) \\ &= fg(e)^{op} = (I \otimes_{B^{op}} e^\dagger)(f \otimes_{B^{op}} g) = (I \otimes_{B^{op}} e^\dagger)(x)\end{aligned}$$

and

$$\begin{aligned}(I \otimes_{B^{op}} e^\dagger) \circ R^\dagger(x) &= (I \otimes_{B^{op}} e^\dagger) \circ \kappa_{V,V}^{-1} \circ R^* \circ \kappa_{V,V}(x) = (I \otimes_{B^{op}} e^\dagger) \circ \kappa_{V,V}^{-1}(\kappa_{V,V}(x) \circ R) \\ &= (I \otimes_{B^{op}} e^\dagger)(\sum_i \hat{f}_i \otimes_{B^{op}} \kappa_{V,V}(x) \circ R(\hat{v}_i \otimes_B \cdot)) = \sum_i \hat{f}_i \kappa_{V,V}(x) \circ R(\hat{v}_i \otimes_B e)^{op} \\ &= \sum_i \kappa_{V,V}(x) \circ R(\hat{v}_i \otimes_B e)\hat{f}_i(\cdot) = \sum_i \kappa_{V,V}(x) \circ R(\hat{v}_i \otimes_B \hat{f}_i(\cdot)e) = \kappa_{V,V}(x) \circ R(\cdot \otimes_B e) \\ &= \kappa_{V,V}(x)(e \otimes_B \cdot) = g(f(e)\cdot) = f(e)^{op} g = (e^\dagger \otimes_{B^{op}} I)(x)\end{aligned}$$

Hence $(V^*, R^\dagger, \varepsilon, e^\dagger) \in$ **r.f.g.pYB str**$_{B^{op}}$.

Morphism $\kappa : () \to ()^{**}$ is natural in ${}_B\mathcal{M}_B$, and as V is finitely generated projective, κ_V is invertible. Therefore it suffices to prove that κ_V is a morphism in **r.f.g.pYB str**$_B$. To this end, observe first that

$$\kappa_V(e) = f \mapsto f(e)^{op} = e^\dagger$$

and, for all $v \in V$,

$$\varepsilon^\dagger \circ \kappa_V(v) = \kappa_V(v)(\varepsilon)^{op} = \varepsilon(v)^{opop} = \varepsilon(v)$$

Note that $\hat{f}_i \in V^*, \kappa_V(\hat{v}_i) \in V^{**}, i \in I$ is a dual basis of V^*. Therefore, for all $\Gamma \in (V^* \otimes_{B^{op}} V^*)^*$,

$$\kappa_{V^*,V^*}(\Gamma) = \sum_i \kappa_V(\hat{v}_i) \otimes_B \Gamma(\hat{f}_i \otimes_{B^{op}} \cdot)$$

and so, for all $v, v' \in V$,

$$
\begin{aligned}
R^{\dagger\dagger} \circ (\kappa_V \otimes_B \kappa_V)(v \otimes_B v') &= \kappa_{V^*,V^*}^{-1} \circ R^{\dagger *} \circ \kappa_{V^*,V^*}(\kappa_V(v) \otimes_B \kappa_V(v')) \\
&= \kappa_{V^*,V^*}^{-1}(\kappa_{V^*,V^*}(\kappa_V(v) \otimes_B \kappa_V(v')) \circ \kappa_{V,V}^{-1} \circ R^* \circ \kappa_{V,V}) \\
&= \kappa_{V^*,V^*}^{-1}(\kappa_{V^*,V^*}(\kappa_V(v) \otimes_B \kappa_V(v')) \circ \kappa_{V,V}^{-1} \circ (x \mapsto \kappa_{V,V}(x) \circ R) \\
&= \kappa_{V^*,V^*}^{-1}(\kappa_{V^*,V^*}(\kappa_V(v) \otimes_B \kappa_V(v')) \circ (x \mapsto \sum_i \hat{f}_i \otimes_{B^{\mathrm{op}}} \kappa_{V,V}(x) \circ R(\hat{v}_i \otimes_B \cdot)) \\
&= \kappa_{V^*,V^*}^{-1}(x \mapsto \sum_i \kappa_V(v')(\kappa_V(v)(\hat{f}_i)\kappa_{V,V}(x) \circ R(\hat{v}_i \otimes_B \cdot))) \\
&= \kappa_{V^*,V^*}^{-1}(x \mapsto \sum_i \kappa_V(v')(\hat{f}_i(v)^{\mathrm{op}}\kappa_{V,V}(x) \circ R(\hat{v}_i \otimes_B \cdot))) \\
&= \kappa_{V^*,V^*}^{-1}(x \mapsto \kappa_V(v')(\kappa_{V,V}(x) \circ R(v \otimes_B \cdot)) \\
&= \kappa_{V^*,V^*}^{-1}(x \mapsto \kappa_{V,V}(x) \circ R(v \otimes_B v')^{\mathrm{op}}) \\
&= \sum_i \kappa_V(\hat{v}_i) \otimes_B \kappa_{V,V}(\hat{f}_i \otimes_{B^{\mathrm{op}}} \cdot) \circ R(v \otimes_B v')^{\mathrm{op}} \\
&= \sum_i \kappa_V(\hat{v}_i) \otimes_B \kappa_V((\hat{f}_i \otimes_{B^{\mathrm{op}}} I) \circ R(v \otimes_B v')) \\
&= (\kappa_V \otimes_B \kappa_V)(\sum_i \hat{v}_i \otimes_B (\hat{f}_i \otimes_{B^{\mathrm{op}}} I) \circ R(v \otimes_B v')) \\
&= (\kappa_V \otimes_B \kappa_V) \circ R(v \otimes_B v')
\end{aligned}
$$

Therefore, κ_V is a morphism in **r.f.g.pYB str**$_B$ as required. □

Problem 2. *Let $\mathcal{R} \in$ **r.f.g.pRge**$_B$, $\mathcal{C} \in$ **r.f.g.pCrg**$_B$. Then $(F\mathcal{R})^* = G(\mathcal{R}^*)$, $(G\mathcal{C})^* = F(\mathcal{C}^*)$, i.e.,*

$$(\mathcal{R}^*, \phi_{\mathcal{R}}{}^{\dagger}, 0, 1_{\mathcal{R}}{}^{\dagger}) = (\mathcal{R}^*, \psi_{\mathcal{R}^*}, 0, \varepsilon^{\mathcal{R}^*}) \tag{23}$$

$$(\mathcal{C}^*, \psi_{\mathcal{C}}{}^{\dagger}, \varepsilon^{\mathcal{C}}, 0) = (\mathcal{C}^*, \phi_{\mathcal{C}^*}, 1_{\mathcal{C}^*}, 0) \tag{24}$$

Proof. From Lemma 1 we know that $1_{\mathcal{R}}{}^{\dagger} = \varepsilon^{\mathcal{R}^*}$ and $1_{\mathcal{C}^*} = \varepsilon^{\mathcal{C}}$. Furthermore, for all $c, c' \in \mathcal{R}^*$,

$$
\begin{aligned}
\phi_{\mathcal{R}}{}^{\dagger}(c \otimes_{B^{\mathrm{op}}} c') &= \kappa_{\mathcal{R},\mathcal{R}}^{-1} \circ \phi_{\mathcal{R}}{}^* \circ \kappa_{\mathcal{R},\mathcal{R}}(c \otimes_{B^{\mathrm{op}}} c') \\
&= \kappa_{\mathcal{R},\mathcal{R}}^{-1}(r \otimes_B r' \mapsto \kappa_{\mathcal{R},\mathcal{R}}(c \otimes_{B^{\mathrm{op}}} c')(rr' \otimes_B 1_{\mathcal{R}} + 1_{\mathcal{R}} \otimes_B rr' - r \otimes_B r') \\
&= \kappa_{\mathcal{R},\mathcal{R}}^{-1}(r \otimes_B r' \mapsto c'(c(rr')1_{\mathcal{R}}) + c'(c(1_{\mathcal{R}})rr')) - c \otimes_{B^{\mathrm{op}}} c' \\
&= \kappa_{\mathcal{R},\mathcal{R}}^{-1}(r \otimes_B r' \mapsto c'(1_{\mathcal{R}})c(rr') + c'(c(1_{\mathcal{R}})rr')) - c \otimes_{B^{\mathrm{op}}} c' \\
&= \kappa_{\mathcal{R},\mathcal{R}}^{-1}(r \otimes_B r' \mapsto c(rr'))c'(1_{\mathcal{R}}) + c(1_{\mathcal{R}})^{\mathrm{op}}\kappa_{\mathcal{R},\mathcal{R}}^{-1}(r \otimes_B r' \mapsto c'(rr')) - c \otimes_{B^{\mathrm{op}}} c' \\
&= (\Delta^{\mathcal{R}^*} \otimes_{B^{\mathrm{op}}} \varepsilon^{\mathcal{R}^*} + \varepsilon^{\mathcal{R}^*} \otimes_{B^{\mathrm{op}}} \Delta^{\mathcal{R}^*} - I \otimes_{B^{\mathrm{op}}} I)(c \otimes_{B^{\mathrm{op}}} c') \\
&= \psi_{\mathcal{R}^*}(c \otimes_{B^{\mathrm{op}}} c')
\end{aligned}
$$

Similarly, for all $r, r' \in \mathcal{C}^*$, $rr' = \kappa_{\mathcal{C},\mathcal{C}}(r \otimes_{B^{op}} r') \circ \Delta^{\mathcal{C}}$, therefore for all $r, r' \in \mathcal{C}^*$,

$$
\begin{aligned}
\psi_{\mathcal{C}}{}^{\dagger}(r \otimes_{B^{op}} r') &= \kappa_{\mathcal{C},\mathcal{C}}^{-1} \circ \psi_{\mathcal{C}}{}^{*} \circ \kappa_{\mathcal{C},\mathcal{C}}(r \otimes_{B^{op}} r') \\
&= \kappa_{\mathcal{C},\mathcal{C}}^{-1}(\kappa_{\mathcal{C},\mathcal{C}}(r \otimes_{B^{op}} r') \circ (\Delta^{\mathcal{C}} \otimes_B \varepsilon^{\mathcal{C}} + \varepsilon^{\mathcal{C}} \otimes_B \Delta^{\mathcal{C}} - I \otimes_B I)) \\
&= \kappa_{\mathcal{C},\mathcal{C}}^{-1}(c \otimes_B c' \mapsto \kappa_{\mathcal{C},\mathcal{C}}(r \otimes_{B^{op}} r') \circ \Delta^{\mathcal{C}}(c)\varepsilon^{\mathcal{C}}(c') + \kappa_{\mathcal{C},\mathcal{C}}(r \otimes_{B^{op}} r')(\varepsilon^{\mathcal{C}}(c)\Delta^{\mathcal{C}}(c'))) - r \otimes_{B^{op}} r' \\
&= \kappa_{\mathcal{C},\mathcal{C}}^{-1}(c \otimes_B c' \mapsto (rr')(c)\varepsilon^{\mathcal{C}}(c') + (\varepsilon^{\mathcal{C}}(c)^{\mathrm{op}}\kappa_{\mathcal{C},\mathcal{C}}(r \otimes_{B^{op}} r'))(\Delta^{\mathcal{C}}(c'))) - r \otimes_{B^{op}} r' \\
&= \kappa_{\mathcal{C},\mathcal{C}}^{-1}(c \otimes_B c' \mapsto \varepsilon^{\mathcal{C}}((rr')(c)c') + \kappa_{\mathcal{C},\mathcal{C}}(\varepsilon^{\mathcal{C}}(c)^{\mathrm{op}} r \otimes_{B^{op}} r') \circ \Delta^{\mathcal{C}}(c')) - r \otimes_{B^{op}} r' \\
&= \kappa_{\mathcal{C},\mathcal{C}}^{-1}(c \otimes_B c' \mapsto \varepsilon^{\mathcal{C}}((rr')(c)c') + (\varepsilon^{\mathcal{C}}(c)^{\mathrm{op}} rr')(c')) - r \otimes_{B^{op}} r' \\
&= \kappa_{\mathcal{C},\mathcal{C}}^{-1}(c \otimes_B c' \mapsto \varepsilon^{\mathcal{C}}((rr')(c)c') + (rr')(\varepsilon^{\mathcal{C}}(c'))) - r \otimes_{B^{op}} r' \\
&= rr' \otimes_{B^{op}} \varepsilon^{\mathcal{C}} + \varepsilon^{\mathcal{C}} \otimes_{B^{op}} rr' - r \otimes_{B^{op}} r' \\
&= \phi_{\mathcal{C}^*}(r \otimes_{B^{op}} r')
\end{aligned}
$$

This completes the proof. □

Remark 2. Put together the statements of Theorem 3.6, Theorem 3.5, Proposition 3.6 and Proposition 3.7, can be summarized in the following diagram:

$$
\begin{array}{ccc}
\mathbf{r.f.g.pYB\ str}_B & \underset{()^*}{\overset{()^*}{\rightleftarrows}} & \mathbf{r.f.g.pYB\ str}_{B^{op}} \\
F \uparrow & & \uparrow G \\
\mathbf{r.f.g.pRge}_B & \underset{()^*}{\overset{()^*}{\rightleftarrows}} & \mathbf{r.f.g.pCrg}_{B^{op}}.
\end{array}
$$

This means that the duality between right finitely generated projective ring extensions of B and B corings extends to the category $\mathbf{r.f.g.pYB\ str}_B$.

4. Conclusions

We extended the duality between right finitely generated projective ring extensions and right finitely generated projective corings to the category of right finitely generated projective generalized Yang–Baxter structures. This duality and its extension could be seen as a more general construction. For example, at the epistemologic level, the extension of the duality of (co)algebra structures seems to be a model for the relation between interdisciplinarity, pluridisciplinarity and transdisciplinarity (see [4]). It would be interesting to interpret this construction in terms of particle interactions.

The relationships between sub(co)algebras and (co)ideals are well-known, and the term of YB ideal was proposed for the first time in [11]. The following question arises: What are the relationships between sub(co)rings, (co)ideals and generalized Yang–Baxter structures?

We think that there are more connections between the Pontryagin–van Kampen duality and the above extension of the duality of (co)algebra structures.

Acknowledgments

We would like to thank Tomasz Brzeziński for helpful remarks. The first author thanks for a Marie Curie Research Fellowship, HPMF-CT-2002-01782 at Swansea University. The work of BZ was supported by the EPSRC grant GR/S01078/01.

References

1. Morris, S.A. *Pontryagin Duality and the Structure of Locally Compact Abelian Groups*; Cambridge University Press: Cambridge, UK, 1977.
2. Nichita, F.F.; Schack, S.D. The duality between algebras and coalgebras. *Ann. Univ. Ferrara - Sez. VII - Sc. Mat.* **2005**, *51*, 173–181.
3. Brzeziński, T.; Wisbauer, R. *Corings and Comodules*; Cambridge University Press: Cambridge, UK, 2003.
4. Nichita, F.F. Algebraic models for transdisciplinarity. *Transdiscipl. J. Eng. Sci.* **2011**, *10*, 42–46.
5. Abe, E. *Hopf Algebras*; Cambridge University Press: Cambridge, UK, 1977.
6. Dăscălescu, S.; Năstăsescu C.; Raianu, S. *Hopf Algebras: An Introduction*; Marcel Dekker, Inc.: New York, NY, USA, 2000.
7. Sweedler, M.E. *Hopf Algebras*; W.A.Benjamin, Inc.: New York, NY, USA, 1969.
8. Sweedler, M.E. The predual theorem to the Jacobson-Bourbaki theorem. *Trans. Am. Math. Soc.* **1975**, *213*, 391–406.
9. Kassel, C. *Quantum Groups*; Springer Verlag: New York, NY, USA, 1995.
10. Lambe, L.; Radford, D. *Introduction to the Quantum Yang-Baxter Equation and Quantum Groups: An Algebraic Approach*; Kluwer Academic Publishers: Dordrecht, Germany, 1997.
11. Nichita, F.F. Non-Linear Equation, Quantum Groups and Duality Theorems. Ph.D. Thesis, The State University of New York at Buffalo, 2001.
12. Nichita, F.F. Self-inverse Yang-Baxter operators from (co)algebra structures. *J. Algebra* **1999**, *218*, 738–759.

11

Bundles over Quantum Real Weighted Projective Spaces

Tomasz Brzeziński * and Simon A. Fairfax

Department of Mathematics, Swansea University, Singleton Park, Swansea SA2 8PP, UK;
E-Mail: 201102@swansea.ac.uk
* E-Mail: T.Brzezinski@swansea.ac.uk

Abstract: The algebraic approach to bundles in non-commutative geometry and the Definition of quantum real weighted projective spaces are reviewed. Principal $U(1)$-bundles over quantum real weighted projective spaces are constructed. As the spaces in question fall into two separate classes, the *negative* or *odd* class that generalises quantum real projective planes and the *positive* or *even* class that generalises the quantum disc, so do the constructed principal bundles. In the negative case the principal bundle is proven to be non-trivial and associated projective modules are described. In the positive case the principal bundles turn out to be trivial, and so all the associated modules are free. It is also shown that the circle (co)actions on the quantum Seifert manifold that define quantum real weighted projective spaces are almost free.

Keywords: quantum real weighted projective space; principal comodule algebra; noncommutative line bundle

---

1. Introduction

In an algebraic setup an action of a circle on a quantum space corresponds to a coaction of a Hopf algebra of Laurent polynomials in one variable on the noncommutative coordinate algebra of the quantum space. Such a coaction can equivalently be understood as a $\mathbb{Z}$-grading of this coordinate algebra. A typical $\mathbb{Z}$-grading assigns degree ± 1 to every generator of this algebra (different from the identity). The degree zero part forms a subalgebra which in particular cases corresponds to quantum complex or real projective spaces (grading of coordinate algebras of quantum spheres [1] or prolonged quantum spheres [2]). Often this grading is strong, meaning that the product of i, j-graded parts is equal to the $i + j$-part of the total algebra. In geometric terms this reflects the freeness of the circle action.

In two recent papers [3,4] circle actions on three-dimensional (and, briefly, higher dimensional) quantum spaces were revisited. Rather than assigning a uniform grade to each generator, separate generators were given degree by pairwise coprime integers. The zero part of such a grading of the coordinate algebra of the quantum odd-dimensional sphere corresponds to the quantum weighted projective space, while the zero part of such a grading of the algebra of the prolonged even dimensional quantum sphere leads to quantum real weighted projective spaces.

In this paper we focus on two classes of algebras $\mathcal{O}(\mathbb{RP}_q^2(l;-))$ (l a positive integer) and $\mathcal{O}(\mathbb{RP}_q^2(l;+))$ (l an odd positive integer) identified in [3] as fixed points of weighted circle actions on the coordinate algebra $\mathcal{O}(\Sigma_q^3)$ of a non-orientable quantum Seifert manifold described in [2]. Our aim is to construct quantum $U(1)$-principal bundles over the corresponding quantum spaces $\mathbb{RP}_q^2(l;\pm)$ and describe associated line bundles. Recently, the importance of such bundles in non-commutative geometry was once again brought to the fore in [5], where the non-commutative Thom construction was outlined. As a further consequence of the principality of $U(1)$-coactions we also deduce that $\mathbb{RP}_q^2(l;\pm)$ can be understood as quotients of Σ_q^3 by almost free S^1-actions.

We begin in Section 2 by reviewing elements of algebraic approach to classical and quantum bundles. We then proceed to describe algebras $\mathcal{O}(\mathbb{RP}_q^2(l;\pm))$ in Section 3. Section 4 contains main results including construction of principal comodule algebras over $\mathcal{O}(\mathbb{RP}_q^2(l;\pm))$. We observe that constructions albeit very similar in each case yield significantly different results. The principal comodule algebra over $\mathcal{O}(\mathbb{RP}_q^2(l;-))$ is non-trivial while that over $\mathcal{O}(\mathbb{RP}_q^2(l;+))$ turns out to be trivial (this means that all associated bundles are trivial, hence we do not mention them in the text). Whether it is a consequence of our particular construction or there is a deeper (topological or geometric) obstruction to constructing non-trivial principal circle bundles over $\mathbb{RP}_q^2(l;+)$ remains an interesting open question.

Throughout we work with involutive algebras over the field of complex numbers (but the algebraic results remain true for all fields of characteristic 0). All algebras are associative and have identity, we use the standard Hopf algebra notation and terminology and we always assume that the antipode of a Hopf algebra is bijective. All topological spaces are assumed to be Hausdorff.

2. Review of Bundles in Non-Commutative Geometry

The aim of this section is to set out the topological concepts in relation to topological bundles, in particular principal bundles. The classical connection is made for interpreting topological concepts in an algebraic setting, providing a manageable methodology for performing calculations. In particular, the connection between principal bundles in topology and the algebraic Hopf–Galois condition is described. The reader familiar with classical theory of bundles can proceed directly to Definition 8.

2.1. Topological Aspects of Bundles

As a natural starting point, bundles are defined and topological properties are described. The principal map is defined and shown that injectivity is equivalent to the freeness condition. The image of the canonical map is deduced and necessary conditions are imposed to ensure the bijectivity of this map. The detailed account of the material presented in this section can be found in [6].

Definition 1. A *bundle* is a triple (E, π, M) where E and M are topological spaces and $\pi : E \to M$ is a continuous surjective map. Here M is called the *base space*, E the *total space* and π the *projection* of the bundle.

For each $m \in M$, the *fibre* over m is the topological space $\pi^{-1}(m)$, *i.e.*, the points on the total space which are projected, under π, onto the point m in the base space. A bundle whose fibres are homeomorphic which satisfies a condition known as local triviality are known as fibre bundles. This is formally expressed in the next Definition.

Definition 2. A *fibre bundle* is a triple (E, π, M, F) where (E, π, M) is bundle and F is a topological space such that $\pi^{-1}(m)$ are homeomorphic to F for each $m \in M$. Furthermore, π satisfies the local triviality condition.

The local triviality condition is satisfied if for each $x \in E$, there is an open neighbourhood $U \subset B$ such that $\pi^{-1}(U)$ is homeomorphic to the product space $U \times F$, in such a way that π carries over to the projection onto the first factor. That is the following diagram commutes:

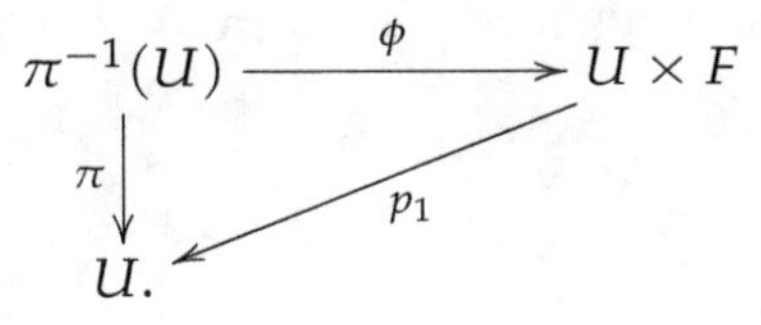

The map p_1 is the natural projection $U \times F \to U$ and $\phi : \pi^{-1}(U) \to U \times F$ is a homeomorphism.

Example 1. An Example of a fibre bundle which is non-trivial, *i.e.*, not a global product space, is the Möbius strip. It has a circle that runs lengthwise through the centre of the strip as a base B and a line segment running vertically for the fibre F. The line segments are in fact copies of the real line, hence each $\pi^{-1}(m)$ is homeomorphic to $\mathbb{R}$ and the Mobius strip is a fibre bundle.

Let X be a topological space which is compact and satisfies the Hausdorff property and G a compact topological group. Suppose there is a right action $\triangleleft : X \times G \to X$ of G on X and write $x \triangleleft g = xg$.

Definition 3. An action of G on X is said to be *free* if $xg = x$ for any $x \in X$ implies that $g = e$, the group identity.

With an eye on algebraic formulation of freeness, the *principal map* $F^G : X \times G \to X \times X$ is defined as $(x, g) \mapsto (x, xg)$.

Proposition 1. *G acts freely on X if and only if F^G is injective.*

Proof. "$\Longleftarrow$" Suppose the action is free, hence $xg = x$ implies that $g = e$. If $(x, xg) = (x', x'g')$, then $x = x'$ and $xg = xg'$. Applying the action of g'^{-1} to both sides of $xg = xg'$ we get $x(gg'^{-1}) = x$, which implies $gg'^{-1} = e$ by the freeness property, concluding $g = g'$ and F^G is injective as required.

"$\Longrightarrow$" Suppose F^G is injective, so $F^G(x, g) = F^G(x', g')$ or $(x, xg) = (x', x'g')$ implies $x = x'$ and $g = g'$. Since $x = xe$ from the properties of the action, if $x = xg$ then $g = e$ from the injectivity property. □

Since G acts on X we can define the quotient space X/G,

$$Y = X/G := \{[x] : x \in X\}, \qquad \text{where} \qquad [x] = xG = \{xg : g \in G\}$$

The sets xG are called the *orbits* of the points x. They are defined as the set of elements in X to which x can be moved by the action of elements of G. The set of orbits of X under the action of G forms a partition of X, hence we can define the equivalence relation on X as,

$$x \sim y \Longleftrightarrow \exists g \in G \text{ such that } xg = y$$

The equivalence relation is the same as saying x and y are in the same orbit, *i.e.*, $xG = yG$. Given any quotient space, then there is a canonical surjective map

$$\pi : X \to Y = X/G, \qquad x \mapsto xG = [x]$$

which maps elements in X to their orbits. We define the pull-back along this map π to be the set

$$X \times_Y X := \{(x, y) \in X \times X : \pi(x) = \pi(y)\}$$

As described above, the image of the principal map F^G contains elements of X in the first leg and the action of $g \in G$ on x in the second leg. To put it another way, the image records elements of $x \in X$ in the first leg and all the elements in the same orbit as this x in the second leg. Hence we can identify the image of the canonical map as the pull back along π, namely $X \times_Y X$. This is formally proved as a part of the following Proposition.

Proposition 2. *G acts freely on X if and only if the map*

$$F_X^G : X \times G \to X \times_Y X, \qquad (x,g) \mapsto (x,xg)$$

is bijective.

Proof. First note that the map F_X^G is well-defined since the elements x and xg are in the same orbit and hence map to the same equivalence class under π. Using Proposition 1 we can deduce that the injectivity of F_X^G is equivalent to the freeness of the action. Hence if we can show that F_X^G is surjective the proof is complete.

Take $(x,y) \in X \times_Y X$. This means $\pi(x) = \pi(y)$, which implies x and y are in the same equivalence class, which in turn means they are in the same orbit. We can therefore deduce that $y = xg$ for some $g \in G$. So, $(x,y) = (x,xg) = F_X^G(x,g)$ implying $(x,y) \in \mathrm{Im}F_X^G$. Hence $\mathrm{Im}F_X^G = X \times_Y X$ completing the proof. □

Definition 4. An action of G on X is said to be *principal* if the map F^G is both injective and continuous (and such that the inverse image of a compact subset is compact in a case of locally compact spaces).

Since the injectivity and freeness condition are equivalent, we can interpret principal actions as both free and continuous actions. We can also deduce that these types of actions give rise to homeomorphisms F_X^G from $X \times G$ onto the space $X \times_{X/G} X$. Principal actions lead to the concept of topological principle bundles.

Definition 5. A *principal bundle* is a quadruple (X,π,M,G) such that

(a) (X,π,M) is a bundle and G is a topological group acting continuously on X with action $\triangleleft : X \times G \to X$, $x \triangleleft g = xg$;

(b) the action $\triangleleft$ is principal;

(c) $\pi(x) = \pi(y) \iff \exists g \in G$ such that $y = xg$;

(d) the induced map $X/G \to M$ is a homeomorphism.

The first two properties tell us that principal bundles are bundles admitting a principal action of a group G on the total space X, *i.e.*, principal bundles correspond to principal actions. By Definition (4), principal actions occur when the principal map is both injective and continuous, or equivalently, when the action is free and continuous. The third property ensures that the fibres of the bundle correspond to the orbits coming from the action and the final property implies that the quotient space can topologically be viewed as the base space of the bundle.

Example 2. Suppose X is a topological space and G a topological group which acts on X from the right. The triple $(X,\pi,X/G)$ where X/G is the orbit space and π the natural projection is a bundle. A principal action of G on X makes the quadruple $(X,\pi,X/G,G)$ a principal bundle.

We describe a principal bundle (X,π,Y,G) as a G-principal bundle over (X,π,Y), or X as a G-principal bundle over Y.

Definition 6. A *vector bundle* is a bundle (E,π,M) where each fibre $\pi^{-1}(m)$ is endowed with a vector space structure such that addition and scalar multiplication are continuous maps.

Any vector bundle can be understood as a bundle associated to a principal bundle in the following way. Consider a G-principal bundle (X,π,Y,G) and let V be a representation space of G, *i.e.*, a (topological) vector space with a (continuous) left G-action $\triangleright : G \times V \to V$, $(g, v) \mapsto g \triangleright v$. Then G acts from the right on $X \times V$ by

$$(x, v) \triangleleft g := (xg, g^{-1} \triangleright v), \quad \text{for all } x \in X, v \in V \text{ and } g \in G$$

We can define $E = (X \times V)/G$ and a surjective (continuous map) $\pi_E : E \to Y$, $(x, v) \triangleleft G \mapsto \pi(x)$ and thus have a fibre bundle (E, π_E, Y, V). In the case where V is a vector space, we assume that G acts linearly on V.

Definition 7. A *section* of a bundle (E, π_E, Y) is a continuous map $s : Y \to E$ such that, for all $y \in Y$,

$$\pi_E(s(y)) = y$$

i.e., a section is simply a section of the morphism π_E. The set of sections of E is denoted by $\Gamma(E)$.

Proposition 3. *Sections in a fibre bundle (E, π_E, Y, V) associated to a principal G-bundle X are in bijective correspondence with (continuous) maps $f : X \to V$ such that*

$$f(xg) = g^{-1} \triangleright f(x)$$

All such G-equivariant maps are denoted by $\mathrm{Hom}_G(X, V)$.

Proof. Remember that $Y = X/G$. Given a map $f \in \mathrm{Hom}_G(X, V)$, define the section $s_f : Y \to E$, $xG \mapsto (x, f(x)) \triangleleft G$.

Conversely, given $s \in \Gamma(E)$, define $f_s : X \to V$ by assigning to $x \in X$ a unique $v \in V$ such that $s(xG) = (x, v) \triangleleft G$. Note that v is unique, since if $(x, w) = (x, v) \triangleleft g$, then $xg = x$ and $w = g^{-1} \triangleright v$. Freeness implies that $g = e$, hence $w = v$. The map f_s has the required equivariance property, since the element of $(X \times V)/G$ corresponding to xg is $g^{-1} \triangleright v$. □

2.2. Non-Commutative Principal and Associated Bundles

To make the transition from algebraic formulation of principal and associated bundles to non-commutative setup more transparent, we assume that X is a complex affine variety with an action of an affine algebraic group G and set $Y = X/G$ (all with the usual Euclidean topology). Let $\mathcal{O}(X)$, $\mathcal{O}(Y)$ and $\mathcal{O}(G)$ be the corresponding coordinate rings. Put $A = \mathcal{O}(X)$ and $H = \mathcal{O}(G)$ and note the identification $\mathcal{O}(G \times G) \cong \mathcal{O}(G) \otimes \mathcal{O}(G)$. Through this identification, $\mathcal{O}(G \times G)$ is a Hopf algebra with comultiplication: $f \mapsto (\Delta f)$, $(\Delta f)(g, h) = f(gh)$, counit $\varepsilon : \mathcal{O}(G) \to \mathbb{C}$, $\varepsilon(f) = f(e)$, and the antipode $S : H \to H$, $(Sf)(g) = f(g^{-1})$.

Using the fact that G acts on X we can construct a right coaction of H on A by $\varrho^A : A \to A \otimes H$, $\varrho^A(a)(x, g) = a(xg)$. This coaction is an algebra map due to the commutativity of the algebras of functions involved.

We have viewed the spaces of polynomial functions on X and G, next we view the space of functions on Y, $B := \mathcal{O}(Y)$, where $Y = X/G$. B is a subalgebra of A by

$$\pi^* : B \to A, \qquad b \mapsto b \circ \pi$$

where π is the canonical surjection defined above. The map π^* is injective, since $b \neq b'$ in $\mathcal{O}(X/G)$ means there exists at least one orbit $xG = [x]$ such that $b([x]) \neq b'([x])$, but $\pi(x) = [x]$, so $b(\pi(x)) \neq b'(\pi(x))$ which implies $\pi^*(b) \neq \pi^*(b')$. Therefore, we can identify B with $\pi^*(B)$. Furthermore, $a \in \pi^*(B)$ if and only if

$$a(xg) = a(x)$$

for all $x \in X, g \in G$. This is the same as

$$\varrho^A(a)(x,g) = (a \otimes 1)(x,g)$$

for all $x \in X$, $g \in G$, where $1 : G \to \mathbb{C}$ is the unit function $1(g) = 1$ (the identity element of H). Thus we can identify B with the *coinvariants* of the coaction ϱ^A:

$$B = A^{coH} := \{a \in A \mid \varrho^A(a) = a \otimes 1\}$$

Since B is a subalgebra of A, it acts on A via the inclusion map $(ab)(x) = a(x)b(\pi(x))$, $(ba)(x) = b(\pi(x))a(x)$. We can identify $\mathcal{O}(X \times_Y X)$ with $\mathcal{O}(X) \otimes_{\mathcal{O}(Y)} \mathcal{O}(X) = A \otimes_B A$ by the map

$$\theta(a \otimes_B a')(x,y) = a(x)a'(y), \quad (\text{with } \pi(x) = \pi(y))$$

Note that θ is well defined because $\pi(x) = \pi(y)$. Proposition 2 immediately yields

Proposition 4. *The action of G on X is free if and only if $F_X^{G*} : \mathcal{O}(X \times_Y X) \to \mathcal{O}(X \times G)$, $f \mapsto f \circ F_X^G$ is bijective.*

In view of the Definition of the coaction of H on A, we can identify $F_X^{G^*}$ with the *canonical map*

$$\mathsf{can} : a \otimes_B a' \mapsto [(x,g) \mapsto a(x)a'(x.g)] = a\varrho^A(a')$$

Thus the action of G on X is free if and only if this purely algebraic map is bijective. In the classical geometry case we take $A = \mathcal{O}(X)$, $H = \mathcal{O}(G)$ and $B = \mathcal{O}(X/G)$, but in general there is no need to restrict oneself to commutative algebras (of functions on topological spaces). In full generality this leads to the following Definition.

Definition 8. (Hopf–Galois Extensions) Let H be a Hopf algebra and A a right H-comodule algebra with coaction $\varrho^A : A \to A \otimes H$. Let $B = A^{coH} := \{b \in A \mid \varrho^A(b) = b \otimes 1\}$, the coinvariant subalgebra of A. We say that $B \subseteq A$ is a *Hopf–Galois extension* if the left A-module, right H-comodule map

$$\mathsf{can} : A \otimes_B A \to A \otimes H, \qquad a \otimes_B a' \mapsto a\varrho^A(a')$$

is an isomorphism.

Proposition 4 tells us that when viewing bundles from an algebraic perspective, the freeness condition is equivalent to the Hopf–Galois extension property. Hence, the Hopf–Galois extension condition is a necessary condition to ensure a bundle is principal. Not all information about a topological space is encoded in a coordinate algebra, so to make a fuller reflection of the richness of the classical notion of a principal bundle we need to require conditions additional to the Hopf–Galois property.

Definition 9. Let H be a Hopf algebra with bijective antipode and let A be a right H-comodule algebra with coaction $\varrho^A : A \to A \otimes H$. Let B denote the coinvariant subalgebra of A. We say that A is a *principal H-comodule algebra* if:

(a) $B \subseteq A$ is a Hopf–Galois extension;

(b) the multiplication map $B \otimes A \to A$, $b \otimes a \mapsto ba$, splits as a left B-module and right H-comodule map (the equivariant projectivity condition).

As indicated already in [7–9], principal comodule algebras should be understood as principal bundles in noncommutative geometry. In particular, if H is the Hopf algebra associated to a C^*-algebra of functions on a quantum group [10], then the existence of the Haar measure together with the results of [8] mean that condition (a) in Definition 9 implies condition (b) (*i.e.*, the freeness of the coaction implies its principality).

The following characterisation of principal comodule algebras [11,12] gives an effective method for proving the principality of coaction.

Proposition 5. *A right H-comodule algebra A with coaction $\varrho^A : A \to A \otimes H$ is principal if and only if it admits a* strong connection form, *that is if there exists a map $\omega : H \longrightarrow A \otimes A$, such that*

$$\omega(1) = 1 \otimes 1 \tag{1a}$$

$$\mu \circ \omega = \eta \circ \varepsilon \tag{1b}$$

$$(\omega \otimes \mathrm{id}) \circ \Delta = (\mathrm{id} \otimes \varrho) \circ \omega \tag{1c}$$

$$(S \otimes \omega) \circ \Delta = (\sigma \otimes \mathrm{id}) \circ (\varrho \otimes \mathrm{id}) \circ \omega \tag{1d}$$

Here $\mu : A \otimes A \to A$ denotes the multiplication map, $\eta : \mathbb{C} \to A$ is the unit map, $\Delta : H \to H \otimes H$ is the comultiplication, $\varepsilon : H \to \mathbb{C}$ counit and $S : H \to H$ the (bijective) antipode of the Hopf algebra H, and $\sigma : A \otimes H \to H \otimes A$ is the flip.

Proof. If a strong connection form ω exists, then the inverse of the canonical map can (see Definition 8) is the composite

$$A \otimes H \xrightarrow{\mathrm{id} \otimes \omega} A \otimes A \otimes A \xrightarrow{\mu \otimes \mathrm{id}} A \otimes A \longrightarrow A \otimes_B A$$

while the splitting of the multiplication map (see Definition 9 (b)) is given by

$$A \xrightarrow{\varrho^A} A \otimes H \xrightarrow{\mathrm{id} \otimes \omega} A \otimes A \otimes A \xrightarrow{\mu \otimes \mathrm{id}} B \otimes A$$

Conversely, if $B \subseteq A$ is a principal comodule algebra, then ω is the composite

$$H \xrightarrow{\eta \otimes \mathrm{id}} A \otimes H \xrightarrow{\mathsf{can}^{-1}} A \otimes_B A \xrightarrow{\mathrm{id} \otimes s} A \otimes_B B \otimes A \xrightarrow{\cong} A \otimes A$$

where s is the left B-linear right H-colinear splitting of the multiplication $B \otimes A \to A$. □

Example 3. Let A be a right H-comodule algebra. The space of $\mathbb{C}$-linear maps $\mathrm{Hom}(H, A)$ is an algebra with the *convolution product*

$$f \otimes g \mapsto \mu \circ (f \otimes g) \circ \Delta$$

and unit $\eta \circ \varepsilon$. A is said to be *cleft* if there exists a right H-colinear map $j : H \to A$ that has an inverse in the convolution algebra $\mathrm{Hom}(H, A)$ and is normalised so that $j(1) = 1$. Writing j^{-1} for the convolution inverse of j, one easily observes that

$$\omega : H \to A \otimes A, \qquad h \mapsto (j^{-1} \otimes j)(\Delta(h))$$

is a strong connection form. Hence a cleft comodule algebra is an Example of a principal comodule algebra. The map j is called a *cleaving map* or a *normalised total integral*.

In particular, if $j : H \to A$ is an H-colinear algebra map, then it is automatically convolution invertible (as $j^{-1} = j \circ S$) and normalised. A comodule algebra A admitting such a map is termed a *trivial* principal comodule algebra.

Example 4. Let H be a Hopf algebra of the compact quantum group. By the Woronowicz Theorem [10], H admits an invariant Haar measure, *i.e.*, a linear map $\Lambda : H \to \mathbb{C}$ such that, for all $h \in H$,

$$\sum h_{(1)}\Lambda(h_{(2)}) = \varepsilon(h), \qquad \Lambda(1) = 1$$

where $\Delta(h) = \sum h_{(1)} \otimes h_{(2)}$ is the Sweedler notation for the comultiplication. Next, assume that the lifted canonical map:

$$\overline{\mathrm{can}} : A \otimes A \to A \otimes H, \qquad a \otimes a' \mapsto a\varrho(a') \tag{2}$$

is surjective, and write

$$\ell : H \to A \otimes A, \qquad \ell(h) = \sum \ell(h)^{[1]} \otimes \ell(h)^{[2]}$$

for the $\mathbb{C}$-linear map such that $\overline{\mathrm{can}}(\ell(h)) = 1 \otimes h$, for all $h \in H$. Then, by the Schneider Theorem [8], A is a principal H-comodule algebra. Explicitly, a strong connection form is

$$\omega(h) = \sum \Lambda\left(h_{(1)}\ell\left(h_{(2)}\right)^{[1]}{}_{(1)}\right)\Lambda\left(\ell\left(h_{(2)}\right)^{[2]}{}_{(1)}S\left(h_{(3)}\right)\right)\ell(h_{(2)})^{[1]}{}_{(0)} \otimes \ell(h_{(2)})^{[2]}{}_{(0)}$$

where the coaction is denoted by the Sweedler notation $\varrho^A(a) = \sum a_{(0)} \otimes a_{(1)}$; see [13].

Having described non-commutative principal bundles, we can look at the associated vector bundles. First we look at the classical case and try to understand it purely algebraically. Start with a vector bundle (E, π_E, Y, V) associated to a principal G-bundle X. Since V is a vector representation space of G, also the set $\mathrm{Hom}_G(X, V)$ is a vector space. Consequently $\Gamma(E)$ is a vector space. Furthermore, $\mathrm{Hom}_G(X, V)$ is a left module of $B = \mathcal{O}(Y)$ with the action $(bf)(x) = b(\pi_E(x))f(x)$. To understand better the way in which B-module $\Gamma(E)$ is associated to the principal comodule algebra $\mathcal{O}(X)$ we recall the notion of the cotensor product.

Definition 10. Given a Hopf algebra H, right H-comodule A with coaction ϱ^A and left H-comodule V with coaction ${}^V\varrho$, the *cotensor product* is defined as an equaliser:

$$A\Box_H V \longrightarrow A \otimes V \underset{\mathrm{id}\otimes {}^V\varrho}{\overset{\varrho^A\otimes\mathrm{id}}{\rightrightarrows}} A \otimes H \otimes V$$

If A is an H-comodule algebra, and $B = A^{coH}$, then $A\Box_H V$ is a left B-module with the action $b(a\Box v) = ba\Box v$. In particular, in the case of a principal G-bundle X over $Y = X/G$, for any left $\mathcal{O}(G)$-comodule V the cotensor product $\mathcal{O}(X)\Box_{\mathcal{O}(G)}V$ is a left $\mathcal{O}(Y)$-module.

The following Proposition indicates the way in which cotensor products enter description of associated bundles.

Proposition 6. *Assume that the fibre V of a vector bundle (E, π_E, Y, V) associated to a principal G-bundle X is finite dimensional. View V as a left comodule of $\mathcal{O}(G)$ with the coaction ${}^V\varrho : v \mapsto \sum v_{(-1)} \otimes v_{(0)}$ (summation implicit) determined by $\sum v_{(-1)}(g)v_{(0)} = g^{-1} \triangleright v$. Then the left $\mathcal{O}(Y)$-module of sections $\Gamma(E)$ is isomorphic to the left $\mathcal{O}(Y)$-module $\mathcal{O}(X)\Box_{\mathcal{O}(G)}V$.*

Proof. First identify $\Gamma(E)$ with $\mathrm{Hom}_G(X, V)$. Let $\{v_i \in V^*,\ v^i \in V\}$ be a (finite) dual basis. Take $f \in \mathrm{Hom}_G(X, V)$, and define $\theta : \mathrm{Hom}_G(X, V) \to \mathcal{O}(X)\Box_{\mathcal{O}(G)}V$ by $\theta(f) = \sum_i v_i \circ f \otimes v^i$.

In the converse direction, define a left $\mathcal{O}(Y)$-module map

$$\theta^{-1} : \mathcal{O}(X)\Box_{\mathcal{O}(G)}V \to \mathrm{Hom}_G(X, V), \qquad a\Box v \mapsto a(-)v$$

One easily checks that the constructed map are mutual inverses. □

Moving away from commutative algebras of functions on topological spaces one uses Proposition 6 as the motivation for the following Definition.

Definition 11. Let A be a principal H-comodule algebra. Set $B = A^{coH}$ and let V be a left H-comodule. The left B-module $\Gamma = A\Box_H V$ is called a *module associated to the principal comodule algebra A*.

Γ is a projective left B-module, and if V is a finite dimensional vector space, then Γ is a finitely generated projective left B-module. In this case it has the meaning of a module of sections over a non-commutative vector bundle. Furthermore, its class gives an element in the K_0-group of B. If A is a cleft principal comodule algebra, then every associated module is free, since $A \cong B \otimes H$ as a left B-module and right H-comodule, so that

$$\Gamma = A\Box_H V \cong (B \otimes H)\Box_H V \cong B \otimes (H\Box_H V) \cong B \otimes V$$

3. Weighted Circle Actions on Prolonged Spheres.

In this section we recall the Definitions of algebras we study in the sequel.

3.1. Circle Actions and $\mathbb{Z}$-Gradings.

The coordinate algebra of the circle or the group $U(1)$, $\mathcal{O}(S^1) = \mathcal{O}(U(1))$ can be identified with the $*$-algebra $\mathbb{C}[u, u^*]$ of Laurent polynomials in a unitary variable u (unitary means $u^{-1} = u^*$). As a Hopf $*$-algebra $\mathbb{C}[u, u^*]$, is generated by the grouplike element u, i.e.,

$$\Delta(u) = u \otimes u, \qquad \varepsilon(u) = 1, \qquad S(u) = u^*$$

and thus it can be understood as the group algebra $\mathbb{C}\mathbb{Z}$. As a consequence of this interpretation of $\mathbb{C}[u, u^*]$, an algebra A is a $\mathbb{C}[u, u^*]$-comodule algebra if and only if A is a $\mathbb{Z}$-graded algebra,

$$A = \bigoplus_{n\in\mathbb{Z}} A_n, \qquad A_n := \{a \in A \mid \varrho^A(a) = a \otimes u^n\}, \qquad A_m A_n \subseteq A_{m+n}$$

A_0 is the coinvariant subalgebra of A. Since $\mathbb{C}[u, u^*]$ is spanned by grouplike elements, any convolution invertible map $j : \mathbb{C}[u, u^*] \to A$ must assign a unit (invertible element) of A to u^n. Furthermore, colinear maps are simply the $\mathbb{Z}$-degree preserving maps, where $\deg(u) = 1$. Put together, convolution invertible colinear maps $j : \mathbb{C}[u, u^*] \to A$ are in one-to-one correspondence with sequences

$$(a_n \ : \ n \in \mathbb{Z},\ a_n \text{ is a unit in } A,\ \deg(a_n) = n)$$

3.2. The $\mathcal{O}(\Sigma_q^{2n+1})$ and $\mathcal{O}(\mathbb{RP}_q(l_0, ..., l_n))$ Coordinate Algebras

Let q be a real number, $0 < q < 1$. The coordinate algebra $\mathcal{O}(S_q^{2n})$ of the even-dimensional quantum sphere is the unital complex $*$-algebra with generators $z_0, z_1, \ldots, z_n$, subject to the following relations:

$$z_i z_j = q z_j z_i \quad \text{for } i < j, \qquad z_i z_j^* = q z_j^* z_i \quad \text{for } i \neq j \tag{3a}$$

$$z_i z_i^* = z_i^* z_i + (q^{-2} - 1) \sum_{m=i+1}^{n} z_m z_m^*, \qquad \sum_{m=0}^{n} z_m z_m^* = 1, \qquad z_n^* = z_n \tag{3b}$$

$\mathcal{O}(S_q^{2n})$ is a $\mathbb{Z}_2$-graded algebra with $\deg(z_i) = 1$ and so is $\mathbb{C}[u, u^*]$ (with $\deg(u) = 1$). In other words, $\mathcal{O}(S_q^{2n})$ is a right $\mathbb{C}\mathbb{Z}_2$-comodule algebra and $\mathbb{C}[u, u^*]$ is a left $\mathbb{C}\mathbb{Z}_2$-comodule algebra, hence one can consider the cotensor product algebra $\mathcal{O}(\Sigma_q^{2n+1}) := \mathcal{O}(S_q^{2n})\Box_{\mathbb{C}\mathbb{Z}_2}\mathbb{C}[u, u^*]$. It was shown in [2] that, as a unital $*$-algebra, $\mathcal{O}(\Sigma_q^{2n+1})$ has generators $\zeta_0, ..., \zeta_n$ and a central unitary ξ which are related in the following way:

$$\zeta_i \zeta_j = q \zeta_j \zeta_i \quad \text{for } i < j, \qquad \zeta_i \zeta_j^* = q \zeta_j^* \zeta_i \quad \text{for } i \neq j \tag{4a}$$

$$\zeta_i\zeta_i^* = \zeta_i^*\zeta_i + (q^{-2}-1)\sum_{m=i+1}^{n}\zeta_m\zeta_m^*, \qquad \sum_{m=0}^{n}\zeta_m\zeta_m^* = 1, \qquad \zeta_n^* = \zeta_n\xi \tag{4b}$$

For any choice of $n+1$ pairwise coprime numbers $l_0,...,l_n$ one can define the coaction of the Hopf algebra $\mathcal{O}(U(1)) = \mathbb{C}[u,u^*]$ on $\mathcal{O}(\Sigma_q^{2n+1})$ as

$$\varrho_{l_0,...,l_n} : \mathcal{O}(\Sigma_q^{2n+1}) \to \mathcal{O}(\Sigma_q^{2n+1}) \otimes \mathbb{C}[u,u^*], \qquad \zeta_i \mapsto \zeta_i \otimes u^{l_i}, \qquad \xi \mapsto \xi \otimes u^{-2l_n} \tag{5}$$

for $i = 0,1,...,n$. This coaction is then extended to the whole of $\mathcal{O}(\Sigma_q^{2n+1})$ so that $\mathcal{O}(\Sigma_q^{2n+1})$ is a right $\mathbb{C}[u,u^*]$-comodule algebra.

The algebra of coordinate functions on the quantum real weighted projective space is now defined as the subalgebra of $\mathcal{O}(\Sigma_q^{2n+1})$ containing all coinvariant elements, *i.e.*,

$$\mathcal{O}(\mathbb{RP}_q(l_0,...,l_n)) = \mathcal{O}(\Sigma_q^{2n+1})^{\mathcal{O}(U(1))} := \{x \in \mathcal{O}(\Sigma_q^{2n+1}) : \varrho_{l_0,...,l_n}(x) = x \otimes 1\}$$

3.3. The 2D Quantum Real Projective Space $\mathcal{O}(\mathbb{RP}_q(k,l)) \subset \mathcal{O}(\Sigma_q^3)$

In this paper we consider two-dimensional quantum real weighted projective spaces, *i.e.*, the algebras obtained from the coordinate algebra $\mathcal{O}(\Sigma_q^3)$ which is generated by ζ_0, ζ_1 and central unitary ξ such that

$$\zeta_0\zeta_1 = q\zeta_1\zeta_0, \quad \zeta_0\zeta_1^* = q\zeta_1^*\zeta_0 \tag{6a}$$

$$\zeta_0\zeta_0^* = \zeta_0^*\zeta_0 + (q^{-2}-1)\zeta_1^2\xi, \qquad \zeta_0\zeta_0^* + \zeta_1^2\xi = 1, \qquad \zeta_1^* = \zeta_1\xi \tag{6b}$$

The linear basis of $\mathcal{O}(\Sigma_q^3)$ is

$$\{\zeta_0^r\zeta_1^s\xi^t,\ \zeta_0^{*r}\zeta_1^s\xi^t,\ |\ r,s,\in\mathbb{N},\ t\in\mathbb{Z}\} \tag{7}$$

For a pair k,l of coprime positive integers, the coaction $\varrho_{k,l}$ is given on generators by

$$\zeta_0 \mapsto \zeta_0 \otimes u^k, \qquad \zeta_1 \mapsto \zeta_1 \otimes u^l, \qquad \xi \mapsto \xi \otimes u^{-2l} \tag{8}$$

and extended to the whole of $\mathcal{O}(\Sigma_q^3)$ so that the coaction is a $*$-algebra map. We denote the comodule algebra $\mathcal{O}(\Sigma_q^3)$ with coaction $\varrho_{k,l}$ by $\mathcal{O}(\Sigma_q^3(k,l))$.

It turns out that the two dimensional quantum real projective spaces split into two cases depending on not wholly the parameter k but instead whether k is either even or odd, and hence only cases $k=1$ and $k=2$ need to be considered [3]. We describe these cases presently.

3.3.1. . The Odd or Negative Case

For $k=1$, $\mathcal{O}(\mathbb{RP}_q^2(l;-))$ is a polynomial $*$-algebra generated by a, b, c_- which satisfy the relations:

$$a = a^*, \qquad ab = q^{-2l}ba, \qquad ac_- = q^{-4l}c_-a, \qquad b^2 = q^{3l}ac_-, \qquad bc_- = q^{-2l}c_-b \tag{9a}$$

$$bb^* = q^{2l}a\prod_{m=0}^{l-1}(1-q^{2m}a), \qquad b^*b = a\prod_{m=1}^{l}(1-q^{-2m}a) \tag{9b}$$

$$b^*c_- = q^{-l}\prod_{m=1}^{l}(1-q^{-2m}a)b, \qquad c_-b^* = q^lb\prod_{m=0}^{l-1}(1-q^{2m}a) \tag{9c}$$

$$c_-c_-^* = \prod_{m=0}^{2l-1}(1-q^{2m}a), \qquad c_-^*c_- = \prod_{m=1}^{2l}(1-q^{-2m}a) \tag{9d}$$

The embedding of generators of $\mathcal{O}(\mathbb{RP}_q^2(l;-))$ into $\mathcal{O}(\Sigma_q^3)$ or the isomorphism of $\mathcal{O}(\mathbb{RP}_q^2(l;-))$ with the coinvariants of $\mathcal{O}(\Sigma_q^3(1,l))$ is provided by

$$a \mapsto \zeta_1^2\xi, \qquad b \mapsto \zeta_0^l\zeta_1\xi, \qquad c_- \mapsto \zeta_0^{2l}\xi \tag{10}$$

Up to equivalence $\mathcal{O}(\mathbb{RP}_q^2(l;-))$ has the following irreducible $*$-representations. There is a family of one-dimensional representations labelled by $\theta \in [0,1)$ and given by

$$\pi_\theta(a) = 0, \qquad \pi_\theta(b) = 0, \qquad \pi_\theta(c_-) = e^{2\pi i\theta} \tag{11}$$

All other representations are infinite dimensional, labelled by $r = 1, \ldots, l$, and given by

$$\pi_r(a)e_n^r = q^{2(ln+r)}e_n^r, \quad \pi_r(b)e_n^r = q^{ln+r}\prod_{m=1}^{l}\left(1 - q^{2(ln+r-m)}\right)^{1/2} e_{n-1}^r, \quad \pi_r(b)e_0^r = 0 \tag{12a}$$

$$\pi_r(c_-)e_n^r = \prod_{m=1}^{2l}\left(1 - q^{2(ln+r-m)}\right)^{1/2} e_{n-2}^r, \qquad \pi_r(c_-)e_0^r = \pi_r(c_-)e_1^r = 0 \tag{12b}$$

where e_n^r, $n \in \mathbb{N}$, is an orthonormal basis for the representation space $\mathcal{H}_r \cong l^2(\mathbb{N})$.

The C*-algebra of continuous functions on $\mathbb{RP}_q^2(l;-)$, obtained as the completion of these bounded representations, can be identified with the pullback of l-copies of the quantum real projective plane $\mathbb{RP}_q^2$ introduced in [14].

3.3.2. . The Even or Positive Case

For $k = 2$ and hence l odd, $\mathcal{O}(\mathbb{RP}_q^2(l;+))$ is a polynomial $*$-algebra generated by a, c_+ which satisfy the relations:

$$a^* = a, \qquad ac_+ = q^{-2l}c_+a \tag{13a}$$

$$c_+c_+^* = \prod_{m=0}^{l-1}(1 - q^{2m}a), \qquad c_+^*c_+ = \prod_{m=1}^{l}(1 - q^{-2m}a) \tag{13b}$$

The embedding of generators of $\mathcal{O}(\mathbb{RP}_q^2(l;+))$ into $\mathcal{O}(\Sigma_q^3)$ or the isomorphism of $\mathcal{O}(\mathbb{RP}_q^2(l;+))$ with the coinvariants of $\mathcal{O}(\Sigma_q^3(2,l))$ is provided by

$$a \mapsto \zeta_1^2\xi, \qquad c_+ \mapsto \zeta_0^l\xi \tag{14}$$

Similarly to the odd k case, there is a family of one-dimensional representations of $\mathcal{O}(\mathbb{RP}_q^2(l;+))$ labelled by $\theta \in [0,1)$ and given by

$$\pi_\theta(a) = 0, \qquad \pi_\theta(c_+) = e^{2\pi i\theta} \tag{15}$$

All other representations are infinite dimensional, labelled by $r = 1, \ldots, l$, and given by

$$\pi_r(a)e_n^r = q^{2(ln+r)}e_n^r, \quad \pi_r(c_+)e_n^r = \prod_{m=1}^{l}\left(1 - q^{2(ln+r-m)}\right)^{1/2} e_{n-1}^r, \quad \pi_r(c_+)e_0^r = 0 \tag{16}$$

where e_n^r, $n \in \mathbb{N}$ is an orthonormal basis for the representation space $\mathcal{H}_r \cong l^2(\mathbb{N})$.

The C*-algebra $C(\mathbb{RP}_q^2(l;+))$ of continuous functions on $\mathbb{RP}_q^2(l;+)$, obtained as the completion of these bounded representations, can be identified with the pullback of l-copies of the quantum disk D_q introduced in [15]. Furthermore, $C(\mathbb{RP}_q^2(l;+))$ can also be understood as the quantum double suspension of l points in the sense of [16, Definition 6.1].

4. Quantum Real Weighted Projective Spaces and Quantum Principal Bundles

The general aim of this paper is to construct quantum principal bundles with base spaces given by $\mathcal{O}(\mathbb{RP}_q^2(l;\pm))$ and fibre structures given by the circle Hopf algebra $\mathcal{O}(S^1) \cong \mathbb{C}[u,u^*]$. The question arises as to which quantum space (*i.e.*, a $\mathbb{C}[u,u^*]$-comodule algebra with coinvariants isomorphic to

$\mathcal{O}(\mathbb{RP}^2_q(l;\pm)))$ we should consider as the total space within this construction. We look first at the coactions of $\mathbb{C}[u,u^*]$ on $\mathcal{O}(\Sigma^3_q)$ that define $\mathcal{O}(\mathbb{RP}_q(k,l))$, *i.e.*, at the comodule algebras $\mathcal{O}(\Sigma^3_q(k,l))$.

4.1. The (Non-)Principality of $\mathcal{O}(\Sigma^3_q(k,l))$

Theorem 1. *$A = \mathcal{O}(\Sigma^3_q(k,l))$ is a principal comodule algebra if and only if $(k,l) = (1,1)$.*

Proof. As explained in [2] $\mathcal{O}(\Sigma^3_q(1,1))$ is a prolongation of the $\mathbb{C}\mathbb{Z}_2$-comodule algebra $\mathcal{O}(S^2_q)$. The latter is a principal comodule algebra (over the quantum real projective plane $\mathcal{O}(\mathbb{RP}^2_q)$ [14]) and since a prolongation of a principal comodule algebra is a principal comodule algebra [8, Remark 3.11], the coaction $\varrho_{1,1}$ is principal as stated.

In the converse direction, we aim to show that the canonical map is not an isomorphism by showing that the image does not contain $1 \otimes u$, *i.e.*, it cannot be surjective since we know $1 \otimes u$ is in the codomain. We begin by identifying a basis for the algebra $\mathcal{O}(\Sigma^3_q) \otimes \mathcal{O}(\Sigma^3_q)$; observing the relations in Equations (6a) and (6b) it is clear that a basis for $\mathcal{O}(\Sigma^3_q(k,l))$ is given by elements of the form

$$b_1 = b_1(p_1,p_2,p_3) = \zeta_0^{p_1}\zeta_1^{p_2}\xi^{p_3}, \qquad b_2 = b_2(\bar{p}_1,\bar{p}_2,\bar{p}_3) = \zeta_0^{\bar{p}_1}\zeta_1^{\bar{p}_2}\xi^{*\bar{p}_3}$$

$$b_3 = b_3(q_1,q_2,q_3) = \zeta_0^{*q_1}\zeta_1^{q_2}\xi^{q_3}, \qquad b_4 = b_4(\bar{q}_1,\bar{q}_2,\bar{q}_3) = \zeta_0^{*\bar{q}_1}\zeta_1^{\bar{q}_2}\xi^{*\bar{q}_3}$$

noting that all powers are non-negative. Hence a basis for $\mathcal{O}(\Sigma^3_q) \otimes \mathcal{O}(\Sigma^3_q)$ is given by elements of the form $b_i \otimes b_j$, where $i,j \in \{1,2,3,4\}$. Applying the canonial map gives

$$\mathsf{can}(b_i \otimes b_j) = b_i\varrho(b_j) = b_ib_j \otimes u^{deg(b_j)}, \qquad \text{where} \quad i,j \in \{1,2,3,4\} \tag{17}$$

where ϱ means $\varrho_{k,l}$ for simplicity of notation. The next stage is to construct all possible elements in $\mathcal{O}(\Sigma^3_q) \otimes \mathcal{O}(\Sigma^3_q)$ which map to $1 \otimes u$. To obtain the identity in the first leg we must use one of the following relations:

$$\zeta_0^m\zeta_0^{*n} = \begin{cases} \prod_{p=0}^{m-1}(1-q^{2p}\zeta_1^2\xi) & \text{when } m=n \\ \zeta_0^{m-n}\prod_{p=0}^{n-1}(1-q^{2p}\zeta_1^2\xi) & \text{when } m>n \\ \prod_{p=0}^{m-1}(1-q^{2p}\zeta_1^2\xi)\zeta_0^{*n-m} & \text{when } n>m \end{cases} \tag{18a}$$

$$\zeta_0^{*n}\zeta_0^m = \begin{cases} \prod_{p=1}^{m}(1-q^{-2p}\zeta_1^2\xi) & \text{when } m=n \\ \zeta_0^{*n-m}\prod_{p=1}^{m}(1-q^{-2p}\zeta_1^2\xi) & \text{when } n>m \\ \prod_{p=1}^{n}(1-q^{-2p}\zeta_1^2\xi)\zeta^{m-n} & \text{when } n<m \end{cases} \tag{18b}$$

or

$$\xi\xi^* = \xi^*\xi = 1$$

We see that to obtain identity in the first leg we require the powers of ζ_0 and ζ_0^* to be equal. We now construct all possible elements of the domain which map to $1 \otimes u$ after applying the canonical map.

Case 1: use the first relation to obtain $\zeta_0^m\zeta_0^{*m}$ $(m>0)$; this can be done in fours ways. First, using $b_1\varrho(b_3)$, $b_1\varrho(b_4)$, $b_2\varrho(b_3)$ and $b_2\varrho(b_4)$. Now,

$$b_1\varrho(b_3) \sim \zeta_0^{p_1}\zeta_0^{*q_1}\zeta_1^{p_2+q_2}\xi^{p_3+q_3} \otimes u^{-kq_1+lq_2-2lq_3} \Longrightarrow p_1 = q_1 = m, p_2 = q_2 = 0, p_3 = q_3 = 0$$

and

$$-kq_1 + lq_2 - 2lq_3 = 1 \Longrightarrow -mk = 1$$

hence no possible terms. A similar calculation for the three other cases shows that $1 \otimes u$ cannot be obtained as an element of the image of the canonical map in this case.

Case 2: use the second relation to obtain $\zeta_0^{*n}\zeta_0^n$ ($n > 0$); this can be done in four ways $b_3\varrho(b_1)$, $b_3\varrho(b_2)$, $b_3\varrho(b_2)$ and $b_4\varrho(b_2)$. Now,

$$b_3\varrho(b_1) \sim \zeta_0^{*q_1}\zeta_0^{p_1}\zeta_1^{p_2+q_2}\xi^{p_3+q_3} \otimes u^{kp_1+lp_2-2lp_3} \Longrightarrow p_1 = q_1 = n, p_2 = q_2 = 0, p_3 = q_3 = 0$$

and

$$nk = 1 \Longrightarrow n = 1 \text{ and } k = 1$$

Note that $k = 1$ is not a problem provided l is not equal to 1. This is reviewed at the next stage of the proof. The same conclusion is reached in all four cases.

In all possibilities $\zeta_0^{*n}\zeta_0^n$ appears only when $n = 1$, in which case the relation simplifies to $\zeta_0^*\zeta_0 = 1 - q^{-2}\zeta_1^2\xi$, so the next stage involves constructing elements in the domain which map to $\zeta_1^2\xi$. There are eight possibilities altogether to be checked: $b_1\varrho(b_1)$, $b_1\varrho(b_2)$, $b_1\varrho(b_3)$, $b_1\varrho(b_4)$, $b_3\varrho(b_1)$, $b_3\varrho(b_2)$, $b_3\varrho(b_3)$ and $b_3\varrho(b_4)$. The first case gives:

$$b_1\varrho(b_1) \sim \zeta_0^{2p_1}\zeta_1^{2p_2}\xi^{2p_3} \otimes u^{kp_1+lp_2-2lp_3} \Longrightarrow 2p_1 = 0,\ 2p_2 = 2,\ 2p_3 = 1$$

and

$$kp_1 + lp_2 - 2lp_3 = 1 \Longrightarrow p_1 = 0, p_2 = 1, p_3 \text{ has no possible values and } l = 1.$$

Hence $1 \otimes u$ cannot be obtained as an element in the image in this case. Similar calculations for the remaining possibilities show that either $1 \otimes u$ is not in the image of the canonical map, or that if $1 \otimes u$ is in the image then $k = l = 1$.

Case 3: finally, it seems possible that $1 \otimes u$, using the third relation, could be in the image of the canonical map. All possible elements in the domain which could potentially map to this element are constructed and investigated. There are eight possibilities: $b_1\varrho(b_2)$, $b_1\varrho(b_4)$, $b_2\varrho(b_1)$, $b_2\varrho(b_3)$, $b_3\varrho(b_2)$, $b_3\varrho(b_4)$, $b_4\varrho(b_1)$ and $b_4\varrho(b_3)$. The first possibility comes out as

$$b_1\varrho(b_2) \sim \zeta_0^{p_1+\bar{p}_1}\zeta_1^{p_2+\bar{p}_2}\xi^{p_3}\xi^{*\bar{p}_3} \otimes u^{k\bar{p}_1+l\bar{p}_2+2l\bar{p}_3} \Longrightarrow p_1 = \bar{p}_1 = 0, p_2 = \bar{p}_2 = 0, p_3 = \bar{p}_3 = 1$$

Also

$$k\bar{p}_1 + l\bar{p}_2 + 2l\bar{p}_3 = 1 \Longrightarrow 2l = 1$$

which implies there are no terms. The same conclusion can be reached for the remaining relations.

This concludes that $1 \otimes u$, which is contained in $\mathcal{O}(\Sigma_q^3) \otimes \otimes\mathbb{C}[u, u^*]$, is not in the image of the canonical map, proving that this map is not surjective and ultimately not an isomorphism when k and l are both not simultaneously equal to 1, completing the proof that $\mathcal{O}(\Sigma_q^3(k,l))$ is not a principal comodule algebra in this case. □

Theorem 1 tells us that if we use $\mathcal{O}(\Sigma_q^3(k,l))$ as our total space, then we are forced to put $(k,l) = (1,1)$ to ensure that the required Hopf–Galois condition does not fail. A consequence of this would be the generators ζ_0 and ζ_1 would have $\mathbb{Z}$-degree 1. This suggests that the comodule algebra $\mathcal{O}(\Sigma_q^3(k,l))$ is too restrictive as there is no freedom with the weights k or l, and that we should in fact consider a subalgebra of $\mathcal{O}(\Sigma_q^3)$ which admits a $\mathcal{O}(S^1)$-coaction that would offer some choice. Theorem 1 indicates that the desired subalgebra should have generators with grades 1 to ensure the Hopf–Galois condition is satisfied. This process is similar to that followed in [4], where the bundles over the quantum teardrops $\mathbb{WP}_q(1,l)$ have the total spaces provided by the quantum lens spaces and structure groups provided by the circle group $U(1)$. We follow a similar approach in the sense that we view $\mathcal{O}(\Sigma_q^3(k,l))$ as a right H-comodule algebra, where H is the Hopf algebra of a suitable cyclic group.

4.2. The Negative Case $\mathcal{O}(\mathbb{RP}_q^2(l;-))$

4.2.1. . The Principal $\mathcal{O}(U(1))$-Comodule Algebra over $\mathcal{O}(\mathbb{RP}_q^2(l;-))$

Take the group Hopf $*$-algebra $H = \mathbb{C}\mathbb{Z}_l$ which is generated by unitary grouplike element w and satisfies the relation $w^l = 1$. The algebra $\mathcal{O}(\Sigma_q^3)$ is a right $\mathbb{C}\mathbb{Z}_l$-comodule $*$-algebra with coaction

$$\mathcal{O}(\Sigma_q^3) \to \mathcal{O}(\Sigma_q^3) \otimes \mathbb{C}\mathbb{Z}_l, \qquad \zeta_0 \mapsto \zeta_0 \otimes w,\ \zeta_1 \mapsto \zeta_1 \otimes 1,\ \xi \mapsto \xi \otimes 1 \tag{19}$$

Note that the $\mathbb{Z}_l$-degree of the generator ξ is determined by the degree of ζ_1: the relation $\zeta_1^* = \zeta_1\xi$ and that the coaction must be compatible with all relations imply that $\deg(\zeta_1^*) = \deg(\zeta_1) + \deg(\xi)$. Since ζ_1 has degree zero, ξ must also have degree zero.

The next stage of the process is to find the coinvariant elements of $\mathcal{O}(\Sigma_q^3)$ given the coaction defined above.

Proposition 7. *The fixed point subalgebra of the above coaction is isomorphic to the algebra $\mathcal{O}(\Sigma_q^3(l;-))$, generated by x, y and z subject to the following relations*

$$y^* = yz, \qquad xy = q^l yx, \qquad xx^* = \prod_{p=0}^{l-1}(1 - q^{2p}y^2z), \qquad x^*x = \prod_{p=1}^{l}(1 - q^{-2p}y^2z) \tag{20}$$

and z is central unitary. The embedding of $\mathcal{O}(\Sigma_q^3(l;-))$ into $\mathcal{O}(\Sigma_q^3)$ is given by $x \mapsto \zeta_0^l$, $y \mapsto \zeta_1$ and $z \mapsto \xi$

Proof. Clearly ζ_1, ξ, ζ_0^l and ζ_0^{*l} are coinvariant elements of $\mathcal{O}(\Sigma_q^3)$. Apply the coaction to the basis (7) to obtain

$$\zeta_0^r\zeta_1^s\xi^t \mapsto \zeta_0^r\zeta_1^s\xi^t \otimes w^r, \qquad \zeta_0^{*r}\zeta_1^s\xi^t \mapsto \zeta_0^{*r}\zeta_1^s\xi^t \otimes w^{-r}$$

These elements are coinvariant, provided $r = r'l$. Hence every coinvariant element is a polynomial in ζ_1, ξ, ζ_0^l and ζ_0^{*l}. Equations (20) are now easily derived from Equations (6) and (18). □

The algebra $\mathcal{O}(\Sigma_q^3(l;-))$ is a right $\mathcal{O}(U(1))$-comodule coalgebra with coaction defined as

$$\varphi : \mathcal{O}(\Sigma_q^3(l;-)) \to \mathcal{O}(\Sigma_q^3(l;-)) \otimes \mathcal{O}(U(1)), \qquad x \mapsto x \otimes u, \qquad y \mapsto y \otimes u, \qquad z \mapsto z \otimes u^{-2} \tag{21}$$

Note in passing that the second and third relations in Equations (20) tell us that the grade of z must be double the grade of y^* since xx^* and x^*x have degree zero, and so

$$\deg(y^2z) = \deg(y^2) + \deg(z) = 2\deg(y) + \deg(z) = 0 \Longrightarrow \deg(z) = -2\deg(y) = 2\deg(y^*)$$

Proposition 8. *The algebra $\mathcal{O}(\Sigma_q^3(l;-))^{co\mathcal{O}(U(1))}$ of invariant elements under the coaction φ is isomorphic to the $\mathcal{O}((\mathbb{RP}_q(l;-))$.*

Proof. We aim to show that the $*$-subalgebra of $\mathcal{O}(\Sigma_q^3(l;-))$ of elements which are invariant under the coaction is generated by x^2z, xyz and y^2z. The isomorphism of $\mathcal{O}(\Sigma_q^3(l;-))^{co\mathcal{O}(U(1))}$ with $\mathcal{O}((\mathbb{RP}_q(l;-))$ is then obtained by using the embedding of $\mathcal{O}(\Sigma_q^3(l;-))$ in $\mathcal{O}(\Sigma_q^3)$ described in Proposition 7, *i.e.*, $y^2z \mapsto \zeta_1\xi \mapsto a$, $xyz \mapsto \zeta_0^l\zeta_1\xi \mapsto b$ and $x^2z \mapsto \zeta_0^{2l}\xi \mapsto c_-$.

The algebra $\mathcal{O}(\Sigma_q^3(l;-))$ is spanned by elements of the type $x^ry^sz^t$, $x^{*r}y^sz^t$, where $r, s \in \mathbb{N}$ and $t \in \mathbb{Z}$. Applying the coaction φ to these basis elements gives $x^ry^sz^t \mapsto x^ry^sz^t \otimes u^{r+s-2t}$. Hence $x^ry^sz^t$ is φ-invariant if and only if $2t = r + s$. If r is even, then s is even and

$$x^ry^sz^t = x^ry^sz^{(r+s)/2} = (x^2z)^{r/2}(y^2z)^{s/2}$$

If r is odd, then so is s and

$$x^r y^s z^t = x^r y^s z^{(r+s)/2} \sim (x^2 z)^{(r-1)/2} (y^2 z)^{(s-1)/2} (xyz)$$

The case of $x^{*r} y^s z^t$ is dealt with similarly, thus proving that all coinvariants of φ are polynomials in $x^2 z$, xyz, $y^2 z$ and their $*$-conjugates. □

The main result of this section is contained in the following Theorem.

Theorem 2. *$\mathcal{O}(\Sigma_q^3(l;-))$ is a non-cleft principal $\mathcal{O}(U(1))$-comodule algebra over $\mathcal{O}(\mathbb{RP}_q(l;+))$ via the coaction φ.*

Proof. To prove that $\mathcal{O}(\Sigma_q^3(l;-))$ is a principal $\mathcal{O}(U(1))$-comodule algebra over $\mathcal{O}(\mathbb{RP}_q(l;+))$ we employ Proposition 5 and construct a strong connection form as follows.

Define $\omega : \mathcal{O}(U(1)) \to \mathcal{O}(\Sigma_q^3(l;-)) \otimes \mathcal{O}(\Sigma_q^3(l;-))$ recursively as follows.

$$\omega(1) = 1 \otimes 1 \tag{22a}$$

$$\omega(u^n) = x^* \omega(u^{n-1}) x - \sum_{m=1}^{l} (-1)^m q^{-m(m+1)} \binom{l}{m}_{q^{-2}} y^{2m-1} z^m \omega(u^{n-1}) y \tag{22b}$$

$$\omega(u^{-n}) = x \omega(u^{-n+1}) x^* - \sum_{m=1}^{l} (-1)^m q^{m(m-1)} \binom{l}{m}_{q^2} y^{2m-1} z^{m-1} \omega(u^{-n+1}) yz \tag{22c}$$

where $n \in \mathbb{N}$ and, for all $s \in \mathbb{R}$, the *deformed* or *q-binomial* coefficients $\binom{l}{m}_s$ are defined by the following polynomial equality in indeterminate t

$$\prod_{m=1}^{l} (1 + s^{m-1} t) = \sum_{m=0}^{l} s^{m(m-1)/2} \binom{l}{m}_s t^m \tag{23}$$

The map ω has been designed such that normalisation property, Equation (1a), is automatically satisfied. To check Equation (1b) for ω given by Equation (22b) and (22c) takes a bit more work. We use proof by induction, but first have to derive an identity to assist with the calculation. Set $s = q^{-2}$, $t = -q^{-2} y^* y$ in Equation (23) to arrive at

$$\sum_{m=1}^{l} (-1)^m q^{-m(m+1)} \binom{l}{m}_{q^{-2}} y^{*m} y^m = \prod_{m=1}^{l} (1 + q^{-2(m-1)} (-q^{-2} y^* y)) - 1$$

which, using Equations (20), simplifies to

$$\sum_{m=1}^{l} (-1)^m q^{-m(m+1)} \binom{l}{m}_{q^{-2}} y^{2m} z^m = \prod_{m=1}^{l} (1 - q^{-2m} y^2 z) - 1 = x^* x - 1 \tag{24}$$

Now to start the induction process we consider the case $n = 1$. By Equation (24) $(\mu \circ \omega)(u) = 1$ providing the basis. Next, we assume that the relation holds for $n = N$, that is $(\mu \circ \omega)(u^N) = 1$, and consider the case $n = N + 1$,

$$\omega(u^{N+1}) = x^* \omega(u^N) x - \sum_{m=1}^{l} (-1)^m q^{-m(m+1)} \binom{l}{m}_{q^{-2}} y^{2m-1} z^m \omega(u^N) y$$

applying the multiplication map to both sides and using the induction hypothesis,

$$(m \circ \omega)(u^{N+1}) = x^* x - \sum_{m=1}^{l} (-1)^m q^{-m(m+1)} \binom{l}{m}_{q^{-2}} y^{2m} z^m = x^* x - (x^* x - 1) = 1$$

showing Equation (1b) holds for all $u^n \in \mathcal{O}(U(1))$, where $n \in \mathbb{N}$. To show this property holds for each $u^{*n} = u^{-n}$ we adopt the same strategy; this is omitted from the proof as it does not provide further insight, instead repetition of similar arguments.

Equation (1c): this is again proven by induction. Applying $(\mathrm{id} \otimes \varphi)$ to $\omega(u)$ gives

$$\begin{aligned}
x^* \otimes x \otimes u &- \sum_{m=1}^{l} (-1)^m q^{-m(m-1)} \binom{l}{m}_{q^2} y^{2m-1} z^m \otimes y \otimes u \\
&= (x^* \otimes x - \sum_{m=1}^{l} (-1)^m q^{-m(m-1)} \binom{l}{m}_{q^2} y^{2m-1} z^m \otimes y) \otimes u \\
&= \omega(u) \otimes u = (\omega \otimes \mathrm{id}) \circ \Delta(u)
\end{aligned}$$

This shows that Equation (1c) holds for ω given by Equation (22b) when $n = 1$. We now assume the property holds for $n = N - 1$, hence $(id \otimes \varphi) \circ w(u^{N-1}) = (\omega \otimes \mathrm{id}) \circ \Delta(u^{N-1}) = \omega(u^{N-1}) \otimes u^{N-1}$, and consider the case $n = N$.

$$\begin{aligned}
(\mathrm{id} \otimes \varphi)(w(u^N)) &= (\mathrm{id} \otimes \varphi)(x^* \omega(u^{N-1})x - \sum_{m=1}^{l} (-1)^m q^{-m(m-1)} \binom{l}{m}_{q^{-2}} y^{2m-1} z^m \omega(u^{N-1}) y) \\
&= x^*((id \otimes \varphi)(\omega(u^{N-1})x)) \\
&\quad - \sum_{m=1}^{l} (-1)^m q^{-m(m-1)} \binom{l}{m}_{q^{-2}} y^{2m-1} z^m ((id \otimes \varphi)(\omega(u^{N-1})y) \\
&= x^* \omega(u^{N-1}) x \otimes u^N - \sum_{m=1}^{l} (-1)^m q^{-m(m-1)} \binom{l}{m}_{q^{-2}} y^{2m-1} z^m \omega(u^{N-1}) y \otimes u^N \\
&= \omega(u^N) \otimes u^N = (\omega \otimes \mathrm{id}) \circ \Delta(u^N)
\end{aligned}$$

hence Equation (1c) is satisfied for all $u^n \in \mathcal{O}(U(1))$ where $n \in \mathbb{N}$. The case for u^{*n} is proved in a similar manner, as is Equation (1d). Again, the details are omitted as the process is identical. This completes the proof that ω is a strong connection form, hence $\mathcal{O}(\Sigma_q^3(l,-))$ is a principal comodule algebra.

Following the discussion of Section 3.1, to determine whether the constructed comodule algebra is cleft we need to identify invertible elements in $\mathcal{O}(\Sigma_q^3(l,-))$. Since

$$\mathcal{O}(\Sigma_q^3(l,-)) \subset \mathcal{O}(\Sigma_q^3) \cong \mathcal{O}(S_q^2) \square_{\mathbb{C}\mathbb{Z}_2} \mathcal{O}(U(1)) \subset \mathcal{O}(S_q^2) \otimes \mathcal{O}(U(1))$$

and the only invertible elements in the algebraic tensor $\mathcal{O}(S_q^2) \otimes \mathcal{O}(U(1))$ are scalar multiples of $1 \otimes u^n$ for $n \in \mathbb{N}$, we can conclude that the only invertible elements in $\mathcal{O}(S_q^2) \square_{\mathbb{C}\mathbb{Z}_2} \mathcal{O}(U(1))$ are the elements of the form $1 \otimes u^n$. These elements correspond to the elements ξ^n in $\mathcal{O}(\Sigma_q^3)$, which in turn correspond to z^n in $\mathcal{O}(\Sigma_q^3(l,-))$.

Suppose $j : H \to A$ is the cleaving map; to ensure the map is convolution invertible we are forced to put $u \mapsto z^n$. Since u has degree 1 in $H = \mathcal{O}(U(1))$ and z has degree -2 in $\mathcal{O}(\Sigma_q^3(l,-))$, the map j fails to preserve the degrees, hence it is not colinear. Therefore, $\mathcal{O}(\Sigma_q^3(l,-))$ is a non-cleft principal comodule algebra. □

4.2.2. . Almost Freeness of the Coaction $\varrho_{1,l}$

At the classical limit, $q \to 1$, the algebras $\mathcal{O}(\mathbb{RP}_q(l;-))$ represent singular manifolds or orbifolds. It is known that every orbifold can be obtained as a quotient of a manifold by an *almost free* action. The latter means that the action has finite (rather than trivial as in the free case) stabiliser groups. As explained in Section 2, on the algebraic level, freeness is encoded in the bijectivity of the canonical map can, or, more precisely, in the surjectivity of the lifted canonical map $\overline{\mathrm{can}}$ (Equation (2)). The surjectivity of $\overline{\mathrm{can}}$ means the triviality of the cokernel of $\overline{\mathrm{can}}$, thus the size of the cokernel of $\overline{\mathrm{can}}$ can be treated as a measure of the size of the stabiliser groups. This leads to the following notion proposed in [4].

Definition 12. *Let H be a Hopf algebra and let A be a right H-comodule algebra with coaction $\varrho^A : A \to A \otimes H$. We say that the coaction is* almost free *if the cokernel of the (lifted) canonical map*

$$\overline{\mathsf{can}} : A \otimes A \to A \otimes H, \qquad a \otimes a' \mapsto a\varrho^A(a')$$

is finitely generated as a left A-module.

Although the coaction φ defined in the preceding section is free, at the classical limit $q \to 1$ $\mathcal{O}(\Sigma_q^3(l,-))$ represents a singular manifold or an orbifold. On the other hand, at the same limit, $\mathcal{O}(\Sigma_q^3)$ corresponds to a genuine manifold, one of the Seifert three-dimensional non-orientable manifolds; see [17]. It is therefore natural to ask, whether the coaction $\varrho_{1,l}$ of $\mathcal{O}(U(1))$ on $\mathcal{O}(\Sigma_q^3)$ which has $\mathcal{O}(\mathbb{RP}_q(l;-))$ as fixed points is almost free in the sense of Definition 12.

Proposition 9. *The coaction $\varrho_{1,l}$ is almost free.*

Proof. Denote by $\iota_- : \mathcal{O}(\Sigma_q^3(l,-)) \hookrightarrow \mathcal{O}(\Sigma_q^3)$, the $*$-algebra embedding described in Proposition 7. One easily checks that the following diagram

$$\begin{array}{ccc} \mathcal{O}(\Sigma_q^3(l,-)) & \xrightarrow{\quad \iota_- \quad} & \mathcal{O}(\Sigma_q^3) \\ \varphi \big\downarrow & & \big\downarrow \varrho_{1,l} \\ \mathcal{O}(\Sigma_q^3(l,-)) \otimes \mathcal{O}(U(1)) & \xrightarrow{\iota_- \otimes (-)^l} & \mathcal{O}(\Sigma_q^3) \otimes \mathcal{O}(U(1)) \end{array}$$

where $(-)^l : u \to u^l$, is commutative. The principality or freeness of φ proven in Theorem 2 implies that $1 \otimes u^{ml} \in \mathrm{Im}(\overline{\mathsf{can}})$, $m \in \mathbb{Z}$, where $\overline{\mathsf{can}}$ is the (lifted) canonical map corresponding to coaction $\varrho_{1,l}$. This means that $\mathcal{O}(\Sigma_q^3) \otimes \mathbb{C}[u^l, u^{-l}] \subseteq \mathrm{Im}(\overline{\mathsf{can}})$. Therefore, there is a short exact sequence of left $\mathcal{O}(\Sigma_q^3)$-modules

$$(\mathcal{O}(\Sigma_q^3) \otimes \mathbb{C}[u,u^{-1}])/(\mathcal{O}(\Sigma_q^3) \otimes \mathbb{C}[u^l,u^{-l}]) \longrightarrow \mathrm{coker}(\overline{\mathsf{can}}) \longrightarrow 0$$

The left $\mathcal{O}(\Sigma_q^3)$-module $(\mathcal{O}(\Sigma_q^3)) \otimes \mathbb{C}[u,u^{-1}])/(\mathcal{O}(\Sigma_q^3) \otimes \mathbb{C}[u^l,u^{-l}])$ is finitely generated, hence so is $\mathrm{coker}(\overline{\mathsf{can}})$. □

4.2.3. . Associated Modules or Sections of Line Bundles

One can construct modules associated to the principal comodule algebra $\mathcal{O}(\Sigma_q^3(l,-))$ following the procedure outlined at the end of Section 2.2; see Definition 11.

Every one-dimensional comodule of $\mathcal{O}(U(1)) = \mathbb{C}[u,u^*]$ is determined by the grading of a basis element of $\mathbb{C}$, say 1. More precisely, for any integer n, $\mathbb{C}$ is a left $\mathcal{O}(U(1))$-comodule with the coaction

$$\varrho_n : \mathbb{C} \to \mathbb{C}[u,u^*] \otimes \mathbb{C}, \qquad 1 \mapsto u^n \otimes 1$$

Identifying $\mathcal{O}(\Sigma_q^3(l,-)) \otimes \mathbb{C}$ with $\mathcal{O}(\Sigma_q^3(l,-))$ we thus obtain, for each coaction ϱ_n

$$\Gamma[n] := \mathcal{O}(\Sigma_q^3(l,-))\Box_{\mathcal{O}(U(1))}\mathbb{C} \cong \{f \in \Sigma_q^3(l,-) \mid \varphi(f) = f \otimes u^n\} \subset \mathcal{O}(\Sigma_q^3(l,-))$$

In other words, $\Gamma[n]$ consists of all elements of $\mathcal{O}(\Sigma_q^3(l,-))$ of $\mathbb{Z}$-degree n. In particular $\Gamma[0] = \mathcal{O}(\mathbb{RP}_q(l;-))$. Each of the $\Gamma[n]$ is a finitely generated projective left $\mathcal{O}(\mathbb{RP}_q(l;-))$-module, *i.e.*, it represents the module of sections of the non-commutative line bundle over $\mathbb{RP}_q(l;-)$. The idempotent matrix $E[n]$ defining $\Gamma[n]$ can be computed explicitly from a strong connection form ω (see Equations (22) in the proof of Theorem 2) following the procedure described in [11]. Write $\omega(u^n) = \sum_i \omega(u^n)^{[1]}{}_i \otimes \omega(u^n)^{[2]}{}_i$. Then

$$E[n]_{ij} = \omega(u^n)^{[2]}{}_i \omega(u^n)^{[1]}{}_j \in \mathcal{O}(\mathbb{RP}^2_q(l;-)) \tag{25}$$

For Example, for $l = 2$ and $n = 1$, using Equations (22b) and (22a) as well as redistributing numerical coefficients we obtain

$$E[1] = \begin{pmatrix} (1-a)(1-q^2a) & q^{-1}\sqrt{1+q^{-2}}\,b & iq^{-3}ba \\ q^{-1}\sqrt{1+q^{-2}}\,b^* & q^{-2}(1+q^{-2})\,a & iq^{-4}\sqrt{1+q^{-2}}\,a^2 \\ iq^{-3}b^* & iq^{-4}\sqrt{1+q^{-2}}\,a & -q^{-6}a^2 \end{pmatrix} \tag{26}$$

Although the matrix $E[1]$ is not hermitian, the left-upper 2×2 block is hermitian. On the other hand, once $\mathcal{O}(\mathbb{RP}_q(2;-))$ is completed to the C^*-algebra $C(\mathbb{RP}_q(2;-))$ of continuous functions on $\mathbb{RP}_q(2;-)$ (and then identified with the suitable pullback of two algebras of continuous functions over the quantum real projective space; see [3]), then a hermitian projector can be produced out of $E[1]$ by using the Kaplansky formula; see [18, page 88].

The traces of tensor powers of each of the $E[n]$ make up a cycle in the cyclic complex of $\mathcal{O}(\mathbb{RP}_q(l;-))$, whose corresponding class in the cyclic homology $HC_\bullet(\mathcal{O}(\mathbb{RP}_q(l;-)))$ is known as the *Chern character* of $\Gamma[n]$. Again, as an illustration of the usage of an explicit form of a strong connection form, we compute the traces of $E[n]$ for general l.

Lemma 1. *The zero-component of the Chern character of $\Gamma[n]$ is the class of the polynomial c_n in generator a of $\mathcal{O}(\mathbb{RP}_q(l;-))$, given by the following recursive formula. First, $c_0(a) = 1$, and then, for all positive n,*

$$c_n(a) = c_{n-1}\left(q^{2l}a\right)\prod_{p=0}^{l-1}\left(1-q^{2p}a\right) + c_{n-1}(a)\left(1-\prod_{p=1}^{l}\left(1-q^{-2p}a\right)\right) \tag{27a}$$

$$c_{-n}(a) = c_{-n+1}\left(q^{-2l}a\right)\prod_{p=1}^{l}\left(1-q^{-2p}a\right) + c_{-n+1}(a)\left(1-\prod_{p=0}^{l-1}\left(1-q^{2p}a\right)\right) \tag{27b}$$

Proof. We will prove the formula (27a) as (27b) is proven by similar arguments. Recall that $c_n = \operatorname{Tr} E[n]$. By normalisation (22a) of the strong connection ω, obviously $c_0 = 1$. In view of Equation (22b) we obtain the following recursive formula

$$c_n = xc_{n-1}x^* - \sum_{m=1}^{l}(-1)^m q^{-m(m+1)}\binom{l}{m}_{q^{-2}} yc_{n-1}y^{2m-1}z^m \tag{28}$$

In principle, c_n could be a polynomial in a, b and c_-. However, the third of Equations (20) together with Equation (24) and identification of a as y^2z yield

$$c_1 = \prod_{p=0}^{l-1}\left(1-q^{2p}a\right) + \left(1-\prod_{p=1}^{l}\left(1-q^{-2p}a\right)\right) \tag{29}$$

that is a polynomial in a only. As commuting x and y through a polynomial in a in Equation (28) will produce a polynomial in a again, we conclude that each of the c_n is a polynomial in a. The second of Equations (20), the centrality of z and the identification of a as y^2z imply that

$$xc_{n-1}(a) = c_{n-1}(q^{2l}a), \qquad yc_{n-1}(a) = c_{n-1}(a)y$$

and in view of Equations (28) and (29) yield Equation (27a). □

4.3. The Positive Case $\mathcal{O}(\mathbb{RP}_q(l;+))$

4.3.1. . The Principal $\mathcal{O}(U(1))$-Comodule Algebra over $\mathcal{O}(\mathbb{RP}^2_q(l;+))$

In the same light as the negative case we aim to construct quantum principal bundles with base spaces $\mathcal{O}(\mathbb{RP}_q(l;+))$, and proceed by viewing $\mathcal{O}(\Sigma^3_q)$ as a right H'-comodule algebra, where H' is a Hopf-algebra of a finite cyclic group. The aim is to construct the total space $\mathcal{O}(\Sigma^3_q(l,+))$ of the bundle over $\mathcal{O}(\mathbb{RP}_q(l;+))$ as the coinvariant subalgebra of $\mathcal{O}(\Sigma^3_q)$. $\mathcal{O}(\Sigma^3_q(l,+))$ must contain generators $\zeta_1^2\xi$ and $\zeta_0^l\xi$ of $\mathcal{O}(\mathbb{RP}_q(l;+))$. Suppose $H' = \mathbb{C}\mathbb{Z}_m$ and $\Phi : \mathcal{O}(\Sigma^3_q) \to \mathcal{O}(\Sigma^3_q) \otimes H'$ is a coaction. We require Φ to be compatible with the algebraic relations and to give zero $\mathbb{Z}_m$-degree to $\zeta_1^2\xi$ and $\zeta_0^l\xi$ are zero. These requirements yield

$$2\deg(\zeta_1) + \deg(\xi) = 0 \mod m, \qquad l\deg(\zeta_0) + \deg(\xi) = 0 \mod m$$

Bearing in mind that l is odd, the simplest solution to these requirements is provided by $m = 2l$, $\deg(\xi) = 0$, $\deg(\zeta_0) = 2$, $\deg(\zeta_1) = l$. This yields the coaction

$$\Phi : \mathcal{O}(\Sigma^3_q) \to \mathcal{O}(\Sigma^3_q) \otimes \mathbb{C}\mathbb{Z}_{2l} \qquad \zeta_0 \mapsto \zeta_0 \otimes v^2, \quad \zeta_1 \mapsto \zeta_1 \otimes v^l, \quad \xi \mapsto \xi \otimes 1$$

where v ($v^{2l} = 1$) is the unitary generator of $\mathbb{C}\mathbb{Z}_{2l}$. Φ is extended to the whole of $\mathcal{O}(\Sigma^3_q)$ so that Φ is an algebra map, making $\mathcal{O}(\Sigma^3_q)$ a right $\mathbb{C}\mathbb{Z}_{2l}$-comodule algebra.

Proposition 10. *The fixed point subalgebra of the coaction Φ is isomorphic to the $*$-algebra $\mathcal{O}(\Sigma^3_q(l,+))$ generated by x', y' and central unitary z' subject to the following relations:*

$$x'y' = q^{2l}y'x', \qquad y'^* = y'z'^2 \tag{30a}$$

$$x'x'^* = \prod_{p=0}^{l-1}(1 - q^{2p}y'z'), \qquad x'^*x' = \prod_{p=1}^{l}(1 - q^{-2p}y'z') \tag{30b}$$

The isomorphism between $\mathcal{O}(\Sigma^3_q(l,+))$ and the coinvariant subalgebra of $\mathcal{O}(\Sigma^3_q)$ is given by $x' \mapsto \zeta_0^l$, $y' \mapsto \zeta_1^2$ and $z' \mapsto \xi$.

Proof. Clearly ζ_1^2, ξ, ζ_0^l and ζ_0^{*l} are coinvariant elements of $\mathcal{O}(\Sigma^3_q)$. Apply the coaction Φ to the basis (7) to obtain

$$\zeta_0^r\zeta_1^s\xi^t \mapsto \zeta_0^r\zeta_1^s\xi^t \otimes v^{2r+ls}, \qquad \zeta_0^{*r}\zeta_1^s\xi^t \mapsto \zeta_0^{*r}\zeta_1^s\xi^t \otimes v^{-2r+ls}$$

These elements are coinvariant, provided $2r + ls = 2ml$ in the first case or $-2r + ls = 2ml$ in the second. Since l is odd, s must be even and then $r = r'l$, hence the invariant elements must be of the form

$$(\zeta_0^l)^{r'}(\zeta_1^2)^{s/2}\xi^t, \qquad (\zeta_0^{*l})^{r'}(\zeta_1^2)^{s/2}\xi^t$$

as required. Equations (30) are now easily derived from Equations (6) and (18). □

The algebra $\mathcal{O}(\Sigma^3_q(l,+))$ is a right $\mathcal{O}(U(1))$-comodule with coaction defined as,

$$\Omega : \mathcal{O}(\Sigma^3_q(l,+)) \to \mathcal{O}(\Sigma^3_q(l,+)) \otimes \mathcal{O}(U(1)), \qquad x' \mapsto x' \otimes u, \quad y' \mapsto y' \otimes u, \quad z' \mapsto z' \otimes u^{-1} \tag{31}$$

The first relation in Equations (30a) bears no information on the possible gradings of the generators of $\mathcal{O}(\Sigma_q^3(l,+))$, however the second relation in Equations (30a) tells us that the grade of y'^* must be the same as that of z' since,

$$\deg(y'^*) = -\deg(y') = \deg(y') + 2\deg(z')$$

hence,

$$2\deg(y'^*) = 2\deg(z'), \text{ or, } \deg(y'^*) = \deg(z')$$

This is consistent with Equations (30b) since the left hand sides, $x'x'^*$ and x'^*x', have degree zero, as do the right had sides,

$$\deg(y'z') = \deg(y') + \deg(y'^*) = \deg(y') + (-\deg(y')) = 0$$

The coaction Ω is defined setting the grades of x' and y' as 1, and putting the grade of z' as -1 to ensure the coaction is compatible with the relations of the algebra $\mathcal{O}(\Sigma_q^3(l,+))$.

Proposition 11. *The right $\mathcal{O}(U(1))$-comodule algebra $\mathcal{O}(\Sigma_q^3(l,+))$ has $\mathcal{O}(\mathbb{RP}_q(l;+))$ as its subalgebra of coinvariant elements under the coaction Ω.*

Proof. The fixed points of the algebra $\mathcal{O}(\Sigma_q^3(l,+))$ under the coaction Ω are found using the same method as in the odd k case. A basis for the algebra $\mathcal{O}(\Sigma_q^3(l,+))$ is given by $x'^r y'^s z'^t$, $x'^{*r} y'^s z'^t$, where $r, s \in \mathbb{N}$ and $t \in \mathbb{Z}$.

Applying the coaction Ω to the first of these basis elements gives,

$$x'^r y'^s z'^t \mapsto x'^r y'^s z'^t \otimes u^{r+s-t}$$

Hence the invariance of $x'^r y'^s z'^t$ is equivalent to $t = r + s$. Simple substitution and re-arranging gives,

$$x'^r y'^s z'^t = x'^r y'^s z'^{r+s} = (x'z')^r (y'z')^s$$

i.e., $x'^r y'^s z'^t$ is a polynomial in $x'z'$ and $y'z'$. Repeating the process for the second type of basis element gives the $*$-conjugates of $x'z'$ and $y'z'$. Using Proposition 10 we can see that $a = \zeta_1^2 \xi = y'z'$ and $c_+ = \zeta_0^l \xi = x'z'$. □

In contrast to the odd k case, although $\mathcal{O}(\Sigma_q^3(l,+))$ is a principal comodule algebra it yields trivial principal bundle over $\mathcal{O}(\mathbb{RP}_q(l;+))$.

Proposition 12. *The right $\mathcal{O}(U(1))$-comodule algebra $\mathcal{O}(\Sigma_q^3(l,+))$ is trivial.*

Proof. The cleaving map is given by,

$$j : \mathcal{O}(U(1)) \to \mathcal{O}(\Sigma_q^3(l,+)), \qquad j(u) = z'^*$$

which is an algebra map since z'^* is central unitary in $\mathcal{O}(\Sigma_q^3(l,+))$, hence must be convolution invertible. Also, j is a right $\mathcal{O}(U(1))$-comodule map since,

$$(\Omega \circ j)(u) = \Omega(z'^*) = z'^* \otimes u = j(u) \otimes u = (j \otimes \mathrm{id}) \circ \Delta(u)$$

completing the proof. □

Since $\mathcal{O}(\Sigma_q^3(l,+))$ is a trivial principal comodule algebra, all associated $\mathcal{O}(\mathbb{RP}_q^2(l;+))$-modules are free.

4.3.2. . Almost Freeness of the Coaction $\varrho_{2,l}$

As was the case for $\mathcal{O}(\Sigma_q^3(l,-))$, the principality of $\mathcal{O}(\Sigma_q^3(l,+))$ can be used to determine that the $\mathcal{O}(U(1))$-coaction $\varrho_{2,l}$ on $\mathcal{O}(\Sigma_q^3)$ that defines $\mathcal{O}(\mathbb{RP}_q^2(l;+))$ is almost free.

Proposition 13. *The coaction $\varrho_{2,l}$ is almost free.*

Proof. Denote by $\iota_+ : \mathcal{O}(\Sigma_q^3(l,+)) \hookrightarrow \mathcal{O}(\Sigma_q^3)$, the $*$-algebra embedding described in Proposition 10. One easily checks that the following diagram

$$\begin{array}{ccc} \mathcal{O}(\Sigma_q^3(l,+)) & \xrightarrow{\quad \iota_+ \quad} & \mathcal{O}(\Sigma_q^3) \\ {\scriptstyle \Omega}\downarrow & & \downarrow{\scriptstyle \varrho_{2,l}} \\ \mathcal{O}(\Sigma_q^3(l,+)) \otimes \mathcal{O}(U(1)) & \xrightarrow{\iota_+ \otimes (-)^{2l}} & \mathcal{O}(\Sigma_q^3) \otimes \mathcal{O}(U(1)) \end{array}$$

where $(-)^{2l} : u \to u^{2l}$ is commutative. By the arguments analogous to those in the proof of Proposition 9 one concludes that there is a short exact sequence of left $\mathcal{O}(\Sigma_q^3)$-modules

$$(\mathcal{O}(\Sigma_q^3) \otimes \mathbb{C}[u, u^{-1}]) / (\mathcal{O}(\Sigma_q^3) \otimes \mathbb{C}[u^{2l}, u^{-2l}]) \longrightarrow \mathrm{coker}(\overline{\mathsf{can}}) \longrightarrow 0$$

where $\overline{\mathsf{can}}$ is the lifted canonical map corresponding to coaction $\varrho_{2,l}$. The left $\mathcal{O}(\Sigma_q^3)$-module $(\mathcal{O}(\Sigma_q^3) \otimes \mathbb{C}[u, u^{-1}]) / (\mathcal{O}(\Sigma_q^3) \otimes \mathbb{C}[u^{2l}, u^{-2l}])$ is finitely generated, hence so is $\mathrm{coker}(\overline{\mathsf{can}})$. □

5. Conclusions

In this paper we discussed the principality of the $\mathcal{O}(U(1))$-coactions on the coordinate algebra of the quantum Seifert manifold $\mathcal{O}(\Sigma_q^3)$ weighted by coprime integers k and l. We concluded that the coaction is principal if and only if $k = l = 1$, which corresponds to the case of a $U(1)$-bundle over the quantum real projective plane. In all other cases the coactions are almost free. We identified subalgebras of $\mathcal{O}(\Sigma_q^3))$ which admit principal $\mathcal{O}(U(1))$-coactions, whose invariants are isomorphic to coordinate algebras $\mathcal{O}(\mathbb{RP}_q^2(l;\pm))$ of quantum real weighted projective spaces. The structure of these subalgebras depends on the parity of k. For the odd k case, the constructed principal comodule algebra $\mathcal{O}(\Sigma_q^3(l,-))$ is non-trivial, while for the even case, the corresponding principal comodule algebra $\mathcal{O}(\Sigma_q^3(l,+))$ turns out to be trivial. The triviality of $\mathcal{O}(\Sigma_q^3(l,+))$ is a disappointment. Whether a different nontrivial principal $\mathcal{O}(U(1))$-comodule algebra over $\mathcal{O}(\mathbb{RP}_q^2(l;+))$ can be constructed or whether such a possibility is ruled out by deeper geometric, topological or algebraic reasons remains to be seen.

References

1. Soibel'man, Y.S.; Vaksman, L.L. Algebra of functions on the quantum group SU(n + 1), and odd-dimensional quantum spheres. *Algebra i Analiz* **1990**, *2*, 101–120.
2. Brzeziński, T.; Zieliński, B. Quantum principal bundles over quantum real projective spaces. *J. Geom. Phys.* **2012**, *62*, 1097–1107.
3. Brzeziński, T. Circle actions on a quantum Seifert manifold. In *Proceedings of the Corfu Summer Institute 2011 School and Workshops on Elementary Particle Physics and Gravity*, Corfu, Greece, 4–18 September 2011.
4. Brzeziński, T.; Fairfax, S.A. Quantum teardrops. *Comm. Math. Phys.*, in press.
5. Beggs, E.J.; Brzeziński, T. Line bundles and the Thom construction in noncommutative geometry. *J. Noncommut. Geom.*, in press.
6. Baum, P.F.; Hajac, P.M.; Matthes, R.; Szymański, W. Noncommutative geometry approach to principal and associated bundles. 2007, arXiv:math/0701033. Available online: http://arxiv.org/abs/math/0701033 (accessed on 10 September 2012).

7. Brzeziński, T.; Majid, S. Quantum group gauge theory on quantum spaces. *Comm. Math. Phys.* **1993**, *157*, 591–638.
8. Schneider, H.-J. Principal homogeneous spaces for arbitrary Hopf algebras. *Israel J. Math.* **1990**, *72*, 167–195.
9. Hajac, P.M. Strong connections on quantum principal bundles. *Comm. Math. Phys.* **1996**, *182*, 579–617.
10. Woronowicz, S.L. Compact matrix pseudogroups. *Comm. Math. Phys.* **1987**, *111*, 613–665.
11. Brzeziński, T.; Hajac, P.M. The Chern-Galois character. *Comptes Rendus Math. (Acad. Sci. Paris Ser. I)* **2004**, *338*, 113–116.
12. Dąbrowski, L.; Grosse, H.; Hajac, P.M. Strong connections and Chern-Connes pairing in the Hopf-Galois theory. *Comm. Math. Phys.* **2001**, *220*, 301–331.
13. Beggs, E.J.; Brzeziński, T. An explicit formula for a strong connection. *Appl. Categor. Str.* **2008**, *16*, 57–63.
14. Hajac, P.M.; Matthes, R.; Szymański, W. Quantum real projective space, disc and spheres. *Algebr. Represent. Theory* **2003**, *6*, 169–192.
15. Klimek, S.; Leśniewski, A. A two-parameter quantum deformation of the unit disc. *J. Funct. Anal.* **1993**, *115*, 1–23.
16. Hong, J.H.; Szymański, W. Quantum spheres and projective spaces as graph algebras. *Comm. Math. Phys.* **2002**, *232*, 157–188.
17. Scott, P. The geometries of 3-manifolds. *Bull. Lond. Math. Soc.* **1983**, *15*, 401–487.
18. Gracia-Bondia, J.M.; Várilly, J.C.; Figueroa, H. *Elements of Noncommutative Geometry*; Birkhäuser: Boston, MA, USA, 2001.

12

The Sum of a Finite Group of Weights of a Hopf Algebra

Apoorva Khare

Department of Mathematics and Department of Statistics, Stanford University, 390 Serra Mall, Stanford, CA 94305, USA; E-Mail: khare@stanford.edu

Abstract: Motivated by the orthogonality relations for irreducible characters of a finite group, we evaluate the sum of a finite group of linear characters of a Hopf algebra, at all grouplike and skew-primitive elements. We then discuss results for products of skew-primitive elements. Examples include groups, (quantum groups over) Lie algebras, the small quantum groups of Lusztig, and their variations (by Andruskiewitsch and Schneider).

Keywords: Hopf algebra; weights; grouplike; skew-primitive

MSC: 16W30 (primary), 17B10 (secondary)

---

1. Introduction

1.1. Motivation

Suppose G is a finite group, with irreducible characters $\widehat{G} = \{\chi : \mathbb{C}G \to \mathbb{C}\}$, over $\mathbb{C}$. As is well known, they satisfy the *orthogonality relations*. Here is a consequence: if one defines $\theta_\chi := \frac{1}{|G|}\sum_{g\in G}\chi(g^{-1})g \in H = \mathbb{C}G$, then $\nu(\theta_\chi) = 0$ for $\chi \neq \nu \in \widehat{G}$. Similarly, the other orthogonality relation (for columns) implies that $\sum_{\chi\in\widehat{G}}(\dim \rho_\chi)\chi(g)$ is either zero or a factor of $|G|$, depending on whether or not $g = 1$. (Here, ρ_χ is the irreducible representation with character χ.)

In what follows, we work over a commutative integral domain R. Note that there is an analogue of the first orthogonality relation for any R-algebra H that is a free R-module. Namely, given a linear character (*i.e.*, algebra map or *weight*) $\lambda : H \to R$, define a *left λ-integral* of H to be any (nonzero) $\Lambda_L^\lambda \in H$ so that $h\Lambda_L^\lambda = \lambda(h)\Lambda_L^\lambda$ for all $h \in H$. One can similarly define right and two-sided λ-integrals in H. (For instance, the θ_χ's above are two-sided χ-integrals for any weight χ.) Then the following result holds in any algebra:

Lemma 1.1. *If $\lambda \neq \nu$ are weights of H with corresponding nonzero left integrals $\Lambda_L^\lambda, \Lambda_L^\nu$ respectively, then $\nu(\Lambda_L^\lambda) = 0 = \Lambda_L^\lambda\Lambda_L^\nu$.*

Proof. Choose h so that $\lambda(h) \neq \nu(h)$. Then

$$\nu(h)\nu(\Lambda_L^\lambda)\Lambda_L^\nu = \nu(h)\Lambda_L^\lambda\Lambda_L^\nu = h\Lambda_L^\lambda\Lambda_L^\nu = (h\Lambda_L^\lambda)\Lambda_L^\nu = \lambda(h)\Lambda_L^\lambda\Lambda_L^\nu = \lambda(h)\nu(\Lambda_L^\lambda)\Lambda_L^\nu$$

so $\nu(\Lambda_L^\lambda)(\lambda(h) - \nu(h))\Lambda_L^\nu = 0$. Since we are working over an integral domain and within a free module, this implies that $\nu(\Lambda_L^\lambda) = 0$. Moreover, $\Lambda_L^\lambda\Lambda_L^\nu = \nu(\Lambda_L^\lambda)\Lambda_L^\nu = 0$. □

It is now natural to seek a "Hopf-theoretic" analogue for the second orthogonality relation (which might involve only weights, and not all irreducible characters). Note that in general, a Hopf algebra might not have a nontrivial (sub)group of linear characters that is finite: for instance, $U(\mathfrak{g})$ for a complex Lie algebra $\mathfrak{g}$. However, if such subgroups do exist, then we attempt to evaluate the sum of all weights in the subgroup (*i.e.*, the coefficient of each weight is $\dim \rho = 1$), at various elements of H. For example, one may ask: does this sum always vanish at a nontrivial grouplike $g \in H$?

This problem is also interesting from another perspective. Given a Hopf algebra H, it is interesting to seek connections between the representations of H and those of H^* (or, the comodules over H^* and over H respectively). A famous example where both of these come into play is the well-known formula due to Radford [1] concerning the antipode. The setting of this paper is another such case: note that the weights of H are precisely the one-dimensional representations of H^* (or the group-like elements of H^*), while grouplike and skew-primitive elements in a Hopf algebra H correspond to the 1-dimensional and non-semisimple 2-dimensional H-comodules (or H^*-modules) respectively. The present paper explores the interplay between these objects.

1.2. One of the Setups, and Some References

Instead of attempting a summary of the results, which are several and computational, we make some remarks. Given an R-algebra A, let Γ_A denote the set of weights of A. (If H is a Hopf algebra, then Γ_H is a group.) Let $\Pi \subset \Gamma_H$ be a finite group of weights of a Hopf algebra H over a commutative unital integral domain R. One has the notion of *grouplike* elements (*i.e.*, $\Delta(g) = g \otimes g$) and *skew-primitive* elements in H.

The first step is to compute Σ_Π at all grouplike and skew-primitive elements in H, where $\Sigma_\Pi := \sum_{\gamma \in \Pi} \gamma : H \to R$. In a wide variety of examples—including finite-dimensional pointed Hopf algebras [2]—the computations reduce to grouplike elements. In other words, there are several families of algebras, where knowing Σ_Π at grouplike elements effectively tells us Σ_Π at *all* elements.

The next objective is to evaluate Σ_Π at products of skew-primitive elements. Once again, in the spirit of the previous paragraph, there are numerous examples of Hopf algebras generated by grouplike and skew-primitive elements in the literature. The first two examples below are from folklore, and references can be found in [3].

1. By the Cartier–Kostant–Milnor–Moore Theorem (*e.g.*, see [3, Theorem 5.6.5]), every cocommutative connected Hopf algebra H over a field of characteristic zero, is of the form $U(\mathfrak{g})$, where $\mathfrak{g}$ is the set of primitive elements in H. Similarly, every complex cocommutative Hopf algebra is generated by primitive and grouplike elements.
2. If the Hopf algebra is pointed (and over a field), then by the Taft–Wilson Theorem [3, Theorem 5.4.1.1], our results can evaluate Σ_Π on any element of C_1, the first term in the coradical filtration (which is spanned by grouplike and skew-primitive elements).
3. The final example is from a recent paper [2]. The Classification Theorem 0.1 says, in particular, that if H is a finite-dimensional pointed Hopf algebra over an algebraically closed field of characteristic zero, and the grouplike elements form an Abelian group of order coprime to 210, then H is generated by grouplike and skew-primitive elements, and is a variation of a small quantum group of Lusztig.

We conclude this section with one of our results. Say that an element $h \in H$ is *pseudo-primitive* with respect to Π if $\Delta(h) = g \otimes h + h \otimes g'$ for grouplike g, g' satisfying $\gamma(g) = \gamma(g')$ for all $\gamma \in \Pi$.

Theorem 1.2. *Fix $n \in \mathbb{N}$, as well as the R-Hopf algebra H and a finite subgroup of weights Π of H. Suppose $h_1, \ldots, h_n \in H$ are pseudo-primitive with respect to Π, and $\Delta(h_i) = g_i \otimes h_i + h_i \otimes g_i'$ for all i. Define $\mathbf{h} = \prod_i h_i$, and similarly, $\mathbf{g}, \mathbf{g}'$.*

1. *If $\operatorname{char}(R) = 0$ or $\operatorname{char}(R) \nmid |\Pi|$, then $\Sigma_\Pi(\mathbf{h}) = 0$.*
2. *Suppose $0 < p = \operatorname{char}(R)$ divides $|\Pi|$, and Π_p is any p-Sylow subgroup. If $\Pi_p \not\cong (\mathbb{Z}/p\mathbb{Z})^m$ for any $m > 0$, then $\Sigma_\Pi(\mathbf{h}) = 0$.*
3. *(p as above.) Define $\Phi := \Pi/[\Pi, \Pi]$, and by above, suppose $\Phi_p \cong (\mathbb{Z}/p\mathbb{Z})^k$ is a p-Sylow subgroup of Φ. Let Φ' be any Hall complement(ary subgroup); thus $|\Phi'| = |\Phi|/|\Phi_p|$. Then*

$$\Sigma_\Pi(\mathbf{h}) = |[\Pi, \Pi]| \cdot \Sigma_{\Phi'}(\mathbf{g}) \cdot \Sigma_{\Phi_p}(\mathbf{h})$$

4. *If* $\Sigma_{\Phi_p}(\mathbf{h})$ *is nonzero, then* $(p-1)|n$, *and* $0 \leq k \leq n/(p-1)$. *(Moreover, examples exist wherein* $\Sigma_{\Phi_p}(\mathbf{h})$ *can take any value* $r \in R$.*)*

These results occur below as Proposition 5.5, and Theorems 7.8, 7.5, and 7.10 respectively.

1.3. Organization

We quickly explain the organization of this paper. In Section 2, we compute Σ_Π at all grouplike elements in a Hopf algebra. This turns out to be extremely useful in computing Σ_Π in a large class of examples. More precisely, many algebras A that are well-studied in the literature contain a Hopf (sub)algebra H and whose sets of weights are subsets of Γ_H and kill a large subspace of A. These examples include quantum groups, the quantum Virasoro algebra, and finite-dimensional pointed Hopf algebras.

In Section 3, we evaluate Σ_Π at all skew-primitive elements of H. This is followed by a brief remark concerning the "degenerate" example of quiver (co)algebras.

Computing Σ_Π at all products of skew-primitive elements is a difficult problem. We show in the next Section 4 that on occasion, it can be reduced to computing $\Sigma_{\Gamma_H/[\Gamma_H,\Gamma_H]}$. In Section 6, we are able to obtain results when Γ_H itself is Abelian. These computations are useful in working with products of special kinds of skew-primitive elements in other sections.

In the rest of the paper, we work with "pseudo-primitive elements" h_i in order to obtain more detailed results. In Section 5, we show that $\Sigma_\Pi(h_1 \dots h_n) = 0$ whenever $\mathrm{char}(R) \nmid |\Pi|$. In Section 7, we study the case when $\mathrm{char}(R) = p$ divides $|\Pi|$. In this case, there are severe restrictions on the p-Sylow subgroup of Π, in order for $\Sigma_\Pi(\mathbf{h})$ to be nonzero; moreover, we write down a result that helps compute $\Sigma_\Pi(\mathbf{h})$.

We conclude with a detailed study of further examples—Lie algebras, degenerate affine Hecke algebras of reductive type, and then a Hopf algebra generated by grouplike and skew-primitive elements, where $\Sigma_\Pi(\mathbf{h})$ can take on all values.

2. Grouplike Elements and Quantum Groups

2.1. Preliminaries

We first set some notation, and make some definitions.

Definition 2.1. Suppose R is a commutative unital integral domain.

1. *Integers* in R are the image of the group homomorphism $\mathbb{Z} \to R$, sending $1 \mapsto 1$.
2. A *weight* of an R-algebra H is an R-algebra map : $H \to R$. Denote the set of weights by Γ_H. Occasionally we will also use $\Gamma = \Gamma_H$. Given $\nu \in \Gamma_H$, the ν-*weight space* of an H-module V is $V_\nu := \{v \in V : h \cdot v = \nu(h)v \ \forall h \in H\}$.
3. Given a left R-module H, define $H^* := \mathrm{Hom}_{R-\mathrm{mod}}(H, R)$.
4. An R-Hopf algebra H is an R-algebra $(H, \mu = \cdot, \eta)$ (where μ, η are coalgebra maps) that is also an R-coalgebra (H, Δ, ε) (where Δ, ε are algebra maps), further equipped with an antipode S (which is an R-(co)algebra anti-homomorphism).
5. In a Hopf algebra (or a bialgebra), an element h is *grouplike* if $\Delta(h) = h \otimes h$, and *primitive* if $\Delta(h) = 1 \otimes h + h \otimes 1$. Define $G(H)$ (respectively H_{prim}) to be the set of grouplike (respectively primitive) elements in a Hopf algebra H.

There are several standard texts on Hopf algebras; for instance, see [3–6]. In particular, since H is a coalgebra, H^* is also an R-algebra under *convolution* Δ^*: given $\lambda, \nu \in H^*$ and $h \in H$, one defines $\langle \lambda * \nu, h\rangle := \langle \lambda \otimes \nu, \Delta(h)\rangle$. By [4, Theorem 2.1.5] (also see [6, Lemma 4.0.3]), the set Γ_H of weights is now a group under $*$, with inverse given by $\langle \lambda^{-1}, h\rangle := \langle \lambda, S(h)\rangle$.

Note that for an algebra H over a field k, the dual space H^* is not a coalgebra in general. However, define H° to be the set of linear functionals $f : H \to k$ whose kernel contains an ideal of finite

k-codimension. Then by [5, Proposition 1.5.3 and Remark 1.5.9], H° is a coalgebra whose set of grouplike elements is precisely Γ_H.

Standing Assumption 2.2. For this article, H is any Hopf algebra over a commutative unital integral domain R. Fix a finite subgroup of weights $\Pi \subset \Gamma = \Gamma_H$.

In general, given a finite subgroup $\Pi \subset \Gamma_H$ for any Hopf algebra H, the element $\Sigma_\Pi := \sum_{\gamma \in \Pi} \gamma$ is a functional in H^*, and if R (*i.e.*, its quotient field) has characteristic zero, then Σ_Π does not kill the scalars $\eta(R)$. What, then, is its kernel? How about if Π is cyclic, or all of Γ_H (this, only if H is R-free, and finite-dimensional over the quotient field of R)?

Lemma 2.3. *Suppose H is a Hopf algebra, and $\Pi \subset \Gamma_H$ is a finite subgroup of weights.*

1. $\Sigma_\Pi(1) = 0$ *if and only if* $\operatorname{char}(R)$ *divides* $|\Pi|$.
2. $[H, H] \subset \ker \Sigma_\Pi$.
3. $\Sigma_\Pi(\operatorname{ad} h(h')) = \varepsilon(h)\Sigma_\Pi(h')$ *for all $h, h' \in H$. In particular, if $h \in \ker \varepsilon$, then* $\operatorname{im}(\operatorname{ad} h) \subset \ker \Sigma_\Pi$.

Here, ad stands for the usual *adjoint action* of H. In other words, $\operatorname{ad} h(h') := \sum h_{(1)} h' S(h_{(2)})$, where we use *Sweedler notation*: $\Delta(h) = \sum h_{(1)} \otimes h_{(2)}$.

Proof. The first part is easy, and the other two follow because the statements hold if Σ_Π is replaced by any (algebra map) $\gamma \in \Gamma_H$. □

The goal of this section and the next is to evaluate Σ_Π at all grouplike and skew-primitive elements in H. For these computations, a key fact to note is that for all $\lambda \in \Pi$, the following holds in the R-algebra H^*:

$$\lambda * \Sigma_\Pi = \sum_{\nu \in \Pi} \lambda * \nu = \Sigma_\Pi = \cdots = \Sigma_\Pi * \lambda \tag{2.4}$$

Remark 2.5. We occasionally compute Σ_Π with H an R-algebra (that is not a Hopf algebra), where $\Pi \subset \Gamma_H$ has a group structure on it. As seen in Proposition 2.13 below, there is an underlying Hopf algebra in some cases.

Definition 2.6. Suppose we have a subset $\Theta \subset \Gamma = \Gamma_H$, and $\lambda \in \Pi$.

1. For $g \in G(H)$, set $\Gamma_g := \{\gamma \in \Gamma : \gamma(g) = 1\}$, and $\Theta_g := \Gamma_g \cap \Theta$.
2. $G_\Theta(H)$ is the set (actually, normal subgroup) of grouplike elements $g \in G(H)$ so that $\gamma(g) = 1$ for all $\gamma \in \Theta$.
3. For finite Θ, the functional $\Sigma_\Theta \in H^*$ is given by $\Sigma_\Theta := \sum_{\gamma \in \Theta} \gamma$. Also set $\Sigma_\varnothing := 0$.
4. $n_\lambda := o_\Pi(\lambda) = |\langle \lambda \rangle|$ is the order of λ in Π.

Remark 2.7.

1. For instance, $G_{\{\varepsilon\}}(H) = G(H)$, and $G(H) \cap [1 + \operatorname{im}(\operatorname{id} - S^2)] \subset G_\Gamma(H)$ because every $\gamma \in \Gamma = \Gamma_H$ equals $(\gamma^{-1})^{-1}$. This follows since from earlier in this section, Γ_H is a group with unit ε and inverse given by $\gamma \mapsto \gamma \circ S$.
2. For any $g \in G(H)$ and $\Theta \subset \Gamma_H$, $\Theta_g = \Theta$ if and only if $g \in G_\Theta(H)$.
3. Θ_g is a subgroup if Θ is.

2.2. Grouplike Elements

We first determine how Σ_Π acts on grouplike elements, and answer the motivating question above of finding a Hopf-theoretic analogue of the second orthogonality relations for group characters.

Proposition 2.8 ("Orthogonality" at grouplike elements). *If $g \in G(H) \setminus G_\Pi(H)$, then $\Sigma_\Pi(g) = 0$. If $g \in G_\Pi(H)$, then $\Sigma_\Pi(g) = |\Pi|$.*

Proof. To show the first part, apply Equation (2.4) to g, with $\lambda \notin \Gamma_g$. The second part is obvious. □

We now introduce some notation, which is used in discussing several examples.

Definition 2.9. Let R be a commutative unital integral domain, and l be a nonnegative integer.

1. $\sqrt{1}$ (respectively $\sqrt[l]{1}$) is the set of (l^{th}) roots of unity in R. (Thus, $\sqrt[0]{1} = R^\times$, $\sqrt[1]{1} = \{1\}$, and $\sqrt{1} = \cup_{l>0}\sqrt[l]{1}$.)
2. Given $q \in R^\times$, $\text{char}(q)$ is the smallest positive integer m so that $q^m = 1$, and zero if no such m exists.
3. The group $\mathfrak{G}_{n,l}$ is the Abelian group generated by $\{K_i : 1 \leq i \leq n\}$, with relations $K_iK_j = K_jK_i, K_i^l = 1$.
4. The group $\mathfrak{G}^*_{n,l}$ is defined to be $(\sqrt{1} \cap \sqrt[l]{1})^n$.

Thus, $\mathfrak{G}_{n,l}$ is free if and only if $l = 0$, $\mathfrak{G}^*_{n,l} = (\sqrt[l]{1})^n \ \forall l > 0$, and $\mathfrak{G}^*_{n,0} = (\sqrt{1})^n$.

In light of the motivation, the first example where we apply the above result is:

Example 2.10 (Group rings)**.** The above result computes Σ_Π on all of H, if H is a group ring. We present a specific example: $G = \mathfrak{G}_{n,l}$ (defined above), for (a fixed) $n \in \mathbb{N}$ and $l \geq 0$. Then $\Gamma = (\sqrt[l]{1})^n$, and any finite order element $\gamma \in \Gamma$ maps each K_i to a root of unity in R. Thus, $\Pi \subset (\sqrt{1})^n \cap \Gamma = \mathfrak{G}^*_{n,l}$.

We now compute $\Sigma_\Pi(g)$ for some $g = \prod_{i=1}^n K_i^{n_i}$, where $n_i \in \mathbb{Z} \ \forall i$. Note that the set $\{\gamma(K_i) : 1 \leq i \leq n, \ \gamma \in \Pi\}$ is a finite set of roots of unity; hence the subgroup of $\sqrt{1} \subset R^\times$ that it generates is cyclic, say $\langle\zeta\rangle$. Thus, $\gamma(K_i) = \zeta^{l_i(\gamma)}$ for some $l_i : \Pi \to \mathbb{Z}$.

The above result now says that $\Sigma_\Pi(g) = 0$ if there exists $\gamma \in \Pi$ so that $|\langle\zeta\rangle|$ does not divide $\sum_{i=1}^n n_i l_i(\gamma)$, and $|\Pi|$ otherwise. (Of course, one can also apply the above result directly to $g = \prod_i K_i^{n_i}$.)

Remark 2.11. From the above proposition, finding out if $g \in G_\Pi(H)$ is an important step. However, since $\beta(g) = 1 \ \forall \beta \in [\Pi, \Pi]$, it suffices to compute if $\lambda(g) = 1$, where the λ's are the lifts of a set of generators of the (finite) Abelian group $\Pi/[\Pi, \Pi]$. We see more on this in Section 4 below.

2.3. *Application to Quantum Groups and Related Examples*

We now mention some more examples where the above result applies: Hopf algebras that quantize semisimple Lie algebras, their Borel subalgebras, and polynomial algebras (*i.e.*, coordinate rings of affine spaces/Abelian Lie algebras). There are yet other algebras mentioned below, which are not Hopf algebras but can be treated similarly.

To discuss these examples, some more basic results are needed; here is the setup for them. Suppose an R-algebra A contains a Hopf subalgebra H, so that A is an ad H-module (with possible weight spaces A_ν). Then one has the following result:

Lemma 2.12. *Suppose $\mu \in \Gamma_A$ (i.e., $\mu : A \to R$ is an R-algebra map). If $\nu \in \Gamma_H$ and $\nu \neq \varepsilon$, then $\mu \equiv 0$ on A_ν.*

Proof. Given $\nu \neq \varepsilon$, choose $h \in H$ such that $\nu(h) \neq \varepsilon(h)$. Now given $a_\nu \in A_\nu$, apply μ to the equation: $\text{ad}\, h(a_\nu) = \nu(h)a_\nu$. Simplifying this yields: $\varepsilon(h)\mu(a_\nu) = \nu(h)\mu(a_\nu)$, and since R is an integral domain, $\mu(a_\nu) = 0$. $\square$

Applying this easily yields the following result.

Proposition 2.13. *(A, H as above.) Suppose an R-algebra A contains H and a vector subspace V, that is of the form $V = \oplus_{\nu\neq\varepsilon} V_\nu$ (for the* ad *H-action).*

1. *Every $\mu \in \Gamma_A$ kills AVA.*
2. *If $A = H + AVA$, then $\Gamma_A \subset \Gamma_H$.*
3. *Say $A = H + AVA$, and $\Pi \subset \Gamma_A$ is a finite subgroup of weights of A (from above). If $a \in A$ satisfies $a - \sum_{g\in G(H)} a_g g \in AVA$ (where $a_g \in R \ \forall g$), then $\Sigma_\Pi(a) = |\Pi| \sum_{g \in G_\Pi(H)} a_g$.*

Remark 2.14. Thus, if $\Pi \subset \Gamma_A$ is a group (*i.e.*, a subset with a group structure on it), then $\Sigma_\Pi(AVA) = 0$. Hence, computing Σ_Π at any $a \in A$ essentially reduces to the case of the Hopf subalgebra H. When H is a group algebra, Proposition 2.8 above tells us the answer in this case-assuming that the group operation in Π agrees with the one in Γ_H.

Moreover, even though A is not a Hopf algebra here, note that the computations come from Hopf algebra calculations (for H).

Proof. The first part follows from Lemma 2.12, the third part now follows from Proposition 2.8 and Lemma 2.12, and the second part follows by observing that $\mu : H \to R$ is an algebra map only if $\mu|_G \in \Gamma_G$ and $\mu|_V \equiv 0$. (Additional relations in H may prevent every $\mu \in \Gamma_H$ from being a weight in Γ_A.) $\square$

It is now possible to apply the above theory to some examples; note that they are not always Hopf algebras. In each case, G is of the form $\mathfrak{G}_{n,l}$ for some n, l. Also choose a special element $q \in R^\times$ in each case; then $\text{char}(q)|l$.

Example 2.15 ("Restricted" quantum groups of semisimple Lie algebras). For this example, $R = k$ is a field with char $k \neq 2$, with a special element $q \neq 0, \pm 1$. Suppose $\mathfrak{g}$ is a semisimple Lie algebra over k, together with a fixed Cartan subalgebra and root space decomposition (*e.g.*, using a Chevalley basis, as in [7, Chapter 7]).

One then defines the (Hopf) algebra $U_q(\mathfrak{g})$ as in [8, §4.2,4.3]. In particular, note that it is generated by $\{K_j^{\pm 1}, e_j, f_j : 1 \leq j \leq n\}$ (here, n is the rank of $\mathfrak{g}$, and the α_i are simple roots), modulo the relations:

$$K_i e_j K_i^{-1} = q^{(\alpha_i, \alpha_j)} e_j, \ K_i f_j K_i^{-1} = q^{-(\alpha_i, \alpha_j)} f_j, \quad e_i f_j - f_j e_i = \delta_{ij} \frac{K_i - K_i^{-1}}{q_i - q_i^{-1}}$$

where $q_i = q^{(\alpha_i, \alpha_i)/2}$ for some bilinear form $(\cdot, \cdot)$ on $\mathfrak{h}^*$. (We may also need that q^4 or q^6 is not 1, and possibly also that char $k \neq 3$.) The other relations are that $K_i K_j = K_j K_i$, $K_i^{\pm 1} K_i^{\mp 1} = 1$, and the (two) quantum Serre relations. Define $V := \oplus_{j=1}^n (k e_j \oplus k f_j)$.

Now define the "restricted" quantum group as in [9, Chapter 6]. More precisely, given some fixed $l \geq 0$ so that $q^l = 1$ (whence $\text{char}(q)|l$), define the associative (not necessarily Hopf) algebra $u_{q,l}(\mathfrak{g})$ to be the quotient of $U_q(\mathfrak{g})$ by the relations (for all j) $K_j^l = 1$, and $e_j^l = f_j^l = 0$ if $l > 0$. Note that $u_{q,l}(\mathfrak{g}) = U_q(\mathfrak{g})$ if $l = 0$, and $u_{q,l}(\mathfrak{g})$ is a Hopf algebra if $l = 0$ or $\text{char}(q)$. Moreover, Proposition 2.13 allows us to compute Σ_Π for all l.

For each j, note that e_j, f_j are weight vectors (with respect to the adjoint action of the Abelian group $G := \mathfrak{G}_{n,l}$ generated by all $K_i^{\pm 1}$) with weights $q^{\pm \alpha_j} \neq \varepsilon = q^0$. Hence Lemma 2.12 implies that $\mu(e_j) = \mu(f_j) = 0 \ \forall j, \mu$. Moreover, given the PBW property for $A = U_q(\mathfrak{g})$, we know that $A = k\mathfrak{G}_{n,l} \bigoplus (V_- A + A V_+)$, where V_+, V_- are the spans of the e_j's and f_j's respectively. Hence $\mu \in \Gamma_A \subset \Gamma_G$ by the result above.

Every $\mu \in \Gamma_G$ is compatible with the commuting of the K_i's, the quantum Serre relations (since $\mu|_V \equiv 0$), and the "l^{th} power relations". The only restriction is the last one left, namely: $0 = \mu([e_i, f_i])$, which gives us that $\mu(K_i) = \mu(K_i^{-1})$ for all i, μ. Hence $\mu(K_i) = \pm 1$, so that $\Gamma_A \cong (\mathbb{Z}/2\mathbb{Z})^n \cap (\sqrt[l]{1})^n$, which is of size 2^n or 1, depending on whether l is even or odd. We now compute $\Sigma_\Pi(a)$ using the second part of Proposition 2.13, for any $a \in A$.

Example 2.16 (Restricted quantum groups of Borel subalgebras). We consider the subalgebra $A' = u_{q,l}(\mathfrak{b})$ of $A = u_{q,l}(\mathfrak{g})$, which is generated by $\{K_i^{\pm 1}, e_i : 1 \leq i \leq n\}$; once again, this algebra quantizes the Borel subalgebra $\mathfrak{b}$ of $\mathfrak{g}$ if $l = 0$ (and is a Hopf algebra if $l = 0$ or $\text{char}(q)$). Moreover, every $\mu \in \Gamma_{A'}$ kills each e_i (where we use V_+ for V), and we have $\Gamma_{A'} \subset \Gamma_G$, by Proposition 2.13 above. Moreover, all such maps $\mu \in \Gamma_G$ are admissible (*i.e.*, extend to all of A'), so $\Gamma_{A'} = \Gamma_G \cong (\sqrt[l]{1})^n$.

Now if $\Pi \subset \Gamma_{A'}$ is a finite subgroup, then as above, $\Pi \subset \mathfrak{G}_{n,l}^*$ (note that $R = k$ here), and furthermore, evaluation of Σ_Π once again reduces to the grouplike case.

Example 2.17 (Taft algebras). Given a primitive nth root q of unity, the *nth Taft algebra* is

$$T_n := R\langle x, g \rangle / (gx - qxg, \ g^n - 1, \ x^n)$$

Once again, every weight must kill x and sends g to some power of q. Hence the set Γ_{T_n} of weights is cyclic, whence so is Π. It is now easy to show:

Lemma 2.18. *Every weight kills x, and $\Sigma_\Pi(g^k) = |\Pi|$ if $|\Pi|$ divides k, or 0 otherwise.*

Example 2.19 (Quantization of affine space). We refer to [10]; once again, R is a unital commutative integral domain. The quantum affine space over R (with a fixed element $q \in R^\times$) is the quadratic algebra $T_R(V)/(x_j x_i - q x_i x_j, i < j)$, where $V := \oplus_i R x_i$. This does not have a Hopf algebra structure; however, Hu presents a quantization of $R[x] := \mathrm{Sym}_R(V)$ in [10, §5]-that is, the quantum group associated to a "finite-dimensional" Abelian Lie algebra. Now consider a more general associative (not necessarily Hopf) R-algebra $A = \mathcal{A}_{q,l}(n)$ generated by $\{K_i^{\pm 1}, x_i : 1 \leq i \leq n\}$, with the relations:

$$K_i K_j = K_j K_i, \qquad K_i^l = 1, \qquad K_i^{\pm 1} K_i^{\mp 1} = 1$$
$$K_i x_j K_i^{-1} = \theta_{ij} q^{\delta_{ij}} x_j, \qquad x_i x_j = \theta_{ij} x_j x_i$$

where $\mathrm{char}(q)|l$, and θ_{ij} equals q (respectively $1, q^{-1}$) if $i > j$ (respectively $i = j$, $i < j$). Note that $\mathcal{A}_{q,\mathrm{char}(q)}(n) = \mathcal{A}_q(n)$, the Hopf algebra introduced and studied by Hu, and $\mathcal{A}_{q,l}(n)$ becomes the Hopf algebra $R[x]$ if $q = l = 1$.

We consider the "nontrivial" case $q \neq 1$ (the $q = 1$ case is discussed later). Once again, each x_j is a weight vector with respect to the (free Abelian) group $G = \mathfrak{G}_{n,l}$, and no x_j is in the ε-weight space, so every $\mu \in \Gamma_A$ kills x_j for all j. As in the previous example, $\Gamma_A \cong (\sqrt[l]{1})^n$, and any finite subgroup Π must be contained in $\mathfrak{G}^*_{n,l}$.

Moreover, evaluation of Σ_Π once again reduces to the grouplike case.

Example 2.20 (Quantization of the Virasoro algebra). For this example, assume that R is a field, and $q \in R^\times$ is not a root of unity. Now refer to [11, Page 100] for the definitions; the Hopf algebra in question is the R-algebra $\mathcal{U}_q$ generated by $\mathcal{T}, \mathcal{T}^{-1}, c, e_m (m \in \mathbb{Z})$ with relations:

$$\begin{aligned}
\mathcal{T}\mathcal{T}^{-1} &= \mathcal{T}^{-1}\mathcal{T} = 1 \\
q^{2m}\mathcal{T}^m c &= c\mathcal{T}^m \\
\mathcal{T}^m e_n &= q^{-2(n+1)m} e_n \mathcal{T}^m \\
q^{2m} e_m c &= c e_m \\
q^{m-n} e_m e_n - q^{n-m} e_n e_m &= [m-n] e_{m+n} + \delta_{m+n,0} \frac{[m-1][m][m+1]}{[2][3]\langle m \rangle} c
\end{aligned}$$

where $[m] := \dfrac{q^m - q^{-m}}{q - q^{-1}}$ and $\langle m \rangle := q^m + q^{-m}$ for all $m \in \mathbb{Z}$.

One can now compute the group of weights, as well as $\Sigma_\Pi(h)$ for any monomial word h in the above alphabet. The following is proved using the defining relations above.

Proposition 2.21. *Setup as above.*

1. *The group of weights is $\Gamma_{\mathcal{U}_q} = (R^\times, \cdot)$ (so every finite subgroup Π is cyclic). A weight $r \in R^\times$ kills c and all e_n, and sends $\mathcal{T}$ to r.*
2. *$\Sigma_\Pi(h) = 0$ if the monomial word h contains c or any e_n. Moreover, $\Sigma_\Pi(\mathcal{T}^m) = 0$ unless $|\Pi|$ divides m, in which case $\Sigma_\Pi(\mathcal{T}^m) = |\Pi|$.*

Example 2.22 (Quantum linear groups). For the definitions, we refer to [12, §2]. The *quantum general* (respectively *special*) *linear group* $GL_q(n) = R_q[GL_n]$ (respectively $SL_q(n) = R_q[SL_n]$) is the localization of the algebra $\mathcal{B}$ (defined presently) at the central *quantum determinant*

$$\det{}_q := \sum_{\pi \in S_n} (-q)^{l(\pi)} \prod_{i=1}^{n} u_{i,\pi(i)}$$

(respectively the quotient of $\mathcal{B}$ by the relation $\det_q = 1$). Here, the algebra $\mathcal{B} = R_q[\mathfrak{gl}(n)]$ is generated by $\{u_{ij} : 1 \leq i,j \leq n\}$, with relations

$$u_{ik}u_{il} = qu_{il}u_{ik}, \qquad u_{ik}u_{jk} = qu_{jk}u_{ik}$$
$$u_{il}u_{jk} = u_{jk}u_{il}, \qquad u_{ik}u_{jl} - u_{jl}u_{ik} = (q - q^{-1})u_{il}u_{jk}$$

where $q \in R^\times$, and $i < j, k < l$.

As above, it is possible to compute the group of weights Γ for both families of algebras, in the "nontrivial" case $q \neq \pm 1$. In either case, note that $\det_q \neq 0$. Given any permutation $\pi \in S_n$, suppose there exist $i < j$ such that $\pi(i) > \pi(j)$. Then

$$(q - q^{-1})\mu(u_{i,\pi(i)}u_{j,\pi(j)}) = \mu([u_{i,\pi(j)}, u_{j,\pi(i)}]) = 0$$

whence $\mu(u_{i,\pi(i)}u_{j,\pi(j)}) = 0$ for all $\mu \in \Gamma$.

The only permutations for which this does not happen is $\{\pi \in S_n : i < j \Rightarrow \pi(i) < \pi(j)\} = \{\mathrm{id}\}$. Hence $\mu(\det_q) = \prod_i \mu(u_{ii}) \neq 0$, whence no u_{ii} is killed by any μ. But now for $i < l$,

$$\mu(u_{ii}u_{il}) = q\mu(u_{il}u_{ii}), \qquad \mu(u_{ii}u_{li}) = q\mu(u_{li}u_{ii})$$

whence $\mu(u_{ij}) = 0$ for all $i \neq j, \mu \in \Gamma$. In particular, since $\Delta(u_{ij}) = \sum_{k=1}^n u_{ik} \otimes u_{kj}$ for (all i,j and) both $GL_q(n)$ and $SL_q(n)$, hence $\Gamma_{GL} \cong (R^\times)^n$, and $\Gamma_{SL} \cong (R^\times)^{n-1}$ (both under coordinate-wise multiplication), since $\mu(u_{nn}) = \prod_{i=1}^{n-1} \mu(u_{ii})^{-1}$ in the latter case.

Finally, computing Σ_Π now reduces to the above results and the first example (of the free group $G = \mathfrak{G}_{n,0}$). This is because any $h \in GL_q(n)$ or $SL_q(n)$ can be reduced to a sum of monomial words, and such a word is not killed by any μ if and only if there is no contribution from any $u_{il}, i \neq l$.

Example 2.23 (Hopf regular triangular algebras)**.** These were defined in [13] (in the special case $\Gamma = 1$).

Definition 2.24. An associative k-algebra A (over a ground field k) is a *Hopf RTA* (or *HRTA*), if:

1. The multiplication map $B_- \otimes_k H \otimes_k B_+ \to A$ is an isomorphism, for some (fixed) associative unital k-subalgebras $H, B_\pm$ of A, and H is, in addition, a commutative Hopf algebra.
2. The set $G := \mathrm{Hom}_{k-alg}(H,k)$ contains a free Abelian group with finite basis Δ, so that $B_\pm = \bigoplus_{\lambda \in \pm\mathbb{Z}_{\geqslant 0}\Delta} (B_\pm)_\lambda$. Each summand here is a finite-dimensional weight space for the (usual) adjoint action of H, and $(B_\pm)_0 = k$.
3. There exists an anti-involution i of A, so that $i|_H = \mathrm{id}\,|_H$.

This is a large family of algebras that are widely studied in representation theory. Examples (when $\mathrm{char}(k) = 0$) are $U(\mathfrak{g})$ for $\mathfrak{g}$ a semisimple, symmetrizable Kac–Moody, centerless Virasoro, or (centerless) extended Heisenberg Lie algebra. Other examples include quantum groups $U_q(\mathfrak{g})$ or (quantized) infinitesimal Hecke algebras over $\mathfrak{sl}_2$.

In all these examples, the computations reduce to H:

Lemma 2.25. $\Gamma_A \subset \Gamma_H$.

The same happens if one works with *skew group rings* over A, for example, wreath products $S_n \wr A := A^{\otimes n} \rtimes S_n$; in this case H is replaced in the above lemma, by the Hopf subalgebra $H^{\otimes n} \rtimes S_n$ of $S_n \wr A$.

Proof. Use Proposition 2.13, with $V = N_- + N_+$, where $N_\pm$ are the augmentation ideals in $B_\pm$. Hence $\Gamma_A \subset \Gamma_H$ (note that each weight must also kill $[A,A] \cap H$). $\square$

Example 2.26 (Finite-dimensional pointed Hopf algebras)**.** Assume that:

1. $R = \overline{k(R)}$ is an algebraically closed field of characteristic zero.

2. H is a finite-dimensional pointed Hopf algebra over R.
3. $G(H)$ is a (finite) Abelian group of order coprime to 210.

Then by the Classification Theorem 0.1 of [2], H is generated by $G(H)$ and some skew-primitive (defined in the next section) generators $\{x_i\}$ that satisfy $gx_ig^{-1} = \chi_i(g)x_i$ for all i and all $g \in G(H)$. Since $\chi_i \neq \varepsilon\ \forall i$ by [2, Equation (0.1)], hence Proposition 2.13 again reduces the computations here, to the grouplike case: $\Gamma_H \subset \Gamma_{G(H)}$.

3. Skew-Primitive Elements

As the previous example 2.26 suggests, large families of Hopf algebras that are the subject of much study in the literature are generated by grouplike and "skew-primitive" elements. Hence in the rest of this paper, we address the question of computing Σ_Π at (monomial words in) such elements.

Definition 3.1. An element $h \in H$ is *skew-primitive* if $\Delta(h) = g \otimes h + h \otimes g'$ for grouplike $g, g' \in G(H)$. Denote the set of such elements by $H_{g,g'}$. Then $g - g' \in H_{g,g'} \cap H_{g',g}$, and $H_{1,1} = H_{prim}$ (recall Definition 2.1).

An element h is *pseudo-primitive* (respectively *almost primitive*) with respect to Π if, moreover, $g^{-1}g' \in G_\Pi(H)$ (respectively $g, g' \in G_\Pi(H)$). In future, we may not specify the finite subgroup Π of Γ_H, because it is part of the given data.

Thus, {skew-primitive} $\supset$ {pseudo-primitive} $\supset$ {almost primitive} $\supset$ {primitive}.

Remark 3.2. Note that grouplike and skew-primitive elements in H correspond to 1-dimensional and non-semisimple 2-dimensional H-comodules, respectively. To see this, suppose $V_0 = Rv_0$ and $g \in H$. Define $\rho(v_0) := v_0 \otimes g$. Then ρ induces an H-comodule structure on V_0 if and only if g is grouplike. Similarly, given $V_1 = Rv_0 \oplus Rv_1$ and $g, g', h \in H$, define:

$$\rho(v_0) := v_0 \otimes g, \qquad \rho(v_1) = v_0 \otimes h + v_1 \otimes g'$$

Then ρ induces an H-comodule structure on V_0 if and only if g, g' are grouplike and $h \in H_{g,g'}$ is (g, g')-skew-primitive.

Lemma 3.3. *Suppose $\Pi \subset \Gamma_H$ is a finite subgroup of weights, and $h \in H_{g,g'}$ is as above.*

1. *The set $\{\gamma \in \Gamma_H : \gamma(h) = 0\}$ is a subgroup of Γ_H.*
2. *$\varepsilon(h) = 0$ and $S(h) = -g^{-1}h(g')^{-1} \in H_{(g')^{-1},g^{-1}}$.*
3. *If g_0 is any grouplike element, then g_0h and hg_0 are also skew-primitive. Moreover, $g_0 - g_0^{-1} \in H_{g_0,g_0^{-1}} \cap H_{g_0^{-1},g_0}$.*
4. *For all $n \geq 0$, one also has:*
$$\Delta^{(n)}(h) = \sum_{i=0}^{n} g^{\otimes i} \otimes h \otimes (g')^{\otimes(n-i)} \tag{3.4}$$
5. *For any $\gamma \in \Pi$, either $\gamma(g) \neq \gamma(g')$, or $\gamma(h) = 0$, or $\text{char}(R)|n_\gamma$.*
6. *h is pseudo-primitive if and only if $g^{-1}h, hg^{-1}, (g')^{-1}h, h(g')^{-1}$ are almost primitive. If $g_0 \in G(H)$ and h is pseudo-primitive, then so are g_0h and hg_0.*

Proof. The first part is an easy verification. The second part follows from the statements

$$(\text{id} * \varepsilon)(h) := \mu(\text{id} \otimes \varepsilon)\Delta(h) = h, \qquad (\text{id} * S)(h) := \mu(\text{id} \otimes S)\Delta(h) = \varepsilon(h)$$

The third, fourth and last parts are now easy to verify. For the fifth part, suppose $\gamma(g) = \gamma(g')$ and γ has order $n_\gamma \geq 1$ in Π. Then compute using Equation (3.4) above:

$$0 = \varepsilon(h) = \gamma^{*n_\gamma}(h) = n_\gamma \gamma(h)\gamma(g)^{n_\gamma - 1}$$

The result follows, since $\gamma(g)$ is a unit and R is an integral domain. $\square$

Our main result here is to compute $\Sigma_\Pi(g_1 h g_2)$ for any $g_1, g_2 \in G(H)$, or equivalently by the lemma above, $\Sigma_\Pi(h)$ for all skew-primitive h. Equation (3.4) implies that if $\gamma \in \Gamma_H$, and $\gamma(g) \neq \gamma(g')$, then

$$\gamma^n(h) = \gamma^{\otimes n}(\Delta^{(n-1)}(h)) = \gamma(h) \cdot \frac{\gamma(g)^n - \gamma(g')^n}{\gamma(g) - \gamma(g')} \tag{3.5}$$

3.1. The Main Result

In all that follows below, assume that h is skew-primitive, with $\Delta(h) = g \otimes h + h \otimes g'$.

Theorem 3.6. *A skew primitive $h \in H_{g,g'}$ satisfies at least one of the following three conditions:*

1. *If there is $\lambda \in \Pi$ so that $\lambda(g), \lambda(g') \neq 1$, then $\Sigma_\Pi(h) = 0$. If no such λ exists, then one of g, g' is in $G_\Pi(H)$.*
2. *Suppose only one of g, g' is in $G_\Pi(H)$, so that there exists $\lambda \in \Pi$ with exactly one of $\lambda(g), \lambda(g')$ equal to 1. Then $\Sigma_\Pi(h) = \dfrac{|\Pi|\lambda(h)}{1 - \lambda(gg')}$.*
3. *If $\lambda(gg') = 1$ for all $\lambda \in \Pi$, then $\Sigma_\Pi(h) = \sum_{\gamma \neq \varepsilon = \gamma^2} \gamma(h)$, and $2\Sigma_\Pi(h) = 0$.*

Remark 3.7.

1. Thus, the expression $\lambda(h)/(1 - \lambda(gg')) = \Sigma_\Pi(h) \in k(R)$ is independent of λ (as long as $\lambda(gg') \neq 1$), for such h. As the proof indicates, $1 - \lambda(gg')$ should really be thought of as $1 - \lambda(g)$ or $1 - \lambda(g')$ (depending on which of g' and g is in $G_\Pi(H)$).
   Moreover, Equation (3.5) implies, whenever $\gamma^n(g) \neq \gamma^n(g')$, that
   $$\frac{\gamma^n(h)}{\gamma^n(g) - \gamma^n(g')} = \frac{\gamma(h)}{\gamma(g) - \gamma(g')}$$
   It is not hard to show that both of these are manifestations of the following easy fact:
   Lemma 3.8. *Given $h \in H_{g,g'}$, define $N_h := \{\gamma \in H^* : \gamma(g) \neq \gamma(g')\}$. Suppose $\mu, \lambda \in N_h$. Then $(\mu * \lambda)(h) = (\lambda * \mu)(h)$ if and only if $f_h(\mu) = f_h(\lambda)$, where $f_h : N_h \to k(R)$ is given by*
   $$f_h(\gamma) := \frac{\gamma(h)}{\gamma(g) - \gamma(g')}$$
   (In other words, weights commute at h precisely when they lie on the same "level surface" for f_h.)
2. Also note that if the first two parts fail to hold, then both $g, g' \in G_\Pi(H)$, and the final part holds. Thus, the above theorem computes $\Sigma_\Pi(h)$ for all skew-primitive h, if $\mathrm{char}(R) \neq 2$ or not both of g, g' are in $G_\Pi(H)$. We address the case when $\mathrm{char}(R) = 2$ and $g, g' \in G_\Pi(H)$, in the next subsection.

Proof of Theorem 3.6.

1. Apply Equation (2.4) to h, to get $\Sigma_\Pi(h) = \lambda(g)\Sigma_\Pi(h) + \lambda(h)\Sigma_\Pi(g')$. Since $g' \notin G_\Pi(H)$, hence the second term vanishes, and we are left with $(1 - \lambda(g))\Sigma_\Pi(h) = 0$. But $\lambda(g) \neq 1$.
   Next, if no such $\lambda \in \Pi$ exists, then $\Pi = \Pi_g \cup \Pi_{g'}$, where $\Pi_g, \Pi_{g'}$ were defined before Proposition 2.8. We claim that one of the two sets is contained in the other whence one of g, g' is in $G_\Pi(H)$, because, if not, then one can choose $\gamma, \gamma' \in \Pi$ so that neither $\gamma(g)$ nor $\gamma'(g')$ equals 1 (whence $\gamma(g') = 1 = \gamma'(g)$). Then one verifies that $\gamma\gamma' \notin \Pi_g \cup \Pi_{g'}$, which is a contradiction. Thus one of $\Pi_g \setminus \Pi_{g'}$, $\Pi_{g'} \setminus \Pi_g$ is empty.
2. Suppose $g' \in G_\Pi(H), \lambda(g) \neq 1$ for some λ (the other case is similar). Now apply Equation (2.4) and Proposition 2.8, and compute:
   $$\Sigma_\Pi(h) = \lambda(h)\frac{\Sigma_\Pi(g) - \Sigma_\Pi(g')}{\lambda(g) - \lambda(g')} = \lambda(h)\frac{0 - |\Pi|}{\lambda(g) \cdot 1 - 1} = \frac{\lambda(h)|\Pi|}{1 - \lambda(g)\lambda(g')}$$

3. If $\lambda(gg') = 1$ for any $\lambda \in \Pi$, then: $\lambda^{-1}(h) = -\lambda(h)$. Thus $\lambda + \lambda^{-1}$ kills h, and the first equation now follows because $\varepsilon(h) = 0$. The second is also easy: $2\Sigma_{\Pi}(h) = \sum_{\gamma \in \Pi}(\gamma(h) + \gamma^{-1}(h)) = 0$.

□

3.2. The Characteristic 2 Case

The only case that Theorem 3.6 does not address is when $\mathrm{char}(R) = 2$ and $g, g' \in G_{\Pi}(H)$. We now address this case.

Proposition 3.9. *Suppose Π is as above, $\mathrm{char}(R) = 2$, and $h \in H_{g,g'}$ is almost primitive with respect to Π.*

1. *If Π has odd order, then $\Sigma_{\Pi}(h) = 0$.*
2. *If 4 divides $|\Pi|$, then $\Sigma_{\Pi}(h) = 0$.*
3. *If Π has even order but $4 \nmid |\Pi|$, then $\Sigma_{\Pi}(h) = \gamma(h)$ for any $\gamma \in \Pi$ of order exactly 2. This may assume any nonzero value in R.*

We omit the proof, since this result is a special case of more general results in general (positive) characteristic, which we state and prove later. See Theorems 7.8 and 7.10, as well as Remark 7.11.

3.3. A Degenerate Example: Quiver (Co)algebras

We conclude this section with an example that is not a Hopf algebra, but an algebra with coproduct, and is generated by grouplike and skew-primitive elements. (Thus, Π is no longer necessarily a group of weights, but a semigroup.)

Consider a quiver $Q = (Q_0, Q_1)$, which is a directed graph with vertex set Q_0 and edges Q_1. Thus, there exist source and target maps $s, t : Q_1 \to Q_0$ such that every edge $e \in Q_1$ starts at $s(e)$ and ends at $t(e)$. A path in Q is a finite sequence of edges $a_1 \cdots a_n$ such that $t(a_i) = s(a_{i+1})$. We also write $s(p) := s(a_1)$ and $t(p) := t(a_n)$, and the length of the path is said to be n. Vertices $v \in Q_0$ are paths of length zero, and one writes $s(v) = t(v) := v$.

There are two structures on the free R-module RQ with basis consisting of all paths in Q. The *path algebra* is defined by setting the product of two paths $a_1 \cdots a_n$ and $b_1 \cdots b_m$ to be their concatenation $a_1 \cdots a_n b_1 \cdots b_m$ if $t(a_n) = s(b_m)$; otherwise the product is zero. Then RQ is an associative R-algebra that contains enough idempotents $\{v : v \in Q_0\}$. However, RQ contains a unit if and only if Q_0 is finite, in which case the unit is $\sum_{v \in Q_0} v$.

Another structure on RQ is given by defining $\Delta : RQ \to RQ \otimes RQ$ via: $\Delta(p) := \sum_{(q,r):qr=p} q \otimes r$. Also define $\varepsilon : RQ \to R$ via: $\varepsilon(v) := 1$ for all $v \in Q_0$ and $\varepsilon(p) := 0$ for all paths p of positive length. This structure makes RQ into the *path coalgebra*.

The path (co)algebra has been the subject of much study in the literature, and it is natural to ask for which quivers Q are these two structures on RQ compatible. It is not hard to show that the answer is: very few.

Lemma 3.10. *The coproduct Δ is multiplicative if and only if there are no paths of length ≥ 2.*

Thus, the quiver bialgebra is a "degenerate" example. Now suppose Q has no paths of length ≥ 2 (and hence, no self-loops). Then Δ is indeed multiplicative. Nevertheless, if $1 < |Q_0| < \infty$, then the unit in RQ is not grouplike. It is now possible to show the following.

Proposition 3.11. *There exists a bijection from $Q_0 \coprod \{0\}$ to the semigroup Γ_{RQ}, sending v to λ_v that sends v to 1 and all other paths to 0.*

Proof. Given $v \neq v' \in Q_0$ and $\lambda \in \Gamma_{RQ}$,

$$\lambda(v) = \lambda(v^2) = \lambda(v)^2, \qquad \lambda(v)\lambda(v') = \lambda(vv') = 0$$

Since R is a unital integral domain, this implies that at most one $\lambda(v)$ is nonzero—and then it equals 1. Moreover, if p is any path of length 1 in Q_1, then at least one of its vertices is killed by λ, whence so is p. This supplies the desired bijection. In particular, $\Gamma_{RQ} \cong Q_0 \coprod \{0\}$ as semigroups, with composition given by:

$$\lambda_v * \lambda_{v'} = \delta_{v,v'}\lambda_v, \qquad \lambda_v * 0 = 0 * \lambda_{v'} = 0, \qquad \forall v, v' \in Q_0$$

□

In fact, the path algebra and path coalgebra are dual to one another, morally speaking. More precisely, each of them can be "recovered" inside the dual space of the other; see [14] for more details.

4. Subgroups and Subquotients of Groups of Weights

4.1. Subgroups Associated to Arbitrary Elements

Recall that the goal of this article is to compute Σ_Π at any element h in a Hopf algebra H. We start this section with the following constructions.

Definition 4.1. Suppose R is a commutative unital integral domain.

1. Suppose H is an R-algebra, such that $\Gamma = \Gamma_H$ has a group structure $*$ on it. Define $\Gamma_h \subset \Gamma$ to be the subgroup of Γ that "stabilizes" h. In other words,

$$\Gamma_h := \{\gamma \in \Gamma : (\beta * \gamma * \delta)(h) = (\beta * \delta)(h)\ \forall \beta, \delta \in \Gamma\}$$

2. Given a coalgebra H, and $h \in H$, define C_h to be the R-subcoalgebra generated by h in H.
3. Given a Hopf algebra H, define Γ'_h to be the *fixed weight monoid* of h, given by $\Gamma'_h := \{\gamma \in \Gamma : \gamma|_{C_h} = \varepsilon|_{C_h}\}$.

In particular, $\gamma(h) = \varepsilon(h)$ if $\gamma \in \Gamma_h$.

A later subsection will discuss how this allows us to consider subquotients of Γ; but first, here are some observations involving these subgroups.

Proposition 4.2. *Suppose H is a Hopf algebra, and $h \in H$.*

1. *For all h, Γ_h is a normal subgroup of Γ, and $\Gamma'_h \subset \Gamma_h$ is a monoid closed under Γ-conjugation.*
2. *Given $\{h_i : i \in I\} \subset H$, and $h \in \langle h_i \rangle$ (i.e., in the subalgebra generated by the h_i's), $\Gamma_h \supset \bigcap_{i\in I} \Gamma_{h_i}$, and similarly for the Γ's.*
3. *Given any $h_i \in H$ (finitely many), suppose $\Pi = \times_i \Pi_i$, with $\Pi_i \subset \Gamma'_{h_j}$ whenever $i \neq j$. Then $\Sigma_\Pi(\mathbf{h}) = \prod_i \Sigma_{\Pi_i}(h_i)$.*

Proof. The first and third parts are straightforward computations. For the second part, for all β, $\delta \in \Gamma$, $\gamma \in \bigcap_i \Gamma_{h_i}$, and polynomials p in the h_i's,

$$(\beta * \gamma * \delta)(p(h_i)) = p((\beta * \gamma * \delta)(h_i)) = p((\beta * \delta)(h_i)) = (\beta * \delta)(p(h_i))$$

The outer equalities hold because weights are algebra maps.

The proof for the Γ's is as follows: if $h = p(h_i)$ as above, then since Δ is multiplicative, hence any $h' \in C_h$ is expressible as a polynomial in elements $h'_j \in \cup_i C_{h_i}$ - say $h' = q(h'_j)$. In particular, if $\gamma \in \Gamma'_{h_i}$ for all i, then

$$\gamma(q(h'_j)) = q(\gamma(h'_j)) = q(\varepsilon(h'_j)) = \varepsilon(q(h'_j))$$

where once again, the outer equalities hold because weights are algebra maps. In other words, $\gamma(h') = \varepsilon(h')$. □

We also mention two examples; the proofs are straightforward.

Lemma 4.3. *If $g \in G(H)$, then this definition of Γ_g coincides with the previous one: $\Gamma_g = \Gamma'_g = \{\gamma \in \Gamma : \gamma(g) = 1\}$. If $h \in H_{g,g'}$, then $\Gamma_h \subset \Gamma_g \cap \Gamma_{g'}$, or $\Gamma_h = \Gamma$. In both cases, $\Gamma'_h = \Gamma_h \cap \Gamma_g \cap \Gamma_{g'}$.*

4.2. Subquotients

We now compute $\Sigma_\Pi(h)$ for more general Π. The following result is used later.

Lemma 4.4. *Fix $h \in H$, and choose any subgroup Γ' of Γ_h that is normal in Γ. Also fix a finite subgroup Π of Γ/Γ'. Now fix any lift $\tilde{\Pi}$ of Π to Γ, and define*

$$\Sigma_\Pi(h) := \sum_{\gamma'' \in \tilde{\Pi}} \gamma''(h), \qquad \Pi^\circ := \{\gamma \in \Gamma : (\gamma + \Gamma') \in \Pi \subset \Gamma/\Gamma'\}$$

1. *$\Sigma_\Pi(h)$ is well-defined.*
2. *If a subgroup $\Gamma'' \subset \Gamma_h$ is normal in Π° (e.g., $\Gamma'' = \Gamma', \Gamma_h \cap \Pi^\circ$), then*

$$\Sigma_{\Pi^\circ}(h) = |\Gamma''| \Sigma_{\Pi^\circ/\Gamma''}(h)$$

Proof. For the first part, choose any other lift Π' of Π. If $\gamma' \in \Pi', \gamma'' \in \tilde{\Pi}$ are lifts of $\gamma \in \Pi$, then $(\gamma')^{-1} * \gamma'' \in \Gamma_h$, so

$$\sum_{\gamma'' \in \tilde{\Pi}} \gamma''(h) = \sum_{\gamma' \in \Pi'} (\gamma' * ((\gamma')^{-1} * \gamma''))(h) = \sum_{\gamma' \in \Pi'} \gamma'(h)$$

by definition of Γ_h. The other part is also easy, since by the first part, $\Sigma_{\Pi^*}(h)$ is also well-defined, where the finite group is $\Pi^* := \Pi^\circ/\Gamma''$. If Π^{**} is any lift to Π° of Π^*, then

$$\Sigma_{\Pi^\circ}(h) = \sum_{\gamma' \in \Pi^{**},\ \beta \in \Gamma''} (\gamma' * \beta)(h) = \sum_{\gamma' \in \Pi^{**},\ \beta \in \Gamma''} \gamma'(h) = |\Gamma''| \sum_{\gamma' \in \Pi^{**}} \gamma'(h)$$

and this equals the desired amount. □

While a special case of the second part is that $\Sigma_{\Pi^\circ}(h) = |\Gamma'| \Sigma_\Pi(h)$, we really use the result when Γ' is itself finite, and we replace Π° by Π. The equation is then used to compute $\Sigma_\Pi(h)$.

4.3. Pseudo-Primitive Elements

For the rest of this paper, H is an R-Hopf algebra, unless stated otherwise. If h is grouplike or (pseudo-)primitive, then it is easy to see that $\Sigma_\Pi(h) = |[\Pi,\Pi]| \Sigma_{\Pi_{ab}}(h)$, where $\Pi_{ab} := \Pi/[\Pi,\Pi]$ (we show the pseudo-primitive case presently). Thus, in the grouplike case, the question of whether or not $h \in G_\Pi(H)$ reduces to evaluating (any lift of a set of) generators of the finite Abelian group Π_{ab}, at h.

Proposition 4.5. *Suppose $h \in H_{g,g'}$ is pseudo-primitive with respect to Π.*

1. *Then $(\gamma * \nu)(h) = (\nu * \gamma)(h) = \nu(g)\gamma(h) + \gamma(g)\nu(h)$ for all $\gamma, \nu \in \Pi$.*
2. *$\Gamma'_h \supset [\Pi, \Pi]$.*
3. *For any $m \geq 0$, $\gamma^{*m}(h) = m\gamma(g)^{m-1}\gamma(h)$ if $\gamma \in \Pi$. In particular, if $\operatorname{char}(R) = p$ is prime, then $\gamma^{*p}(h) = 0\ \forall \gamma \in \Pi$.*

Proof. The first and last parts are by definition and induction respectively. As for the second part, one shows the following computation for any skew-primitive $h \in H_{g,g'}$, and $\beta, \beta' \in \Gamma_H$:

$$\begin{aligned}(\beta * \beta' * \beta^{-1} * (\beta')^{-1})(h) &= \beta(h(g')^{-1})(1 - \beta'(g(g')^{-1})) \\ &\quad + \beta'(h(g')^{-1})(\beta(g(g')^{-1}) - 1)\end{aligned}$$

using Lemma 3.3. Since h is pseudo-primitive with respect to Π, this shows that every generator (and hence element) λ of $[\Pi, \Pi]$ satisfies: $\lambda(g) = \lambda(g') = 1$ and $\lambda(h) = 0$. Since $C_h = Rh + Rg + Rg'$, these imply that $\lambda \in \Gamma'_h$. $\square$

An easy consequence of Propositions 4.2 and 4.5 is

Corollary 4.6. *If $h \in H$ is (in the subalgebra) generated by grouplike and pseudo-primitive elements (with respect to Π), then $\Gamma'_h \supset [\Pi, \Pi]$.*

Also note that given some $h \in H$, one can compute $\Sigma_\Pi(h)$ for more general Π, and hence the results in this paper can be generalized; however, we stay with the original setup when $\Pi \subset \Gamma$ (*i.e.*, $\Gamma' = \{\varepsilon\}$). This (general) case is noteworthy, however, because it is used below.

We conclude by specifying more precisely what is meant by $\Sigma_\Pi(h_1 \dots h_n)$ for "pseudo-primitive" h_i's, when Π is a subquotient of Γ as above. In this case, start with some skew-primitive h_i's, then let Π be a finite subgroup of Γ/Γ', for some subgroup $\Gamma' \subset \cap_i \Gamma_{h_i}$ that is normal in Γ. Moreover, if $\Delta(h_i) = g_i \otimes h_i + h_i \otimes g'_i$, then assume further that $\gamma(g_i) = \gamma(g'_i)$, for all elements γ of the subgroup Π° (defined above).

This is what is meant in the case of general Π, when we say that $g_i, g'_i \in G_\Pi(H)$-*i.e.*, that the h_i's are pseudo-primitive (with respect to Π). Similarly, to say that the h_i's are almost primitive with respect to Π means that $\gamma(g_i) = \gamma(g'_i) = 1\ \forall \gamma \in \Pi^\circ$.

5. Products of Skew-Primitive Elements

We now mention some results on (finite) products of skew-primitive elements and grouplike elements. From now on, Π denotes a finite subgroup of $\Gamma = \Gamma_H$ and not a general subquotient; however, in the next section, we need to use a subquotient Φ of this Π.

Since the set of skew-primitive elements is closed under multiplication by grouplike elements, any "monomial" in them can be expressed in the form $\mathbf{h} = \prod_i h_i$. The related "grouplike" elements that would figure in the computations are $\mathbf{g} = \prod_i g_i$ and $\mathbf{g}' = \prod_i g'_i$.

Standing Assumption 5.1. For this section and the next two, assume that $h_i \in H_{g_i, g'_i}$ for all (finitely many) i.

First, here are some results that hold in general.

Proposition 5.2. *If $\lambda(h_i) = 0\ \forall i$ for some $\lambda \in \Pi$, and (at least) one of $\lambda(\mathbf{g}), \lambda(\mathbf{g}')$ is not 1, then $\Sigma_\Pi(\mathbf{h}) = 0$.*

Proof. If $\lambda(h_i) = 0$ then $\lambda^m(h_i) = 0$ for all i, m. Now choose a set $\mathcal{B}$ of coset representatives for $\langle\lambda\rangle$ in Π, and assume that $\lambda(\mathbf{g}') \neq 1$ (the other case is similar). Then compute:

$$\Sigma_\Pi(\mathbf{h}) = \sum_{\beta \in \mathcal{B},\ \gamma \in \langle\lambda\rangle} \prod_i (\beta(g_i)\gamma(h_i) + \beta(h_i)\gamma(g'_i))$$

Since $\gamma(h_i) = 0$ for all $\gamma \in \langle\lambda\rangle$ and all i, hence the entire product in the summand collapses, to give $\sum_{\beta \in \mathcal{B}} \beta(\mathbf{h}) \cdot \Sigma_{\langle\lambda\rangle}(\mathbf{g}')$. But now the second factor vanishes by our assumption (and Proposition 2.8). $\square$

Next, if $\Sigma_\Pi(\prod_{i=1}^n h_i)$ is known for all skew-primitive h_i's, then one can evaluate the product of $(n+1)$ such h's in some cases. The following result relates Σ_Π-values of strings to the Σ_Π-values of proper substrings (with skew-primitive "letters"), which are "corrected" by grouplike elements. The proof is that both equations below follow by evaluating Equation (2.4) at $\mathbf{h}$.

Proposition 5.3. *Suppose one of $\mathbf{g}, \mathbf{g}'$ is not in $G_\Pi(H)$. Thus, if $\lambda(\mathbf{g}) \neq 1$ (or respectively $\lambda(\mathbf{g}') \neq 1$) for some $\lambda \in \Pi$, then*

$$\Sigma_\Pi(\mathbf{h}) = \frac{\sum_{\nu \in \Pi} \prod_i (\lambda(g_i)\nu(h_i) + \lambda(h_i)\nu(g'_i)) - \lambda(\mathbf{g})\Sigma_\Pi(\mathbf{h})}{1 - \lambda(\mathbf{g})}$$

or respectively,

$$\Sigma_\Pi(\mathbf{h}) = \frac{\sum_{\nu\in\Pi}\prod_i(\nu(g_i)\lambda(h_i)+\nu(h_i)\lambda(g_i')) - \Sigma_\Pi(\mathbf{h})\lambda(\mathbf{g}')}{1-\lambda(\mathbf{g}')}$$

Note here that both numerators on the right side have an $\Sigma_\Pi(\mathbf{h})$ in them, which cancels the only such term present in the summations. Thus, what one is left with in either case are linear combinations of Σ_Π-values of "corrected" proper substrings, with coefficients of the form $\lambda(\prod g_j h_k)$.

Also note that if the Σ_Π-values of all "corrected" proper substrings are known, and $\mathrm{char}(R) \neq 2$, then the two propositions, one above and one below, can be used, for instance, to compute Σ_Π at all monomials of odd length (in skew-primitive elements).

The statement and proof of the following result are essentially the same as those of the last part of Theorem 3.6 above.

Proposition 5.4. *Suppose* $\mathbf{gg}' \in G_\Pi(H)$, *and* $\mathrm{char}(R) = 2$ *if the number of* h_i*'s is even. Then* $\Sigma_\Pi(\mathbf{h}) = \sum_{\gamma\neq\varepsilon=\gamma^2}\gamma(\mathbf{h})$, *and* $2\Sigma_\Pi(\mathbf{h}) = 0$.

This is because once again, one shows that $\lambda^{-1}(\mathbf{h}) = -\lambda(\mathbf{h})\ \forall\lambda$.

The next result in this subsection is true for almost all values of char R. The proof is immediate from the penultimate part of Lemma 3.3 above.

Proposition 5.5. *If* $\mathrm{char}(R) \nmid |\Pi|$ *and* $g_i^{-1}g_i' \in G_\Pi(H)$ (i.e., h_i *is pseudo-primitive) for some* i, *then* $\gamma(h_i) = 0\ \forall\gamma\in\Pi$. *In particular,* $\Sigma_\Pi(\mathbf{h}) = 0$.

We conclude this section with one last result—in characteristic p.

Theorem 5.6. *Suppose* $\mathrm{char}(R) = p > 0$, *and* $h_i \in H_{g_i,g_i'}$ *for all* i. *Choose and fix a* p*-Sylow subgroup* Π_p *of* Π.

1. *Then each* h_i *is almost primitive with respect to* Π_p.
2. *If* Π_p *contains an element of order* p^2, *then* $\Sigma_\Pi(\mathbf{h}) = \Sigma_{\Pi_p}(\mathbf{h}) = 0$.

It is also shown later that $\Sigma_{\Pi_p}(\mathbf{h}) = 0$ whenever $\Pi_p \not\cong (\mathbb{Z}/p\mathbb{Z})^k$ for any $k > 0$.

Proof.

1. If $z^p = 1$ in R, then $(1-z)^p = 1 - z^p \mod p = 0$ in R. Since R is an integral domain, $z = 1$. Now assume that $|\Pi_p| = p^f$. Then $\gamma^{*p^f} = \varepsilon$ for each $\gamma\in\Pi_p$, whence $\gamma(g_i)^{p^f} = \gamma(g_i')^{p^f} = 1$ for all i. Successively set $z = \gamma(g_i)^{p^t}$, for $t = f-1, f-2, \ldots, 1, 0$. Hence $\gamma(g_i) = 1$ for all i; the other case is the same.
2. Next, if λ has order p^2, then choose a set $\mathcal{B}$ of coset representatives for $\langle\lambda\rangle$ in Π, and compute using Proposition 4.5 above (since all h_i's are pseudo/almost primitive with respect to Π_p, hence for $\langle\lambda\rangle$):

$$\Sigma_\Pi(\mathbf{h}) = \sum_{\beta\in\mathcal{B}}\sum_{j=0}^{p^2-1}\prod_{i=1}^{n}(\beta*\lambda^{*j})(h_i) = \sum_{\beta\in\mathcal{B}}\sum_{j=0}^{p^2-1}\prod_{i=1}^{n}\left(\beta(h_i)\lambda^j(g_i) + j\beta(g_i)\lambda(g_i)^{j-1}\lambda(h_i)\right)$$

Call the factor in the product $a_{i,j}$ (it really is $a_{\beta,i,j}$). Now observe that $a_{i,p+j} = \lambda^p(g_i)a_{i,j}$ for all i,j, whence $a_{i,kp+j} = \lambda^{kp}(g_i)a_{i,j}$. Therefore for any $\beta\in\mathcal{B}$, one can take $\sum_{k=0}^{p-1}\lambda(\mathbf{g})^{kp}$ out of the summand. But $\lambda(\mathbf{g})^p = 1$, so every β-summand vanishes.

□

6. Special Case: Abelian Group of Weights

In this section, the focus is on evaluating $\Sigma_\Pi(\mathbf{h})$ in the special case where the group Π of weights is Abelian.

Definition 6.1.

1. For all $n \in \mathbb{N}$, define $[n] := \{1, 2, \ldots, n\}$.
2. Given $I \subset [n]$, define $g_I := \prod_{i \in I} g_i$, and similarly define g'_I, h_I.
3. Define Π_p to be any fixed p-Sylow subgroup of Π if $\operatorname{char}(R) = p > 0$, and $\{\varepsilon\}$ otherwise. Also choose and fix a "complementary" subgroup Π' to Π_p in Π (if Π is Abelian), *i.e.*, $|\Pi_p| \cdot |\Pi'| = |\Pi|$. (And if $\operatorname{char}(R) = 0$, set $\Pi' := \Pi$.)

We now present two results. The first is (nontrivial only) when $\operatorname{char}(R)$ divides the order of Π, and the second (which really is the main result) is when it does not.

Theorem 6.2. *Suppose Π is Abelian; let Π_p, Π' be as above. Let $J \subset [n]$ be the set of i's such that h_i is pseudo-primitive with respect to Π. Then*

$$\Sigma_\Pi(\mathbf{h}) = \Sigma_{\Pi'}(g_J h_{[n]\setminus J}) \cdot \Sigma_{\Pi_p}(h_J)$$

In particular, if $\operatorname{char}(R) = 0$, or $0 < \operatorname{char}(R) \nmid |\Pi|$, then $\Sigma_{\Pi_p}(h_J) = \varepsilon(h_J) = 0$, whence $\Sigma_\Pi(\mathbf{h}) = 0$ too.

Proof. First note that $\lambda(g_i) = \lambda(g'_i) = 1$ for all $\lambda \in \Pi_p$ and all i, by Theorem 5.6 above. Since Π is Abelian, every $\gamma \in \Pi$ is uniquely expressible as $\gamma = \beta * \lambda$ with $\beta \in \Pi'$, $\lambda \in \Pi_p$. We now compute $\gamma(h_i)$ in both cases: $i \in J$ and $i \notin J$.

First consider the case when $i \in J$. Then $\beta(h_i) = 0$ for all $\beta \in \Pi'$, by Proposition 5.5. Thus, $\gamma(h_i) = \beta(g_i)\lambda(h_i) + \beta(h_i)\lambda(g_i) = \beta(g_i)\lambda(h_i)$.

Now suppose $i \notin J$. Choose $\gamma \in \Pi$ so that $\gamma(g_i) \neq \gamma(g'_i)$. Then $(\gamma * \lambda)(h_i) = (\lambda * \gamma)(h_i)$, which leads (upon simplifying) to

$$\lambda(h_i)(\gamma(g_i) - \gamma(g'_i)) = \gamma(h_i)(\lambda(g_i) - \lambda(g'_i))$$

But $\lambda(g_i) = \lambda(g'_i)$, and $\gamma(g_i) \neq \gamma(g'_i)$, so $\lambda(h_i) = 0$ for all $\lambda \in \Pi_p$. Hence:

$$\gamma(h_i) = (\beta * \lambda)(h_i) = \beta(g_i)\lambda(h_i) + \beta(h_i)\lambda(g_i) = \beta(h_i)$$

The proof can now be completed:

$$\begin{aligned} \Sigma_\Pi(\mathbf{h}) &= \sum_{\beta \in \Pi',\, \lambda \in \Pi_p} \prod_{i \in J} \beta(g_i)\lambda(h_i) \cdot \prod_{i \notin J} \beta(h_i) \\ &= \sum_{\beta \in \Pi',\, \lambda \in \Pi_p} \beta(g_J h_{[n]\setminus J})\lambda(h_J) = \Sigma_{\Pi'}(g_J h_{[n]\setminus J}) \cdot \Sigma_{\Pi_p}(h_J) \end{aligned}$$

as claimed. □

We compute $\Sigma_{\Pi_p}(\mathbf{h})$ in a later section. For now, we mention how to compute the other factor.

Theorem 6.3. *Suppose* $\operatorname{char}(R) \nmid |\Pi|$, *and no h_i is pseudo-primitive with respect to (the Abelian group of weights) Π. For each i, let $f_i \in k(R)$ denote $\beta(h_i)/(\beta(g_i) - \beta(g'_i))$ for some $\beta \in N_{h_i} \cap \Pi$. Also define $S = \{I \subset [n] : g_I g'_{[n]\setminus I} \in G_\Pi(H)\}$. Then $\Sigma_\Pi(\mathbf{h}) = (-1)^n |\Pi| \prod_{i=1}^{n} f_i \cdot \sum_{I \in S} (-1)^{|I|}$.*

Note that if $n = 1$, this is a special case of the first two parts of Theorem 3.6 above. Moreover, if some h_i is pseudo-primitive with respect to Π, then $\Sigma_\Pi(\mathbf{h}) = 0$ from Proposition 5.5 above.

Proof. Let us fix some generators $\beta_1, \ldots, \beta_k$ of Π (by the structure theory of finite Abelian groups), so that $\Pi = \bigoplus_{j=1}^k \mathbb{Z}\beta_j$. Then (by assumption), for each i there is at least one j so that $\beta_j(g_i) \neq \beta_j(g_i)^{-1}$.

Now fix i, and compute $\beta(h_i)$ for arbitrary $\beta \in \Pi$. Suppose N'_i indexes the set of β_j's that are in N_{h_i}; then write $\beta = \beta' + \sum_{j \in N'_i} r_j \beta_j$ for some $r_j \geq 0$ and $\beta' \in \bigoplus_{j \notin N'_i} \mathbb{Z}\beta_j$.

Note by Proposition 5.5 for $\Pi \leftrightarrow \bigoplus_{j \notin N_i'} \mathbb{Z}\beta_j$ that $\beta'(h_i) = 0$. So if $\beta'' = (\beta')^{-1}\beta \in \Pi$, then

$$\beta(h_i) = \beta'(g_i)\beta''(h_i) + \beta'(h_i)\beta''(g_i) = \beta'(g_i)\beta''(h_i)$$

It remains to compute the last factor above. This is done using the following claim.
Claim. Say $\beta'' = \sum_{j \in N_i'} r_j\beta_j$. Then $\beta''(h_i) = (\beta''(g_i) - \beta''(g_i'))f_i$, where f_i is defined in the statement of the theorem.
Proof of the theorem, modulo the claim. By the claim,

$$\beta(h_i) = \beta'(g_i)(\beta''(g_i) - \beta''(g_i'))f_i = (\beta(g_i) - \beta(g_i'))f_i$$

(for all i) since $\beta'(g_i) = \beta'(g_i')$ by pseudo-primitivity. Using the notation that $\beta_{\mathbf{r}} = \sum_{j=1}^{k} r_j\beta_j \in \Pi$, one can compute $\Sigma_\Pi(\mathbf{h})$ to be

$$= \sum_{\mathbf{r}} \prod_{i=1}^{n} (\beta_{\mathbf{r}}(g_i) - \beta_{\mathbf{r}}(g_i'))f_i = \prod_{i=1}^{n} f_i \cdot \sum_{\mathbf{r}} \sum_{I \subset [n]} (-1)^{n-|I|}\beta_{\mathbf{r}}\left(\prod_{i \in I} g_i \prod_{j \notin I} g_j'\right)$$
$$= (-1)^n \prod_{i=1}^{n} f_i \cdot \sum_{I \subset [n]} (-1)^{|I|}\Sigma_\Pi(g_I g'_{[n]\setminus I}) = (-1)^n \prod_{i=1}^{n} f_i \cdot \sum_{I \subset [n]} (-1)^{|I|}\delta_{I \in S}|\Pi|$$

by Proposition 2.8, where the last δ is 1 if $I \in S$, and 0 otherwise. □

The proof is completed by showing the claim.

Proof of the claim. By Equation (3.5), and Lemma 3.8,

$$\beta_j^{*r_j}(h_i) = (\beta_j(g_i)^{r_j} - \beta_j(g_i')^{r_j})f_i, \qquad \forall j \in N_i'$$

Suppose without loss of generality that we relabel the set $\{\beta_j : j \in N_i'\}$ as $\{\beta_1, \ldots, \beta_m\}$ (*i.e.*, relabel the generators β_j of Π so that these are before the others). Now compute the expression using the above equation:

$$\beta''(h_i) = \left(\sum_{j=1}^{m} r_j\beta_j\right)(h_i) = \sum_{j=1}^{m} \prod_{l<j} \beta_l(g_i)^{r_l} \cdot (\beta_j(g_i)^{r_j} - \beta_j(g_i')^{r_j})f_i \cdot \prod_{l>j} \beta_l(g_i')^{r_l}$$

and this telescopes to $f_i \cdot \prod_j \beta_j(g_i)^{r_j} - f_i \cdot \prod_j \beta_j(g_i')^{r_j} = (\beta''(g_i) - \beta''(g_i'))f_i$, as claimed. □

7. Products of Pseudo-Primitive Elements: Positive Characteristic

We now mention results for pseudo-primitive elements h_i (and not necessarily Abelian Π) in prime characteristic; note that for almost all characteristics (including zero), Proposition 5.5 above says that $\Sigma_\Pi(\mathbf{h}) = 0$. Before considering the positive case, we need a small result.

Lemma 7.1. *Given $f \in \mathbb{N}$ and a prime $p > 0$, define $\varphi_p(f) = \varphi_p(f) := \sum_{i=0}^{p-1} i^f$. If $f > 0$, then $\varphi(f) \not\equiv 0 \mod p$ if and only if $(p-1)|f$, and in this case, $\varphi(p-1) = p - 1 \equiv -1 \mod p$.*

Proof. Let g be any cyclic generator of (the finite cyclic group) $(\mathbb{Z}/p\mathbb{Z})^\times$. Then $\sum_{i=1}^{p-1} i^f \equiv \sum_{j=1}^{p-1} g^{jf} \mod p$. Now if $(p-1)|f$, then each summand is 1, which yields $p - 1 \mod p$. Otherwise, g^f is not 1, and its powers add up to 0 (by the geometric series formula). □

7.1. Preliminaries

Recall that a pseudo-primitive element is any $h \in H_{g,g'}$ so that $g^{-1}g' \in G_\Pi(H)$. Some terminology is now needed. Note by Hall's theorems that a finite group Φ is solvable if and only if it contains Hall subgroups of all possible orders (*e.g.*, see [15, §11]). So if $|\Phi| = p^k \cdot m$ with $p \nmid m$, let Φ_m be any Hall subgroup of order m.

Definition 7.2. ($p > 0$ a fixed prime.) Given a finite solvable group Φ, denote by Φ_p, Φ' respectively, any p-Sylow subgroup and any Hall subgroup of order $|\Phi|/|\Phi_p|$. (From above, we mean $\Phi' = \Phi_m$.)

For the rest of this section, $\mathrm{char}(R) = p > 0$; also fix n, the number of h_i's. (Recall Assumption 5.1.)

Proposition 7.3. *Suppose, given skew-primitive $h_i \in H_{g_i, g_i'}$ for $1 \leq i \leq n$, that $\Phi \subset \Gamma/\Gamma'$ is a finite solvable subquotient of Γ (as in a previous section) with respect to which every h_i is pseudo-primitive. Then $\Sigma_\Phi(\mathbf{h}) = \Sigma_{\Phi'}(\mathbf{g}) \cdot \Sigma_{\Phi_p}(\mathbf{h})$.*

Proof. The first claim is that the "set-product" $\Phi'\Phi_p := \{\beta * \lambda : \beta \in \Phi', \lambda \in \Phi_p\}$ equals the entire group Φ. Next, if $\beta \in \Phi'$, then $\beta(h) = 0$ for any pseudo-primitive $h \in H_{g,g'}$, because if $|\Phi'| = m \not\equiv 0$ mod p, then $\beta^{*m} = \varepsilon \in \Gamma' \subset \Gamma_h$, whence $0 = \varepsilon(h) = \beta^{*m}(h) = m\beta(g)^{m-1}\beta(h)$. Therefore,

$$\begin{aligned}\Sigma_\Phi(\mathbf{h}) &= \sum_{\beta\in\Phi',\ \lambda\in\Phi_p} \prod_{i=1}^{n} (\beta * \lambda)(h_i) = \sum_{\beta\in\Phi',\ \lambda\in\Phi_p} \prod_{i=1}^{n} \beta(g_i)\lambda(h_i) \\ &= \sum_{\beta\in\Phi'} \prod_{i=1}^{n} \beta(g_i) \cdot \sum_{\lambda\in\Phi_p} \prod_{i=1}^{n} \lambda(h_i) = \Sigma_{\Phi'}(\mathbf{g})\Sigma_{\Phi_p}(\mathbf{h})\end{aligned}$$

as claimed. □

Remark 7.4. The above proposition thus holds for *any* group Φ, such that some p-Sylow subgroup Φ_p has a complete set of coset representatives, none of whom has order divisible by p. Obvious examples are Abelian groups or groups of order $p^a q^b$ for primes $p \neq q$ (but these are solvable by Burnside's Theorem).

Also note that the above sum is independent of the choices of Φ_p, Φ'.

The next result is crucial in computing $\Sigma_\Pi(\mathbf{h})$, and uses subquotients of Π.

Theorem 7.5. *Given a finite subgroup of weights $\Pi \subset \Gamma = \Gamma_H$, suppose $h_i \in H_{g_i, g_i'}$ is pseudo-primitive with respect to Π for all $1 \leq i \leq n$. Define $\Phi = \Pi_{ab} := \Pi/[\Pi, \Pi]$. Then,*

$$\Sigma_\Pi(\mathbf{h}) = |[\Pi, \Pi]| \cdot \Sigma_{\Phi'}(\mathbf{g}) \cdot \Sigma_{\Phi_p}(\mathbf{h}) \tag{7.6}$$

For instance, if every h_i was almost primitive, then $\Sigma_{\Phi'}(\mathbf{g}) = [\Phi : \Phi_p]$.

Proof. At the outset, note that $\Sigma_{\Phi'}(\mathbf{g})$ and $\Sigma_{\Phi_p}(\mathbf{h})$ make sense because of Corollary 4.6 and Proposition 4.2 above. Now, the proof is in two steps; each step uses a previously unused result above.
Step 1. We claim that $\Sigma_\Pi(\mathbf{h}) = |[\Pi, \Pi]|\Sigma_\Phi(\mathbf{h})$. This follows immediately from Lemma 4.4, where h, Γ'', Π° are replaced by $\mathbf{h}, [\Pi, \Pi], \Pi$ respectively.

The only thing to check is that the above replacements are indeed valid. Since $[\Pi, \Pi]$ is normal in Π, it suffices to check that $[\Pi, \Pi] \subset \Gamma_{\mathbf{h}}$. But this follows from Corollary 4.6 and Proposition 4.2 above.
Step 2. The proof is now complete by invoking Proposition 7.3 above. □

We conclude the preliminaries with one last result—for *skew*-primitive elements in general.

Proposition 7.7. *If $\Pi_p \not\cong (\mathbb{Z}/p\mathbb{Z})^k$ for any $k > 0$, then $\Sigma_{\Pi_p}(\mathbf{h}) = 0$.*

Proof. By Theorem 5.6, the h_i's are almost primitive with respect to Π_p. Now invoke Equation (7.6) above, replacing Π by Π_p. Now, if Π_p is not Abelian, then $|[\Pi_p, \Pi_p]| > 1$, hence is a power of p, whence the right-hand side vanishes. Next, if Π_p is Abelian, but contains an element of order p^2, then $\Sigma_{\Pi_p}(\mathbf{h}) = 0$ by Theorem 5.6 again. Therefore $\Pi_p \cong (\mathbb{Z}/p\mathbb{Z})^k$ for some k. If $k = 0$, then $\Pi_p = \{\varepsilon\}$, and $\varepsilon(\mathbf{h}) = 0$. □

7.2. The Main Results-Pseudo-Primitive Elements

The following result now computes $\Sigma_\Pi(\mathbf{h})$ (for pseudo-primitive h_i's) in most cases in prime characteristic that are "non-Abelian". For the "Abelian" case, we appeal to Theorem 7.10 below and mention at the outset that it is only for *almost* primitive (and not merely pseudo-primitive) elements, that we obtain a much clearer picture—as its last part shows.

Theorem 7.8. *Suppose* $\mathrm{char}(R) = p \in \mathbb{N}$*,* Π_p *is any (fixed) p-Sylow subgroup of* Π*, and every* h_i *is pseudo-primitive (with respect to* Π*).*

1. $\Sigma_\Pi(\mathbf{h}) = 0$ *if* Π_p
   (a) *is trivial,*
   (b) *contains an element of order* p^2*, or*
   (c) *intersects* $[\Pi,\Pi]$ *nontrivially.*

   This last part includes the cases when Π_p
   (d) *is not Abelian,*
   (e) *does not map isomorphically onto (some)* Φ_p*, via (the restriction of) the quotient map* $\pi : \Pi \twoheadrightarrow \Phi = \Pi/[\Pi,\Pi]$*, or*
   (f) *has size strictly greater than* Φ_p*.*
2. *Otherwise* $\Pi_p \cong \Phi_p \cong (\mathbb{Z}/p\mathbb{Z})^k$ *for some* $k > 0$*, and then* $\Sigma_\Pi(\mathbf{h}) = |[\Pi,\Pi]| \cdot \Sigma_{\Phi'}(\mathbf{g}) \cdot \Sigma_{\Pi_p}(\mathbf{h})$*.*

Remark 7.9. Any finite Abelian group of exponent p is of the form $(\mathbb{Z}/p\mathbb{Z})^k$, hence one part of the second statement is clear. Moreover, every subquotient of such a group is of the same form. Finally (especially when all of the h_i's are almost primitive with respect to Π), the cases that remain reduce to computing $\Sigma_{\Pi_p}(\mathbf{h})$, and when $\Pi_p \cong (\mathbb{Z}/p\mathbb{Z})^k$; this is addressed below.

Proof. The second part follows from the first part, the remarks above, and Equation (7.6). We now show the first part.

(a) If Π_p is trivial, then $p \nmid |\Pi|$, and we are done by Proposition 5.5.
(b) This has been done in Theorem 5.6 above.
(c) Now suppose that $[\Pi,\Pi] \cap \Pi_p \neq \emptyset$. Then $[\Pi,\Pi]$ contains an element of order p, whence p divides $|[\Pi,\Pi]|$. Now use Equation (7.6).

It remains to show how this last includes the remaining cases.

(d) First, if Π_p is non-Abelian, then $[\Pi_p,\Pi_p]$ is a nontrivial subgroup of the p-group Π_p. In particular, Π_p intersects $[\Pi,\Pi]$.
(e) Next, note that $\pi(\Pi_p)$ is a p-group in Φ, and $|\Pi_p| \geq |\Phi_p|$ (since $|\Phi|$ divides $|\Pi|$). Hence Π_p does not map isomorphically onto (some) Φ_p if and only if π is not one-to-one on Π_p. But then $[\Pi,\Pi]$ intersects Π_p.
(f) Finally, if $|\Pi_p| > |\Phi_p|$, then Π_p cannot map isomorphically onto Φ_p, so we are done by the preceding paragraph.

□

We conclude by analyzing $\Sigma_{\Pi_p}(\mathbf{h})$. Note that the results below that pertain only to $\Sigma_{\Pi_p}(\mathbf{h})$ are applicable in general to all skew-primitive h_i's, by Theorem 5.6 above.

Theorem 7.10. *Suppose* $\mathrm{char}(R) = p \in \mathbb{N}$*,* Π_p *is any (fixed) p-Sylow subgroup of* Π*, and every* $h_i \in H_{g_i,g_i'}$ *is pseudo-primitive (with respect to* Π*) for all* $1 \leq i \leq n$*. Suppose moreover that* $\Pi_p \cong (\mathbb{Z}/p\mathbb{Z})^k$*.*

1. $\gamma(g_i) = 1$ *for all i and* $\gamma \in \Pi_p$*. In particular,* $g_i, \mathbf{g} \in G_{\Pi_p}(H)$*.*
2. *If* $k > n$*, then* $\Sigma_\Pi(\mathbf{h}) = \Sigma_{\Pi_p}(\mathbf{h}) = 0$*.*

3. *If $k = n$, then $\Sigma_{\Pi_p}(\mathbf{h}) = \binom{p}{2}^k \cdot \mathrm{perm}(A)$, where A is the matrix given by $a_{ij} = \gamma_j(h_i)$, the γ_j's form a $\mathbb{Z}/p\mathbb{Z}$-basis of Π_p, and* perm *is the* matrix permanent*:*

$$\mathrm{perm}(A_{n\times n}) = \sum_{\sigma\in S_n}\prod_{i=1}^{n} a_{i,\sigma(i)}$$

In particular, $\Sigma_{\Pi_p}(\mathbf{h}) = 0$ unless $p = 2$, in which case $\Sigma_{\Pi}(\mathbf{h}) = \Sigma_{\Pi_p}(\mathbf{h}) = \det A$.

4. *If $\Sigma_{\Pi_p}(\mathbf{h}) \neq 0$, then $(p-1)|n$ and $0 < k \leq n/(p-1)$, and then $\Sigma_{\Pi_p}(\mathbf{h})$ can take any value $r \in R$. (If $k = n$ and $p = 2$, then $r \neq 0$.)*

Remark 7.11. This result is independent of the chosen p-Sylow subgroup Π_p, as well as the choices of generators γ_j. It generalizes Proposition 3.9 above, in the special case $p = 2$.

The rest of this section is devoted to proving the above result. First, suppose there exists a subgroup $\Pi'_p \cong (\mathbb{Z}/p\mathbb{Z})^k$ of Π (so $\Pi'_p \subset \Pi_p$ in general). Choose a set of coset representatives $\mathcal{B}$ for Π'_p in Π, and write

$$\Sigma_{\Pi}(\mathbf{h}) = \sum_{\beta\in\mathcal{B},\ \gamma\in\Pi'_p}(\gamma*\beta)(\mathbf{h}) = \sum_{\beta\in\mathcal{B},\ \gamma\in\Pi'_p}\prod_{i=1}^{n}(\gamma*\beta)(h_i)$$

Recall that every element of Π'_p is $\gamma_{a,I} := \sum_{j\in I} a_j\gamma_j$ (with γ_i as above), for some subset I of $[k] := \{1,2,\ldots,k\}$, and some $|I|$-tuple $a = (a_j)_{j\in I}$ of elements of $(\mathbb{Z}/p\mathbb{Z})^\times$. Recall, moreover, that we had previously defined g_I, g'_I, h_I for $I \subset [n]$.

We also need the following lemma that is proved by computations using Proposition 4.5.

Lemma 7.12. *If $h \in H_{g,g'}$ is pseudo-primitive with respect to Π, and given $\beta \in \mathcal{B}, \gamma_{a,I} \in \Pi_p$ as above, one has*

$$(\beta*\gamma_{a,I})(h) = \prod_{j=1}^{k}\gamma_j(g)^{a_j}\cdot\left[\beta(h) + \beta(g)\sum_{j=1}^{k}a_j\gamma_j(g^{-1}h)\right]$$

The key observation now is that the only "monomials" that occur in the product $\prod_{i=1}^{n}(\beta*\gamma_{a,I})(h_i)$ are of the form $\beta(\mathbf{g}g_{I_0}^{-1}h_{I_0})\cdot\prod_{j\in I}\gamma_j(h_{I_j})$, where $\coprod_j I_j \coprod I_0 = [n]$, and $I_j \subset I$ for all j. The coefficient of such a monomial in this particular summand is $\prod_{j\in I} a_j^{|I_j|}\gamma_j(\mathbf{g})^{a_j}$ by the lemma above. Moreover, every such monomial occurs at most once inside each $(\beta*\gamma_{a,I})(\mathbf{h})$.

The crucial fact that proves Theorem 7.10 above is the following

Key claim. The coefficient of $\beta(\mathbf{g}g_{I_0}^{-1}h_{I_0})\prod_{j\in I}\gamma_j(h_{I_j})$ in $\sum_{\gamma\in\Pi'_p}(\beta*\gamma)(\mathbf{h})$ equals $p^{k-|I|}\prod_{j\in I}\varphi(|I_j|)$ (where $\varphi(0) := 0$).

Proof of the key claim. (Note that $\varphi(f)$ was defined in Lemma 7.1.) A monomial of the desired form occurs in precisely those (a', I')-summands, so that $I' \supset I$. Moreover, all such summands can be split up into a disjoint union over all $a \in ((\mathbb{Z}/p\mathbb{Z})^\times)^{|I|}$, with each disjoint piece containing all (a', I') so that $I' \supset I$ and the I-component of a' is a.

Such a piece contains exactly $p^{k-|I|}$ elements (and hence exactly that number of copies of the monomial with this selfsame coefficient). Each of these "extra" $[k]\setminus I$ factors contributes a $\beta(h_i)$, which gives $\beta(h_{I_0})$.

Moreover, there is one contribution for each $a \in ((\mathbb{Z}/p\mathbb{Z})^\times)^{|I|}$, and it is $\prod_{j\in I} a_j^{|I_j|}\gamma_j(\mathbf{g})^{a_j}\cdot\beta(\mathbf{g}g_{I_0}^{-1})$, since the argument for the β-factor here is precisely $\prod_{j=1}^{k} g_{I_j}$. Moreover, $\gamma_j(\mathbf{g}) = 1$ by Theorem 5.6 above.

Summing over all possible tuples $a \in ((\mathbb{Z}/p\mathbb{Z})^\times)^{|I|}$, the coefficient (apart from the β-part) is

$$p^{k-|I|}\sum_a\prod_{j\in I}a_j^{|I_j|} = p^{k-|I|}\prod_{j\in I}\sum_{a_j=1}^{p-1}a_j^{|I_j|}$$

and this equals $p^{k-|I|}\prod_{j\in I}\varphi(|I_j|)$ as desired, because the only problem may occur when some $|I_j| = 0$. But then $|I| < k$, so

$$p^{k-|I|}\sum_{a_j=1}^{p-1} a_j^0 = 0 \sum_{a_j=1}^{p-1} a_j^0 = 0 \sum_{a_j=1}^{p} a_j^0 = p^{k-|I|}\varphi(0)$$

□

Proof of Theorem 7.10.

1. This is from Theorem 5.6 above.

Now set $\Pi_p = \Pi'_p$. We first note from the key claim that if I_0 is nonempty, or any I_j is empty, then the coefficient of that particular monomial vanishes— because $\mathrm{char}(R) = p$ and $\varphi(0) = 0$.

2. Suppose $k > n$. Then at least one I_j must be empty in every monomial above, by the Pigeonhole Principle, and we are done.
3. If $k = n$, then the only monomials that have a nonzero contribution to the sum $\Sigma_{\Pi'_p}(\mathbf{h})$ must correspond to empty I_0 and singleton I_j's (since $\coprod_{j=1}^{k} I_j = [n] = [k]$). In other words, $\sigma \in S_n$: $j \mapsto i_j\ \forall j$. Moreover, the coefficient of such a monomial is $p^0\prod_{j=1}^{n}\varphi(1)$, and these monomials all add up to give the matrix permanent, as claimed. The rest of the statements are now easy to see.
4. In this part, we are only concerned with $\Sigma_{\Pi_p}(\mathbf{h})$, so that β does not contribute here either (so $I_0 = \varnothing$ and $[n] = \coprod_{j\in I} I_j$).
   From the key claim and Lemma 7.1 above, observe that if some monomial has a nonzero contribution, then $(p-1)$ divides $|I_j|$ for all j, and $I = [k]$. In particular, $(p-1)$ divides $\sum_{j\in I}|I_j| = n$, and
   $$n = \sum_{j\in I}|I_j| = \sum_{j=1}^{k}|I_j| \geq \sum_{j=1}^{k}(p-1) = k(p-1)$$
   whence $k \leq n/(p-1)$. Moreover, $\Sigma_{\Pi_p}(\mathbf{h}) = \varepsilon(\mathbf{h}) = 0$ if $k = 0$.
   It remains to present, for each $0 < k \leq n/(p-1)$ and (nonzero) $r \in R$, an example of $(H, \Pi = \Pi_p)$, so that $\Sigma_{\Pi}(\mathbf{h}) = \Sigma_{\Pi_p}(\mathbf{h}) = r$. This example is analyzed in the next section.

□

8. Example: Lie Algebras

Suppose $H = U(\mathfrak{g})$ for some Lie algebra $\mathfrak{g}$ (say over $\mathbb{C}$). Then any weight $\mu \in \Gamma_H$ kills $[\mathfrak{g},\mathfrak{g}]$, hence belongs to $(\mathfrak{g}/[\mathfrak{g},\mathfrak{g}])^*$. Let us denote $\mathfrak{g}_{ab} := \mathfrak{g}/[\mathfrak{g},\mathfrak{g}]$. Conversely, any element of the set above, is a weight of H, using multiplicativity and evaluating it at the projection down to the quotient $\mathfrak{g}_{ab}$. Thus, Γ_H is the dual space (under addition) of the abelianization $\mathfrak{g}_{ab}$ of $\mathfrak{g}$. Hence we now examine what happens in the case of an (R-free) Abelian Lie algebra $\mathfrak{h}$.

In this case, we have the free R-module $\mathfrak{h} = \oplus_i Rh_i$ with the trivial Lie bracket, and $H = U(\mathfrak{h}) = \mathrm{Sym}(\mathfrak{h})$. Thus, H inherits the usual Hopf algebra structure now (*i.e.*, $\Delta(h_i) = 1\otimes h_i + h_i\otimes 1$, $S(h_i) = -h_i$, $\varepsilon(h_i) = 0\ \forall i$).

First, $(\Gamma_H, *) = (\mathfrak{h}^*, +)$. By Proposition 5.5, if $\mathrm{char}(R) \nmid |\Pi|$, then $\Sigma_{\Pi}(\mathbf{h}) = 0$ for all products $\mathbf{h}$ of primitive elements in H (and hence for all $\mathbf{h}$ in the augmentation ideal $\mathfrak{h}U(\mathfrak{h})$ of H). Thus, the only case left to consider is when $\mathrm{char}(R) = p > 0$. But then $(\mathfrak{h}^*, +)$ is a $\mathbb{Z}/p\mathbb{Z}$-vector space, so every finite subgroup Π is of the form $\Pi = \Pi_p \cong (\mathbb{Z}/p\mathbb{Z})^k$ for some k. Moreover, Theorem 3.6 and (the last part of) Theorem 7.10 provide more results in this case.

We therefore conclude the example (and the proof of the theorem above) by analyzing the computation of $\Sigma_{\Pi}(\mathbf{h})$ for $\mathbf{h} = h_1 \dots h_n$. For any (nonzero) $r \in R$, we produce such a finite subgroup $\Pi = \Pi_p \cong (\mathbb{Z}/p\mathbb{Z})^k$, so that $0 < k \leq n/(p-1)$ and $\Sigma_{\Pi}(\mathbf{h}) = r$.
Construction: Given k, partition $[n]$ into k disjoint nonempty subsets $[n] = \coprod_{j=1}^{k} I_j$, reordered so that $I_1 = \{1, \dots, n-(k-1)(p-1)\}$, and so that $|I_j| = p-1$ for all $j > 1$. For each $1 \leq j \leq n$, define

$\gamma_j \in \mathfrak{h}^* = \Gamma_H$ as follows: $\gamma_1(h_1) = r$, $\gamma_j(h_i) = 1$ if $i \neq 1 \in I_j$, and $\gamma_j(h_i) = 0$ otherwise. (One verifies that the γ_i's thus defined are indeed linearly independent over $k(R)$, hence over $\mathbb{Z}/p\mathbb{Z}$ as well, but for this, one needs that $r \neq 0$ if $n = k, p = 2$.) Thus for any $K \subset [k]$, $\gamma_K(h_i) := \sum_{j \in K} \gamma_j(h_i)$ vanishes unless $i \in \cup_{j \in K} I_j$.

Now evaluate $\Sigma_\Pi(\mathbf{h}) = \sum_{K \subset [k],a} \prod_{i=1}^n \gamma_{a,K}(h_i)$, where $\Pi := \sum_{i=1}^n \mathbb{Z}\gamma_i = \bigoplus_{i=1}^n (\mathbb{Z}/p\mathbb{Z})\gamma_i$. By the key claim in the previous section, the only monomials $\prod_{j \in I} \gamma_j(h_{I'_j})$ that do not vanish are for $|I| = k$, and with $(p-1)$ divides $|I'_j|$ for all j. Moreover, $\gamma_j(h_i)$ is zero except when $i \in I_j$, so there is only one type of monomial remaining: $\prod_{j \in I} \gamma_j(h_{I_j})$. (Note that this satisfies the earlier condition: $(p-1)$ divides $|I_j|$ for all j.)

Moreover, by the key claim in the preceding section, the coefficient of this monomial, which itself equals $r \cdot \prod_{i=2}^n 1 = r$, is $\prod_{j=1}^k \varphi(|I_j|)$, and by Fermat's Little Theorem, $\varphi(|I_j|) = p - 1 = -1\ \forall j$ (in characteristic p). We conclude that $\Sigma_\Pi(\mathbf{h}) = \Sigma_{\Pi_p}(\mathbf{h}) = (-1)^k r$, whence we are done (start with $r' = (-1)^k r$ to get r).

9. Example: Degenerate Affine Hecke Algebras of Reductive Type with Trivial Parameter

In this section, we apply the general theory above, to a special case, wherein a finite group acts on a vector space (or free R-module in our case), with the group and the module corresponding to the Weyl group and the Cartan subalgebra (actually, its dual space) respectively, of a reductive Lie algebra. We use the $\mathbb{Z}$-basis of simple roots (and any $\mathbb{Z}$-basis for the center), to try and compute the value of $\Sigma_\Pi(\mathbf{h})$.

9.1. Hopf Algebras Acting on Vector Spaces

We will consider special cases of the following class of Hopf algebras. Suppose that a cocommutative R-Hopf algebra H acts on a free R-module V; denote the action by $h \cdot v$ for $h \in H, v \in V$. Then H also acts on V^* by: $\langle h \cdot \lambda, v \rangle := \langle \lambda, S(h) \cdot v \rangle$.

Now consider the R-algebra A generated by the sets H and V, with obvious relations in H, and the extra relations $vv' = v'v$, $\sum h_{(1)} v S(h_{(2)}) =: \operatorname{ad} h(v) = h \cdot v$ for all $h \in H$ and $v, v' \in V$. Note that the relation $\operatorname{ad} h(v) = h \cdot v$ can be rephrased, as the following lemma shows.

Lemma 9.1. *Suppose some R-Hopf algebra H acts on a free R-module V, and an R-algebra B contains H, V. Then the following relations are equivalent (in B) for all $v \in V$:*

1. $\sum h_{(1)} v S(h_{(2)}) = h \cdot v$ *for all* $h \in H$.
2. $hv = \sum (h_{(1)} \cdot v) h_{(2)}$ *for all* $h \in H$.

If H is cocommutative, then both of these are also equivalent to:

3. $vh = \sum h_{(1)} (S(h_{(2)}) \cdot v)$ *for all* $h \in H$.

Moreover, if this holds, then any unital subalgebra M of B that is also an H-submodule (via ad*) is an H-(Hopf-)module algebra under the action*

$$h \cdot m := \operatorname{ad} h(m) = \sum h_{(1)} m S(h_{(2)})\ \forall h \in H,\ m \in M$$

(The proof is straightforward.) For instance, one can take $M = B$ or H—or in the above example of A, consider $M = \operatorname{Sym}_R V$.

It is straightforward (but perhaps tedious) to check that A is a Hopf algebra with the usual operations: on H, they restrict to the Hopf algebra structure of H, and V consists of primitive elements.

By the above lemma, if H is R-free, then the ring A is an R-free R-Hopf algebra, with R-basis given by $\{h \cdot m\}$, where $h \in H$ and m run respectively over some R-basis of H, and all (monomial) words (including the empty word) with alphabet given by an R-basis of V. It has the subalgebras H and $\operatorname{Sym}_R(V)$, and is called the *smash product* $H \ltimes \operatorname{Sym}_R V$ of H and $\operatorname{Sym}_R V$.

We now determine the weights of A. Denote by Γ_H the group of weights of H (under convolution). One can now use Proposition 2.13 to prove:

Proposition 9.2. *The weights Γ_A of A form a group, which is the Cartesian product $\Gamma_H \times V_\varepsilon^*$, with convolution given by*

$$(\nu_1, \lambda_1) * (\nu_2, \lambda_2) = (\nu_1\nu_2, \lambda_1 + \lambda_2) = (\nu_1 *_H \nu_2, \lambda_1 *_V \lambda_2)$$

for $\nu_i \in \Gamma_H$, $\lambda_i \in V_\varepsilon^*$. (Here, V_ε^* is the ε-weight space of the H-module V^*.)

9.2. Degenerate Affine Hecke Algebras

Since we work over any commutative unital integral domain R, we can generate examples over all R if there exists a lattice in V that is fixed by H, and one considers its R-span. Now specialize to the case when $H = RW$ is the group ring of a Weyl group acting on a Cartan subalgebra of the corresponding semisimple Lie algebra. Then one uses the root lattice Q inside $V = \mathfrak{h}^*$.

We work in slightly greater generality. Given a finite-dimensional reductive complex Lie algebra $\mathfrak{g}$, let W be its Weyl group and $\mathfrak{h}$ a fixed chosen Cartan subalgebra. Thus $\mathfrak{h} = \oplus_{i \geq 0}\mathfrak{h}_i$, where for $i > 0$, $\mathfrak{h}_i$ corresponds to a simple component (ideal) of $\mathfrak{g}$, with corresponding base of simple roots Δ_i and Weyl group W_i, say; and $\mathfrak{h}_0$ is the central ideal in $\mathfrak{g}$.

Define $Q_i = \oplus_{\alpha \in \Delta_i}\mathbb{Z}\alpha$, the root lattice inside $\mathfrak{h}_i^*$, and choose and fix some $\mathbb{Z}$-lattice Q_0 inside $\mathfrak{h}_0^*$. Now replace $\mathfrak{h}_i^*$ by $V_i = \mathfrak{h}_i^* := R \otimes_{\mathbb{Z}} Q_i$, and $\mathfrak{h}_i$ by the R-dual of $\mathfrak{h}_i^*$, for all $i \geq 0$. Thus, for the entire Lie algebra, $\Delta = \coprod_{i>0} \Delta_i$ and $W = \times_{i>0} W_i$.

Now define $V = \oplus_{i \geq 0} V_i$, whence the previous subsection applies and one can form the algebra $A = RW \ltimes \mathrm{Sym}_R\, V$. This is the *degenerate affine Hecke algebra with trivial parameter* (the parameter is trivial since $wv - w(v)w$ is always zero), of reductive type. This is a special case of [16, Definition 1.1], where one sets $\eta = 0$.

Before we address the general case, note that there are two types of $\mathfrak{h}_i$'s in here: ones corresponding to simple Lie algebras, which we address first, and the "central part", which is fixed by W (hence so is $\mathfrak{h}_0^*$).

9.3. The Simple Case

The first case to consider is: $V = \mathfrak{h}^* = R \otimes_{\mathbb{Z}} Q$, for a *simple* Lie algebra. Thus Δ is irreducible, and given $A = RW \ltimes \mathrm{Sym}_R(\mathfrak{h}^*)$, $\Gamma_A = \Gamma_W \times \mathfrak{h}^W$ (because the condition in Proposition 9.2 above translates to: $w(\gamma) = \varepsilon(w)\gamma = \gamma$ for all $w \in W, \gamma \in \Gamma_A$). Here, $\Gamma_W = \Gamma_{RW}$.

We now state our main result, using the convention that all roots in the simply laced cases (types A, D, E) are short. The result helps compute Σ_Π at any element of the R-basis $\{g \cdot m\}$ mentioned in an earlier subsection.

Theorem 9.3. *Suppose $\mathfrak{g}$ is a complex simple Lie algebra with simple roots Δ, Weyl group W, $V = \mathfrak{h}^* = R \otimes_{\mathbb{Z}} \mathbb{Z}\Delta$, and $A = RW \ltimes \mathrm{Sym}_R(\mathfrak{h}^*)$. As above, let $\Pi \subset \Gamma_A$ be a finite subgroup of weights. Let $h_1, \ldots, h_n \in \mathfrak{h}^*$.*

1. *If* $\mathrm{char}(R) \neq 2$, *or W is of type G_2, or W has more than one short simple root, then every weight acts as $\varepsilon = 0$ on $\mathfrak{h}^*$. In particular, $\Sigma_\Pi = 0$ on $\mathrm{Sym}_R(\mathfrak{h}^*)$.*
2. *If* $\mathrm{char}(R) = 2$, *then every weight acts as ε on W. Now suppose also that W is not of type G_2, and has only one short simple root α_s, say.*

   If Π has an element of order 4, or h_i has no "α_s-contribution" (i.e., $h_i \in \oplus_{\alpha_s \neq \alpha \in \Delta} R \cdot \alpha$) *for some i, then $\Sigma_\Pi(\mathbf{h}) = 0$.*
3. *If this does not happen,* i.e., $\Pi = (\mathbb{Z}/2\mathbb{Z})^k$ *for some k, and the hypotheses of the previous part hold, then*

$$\Sigma_\Pi(\alpha_s^n) = \sum_{\substack{l_i > 0\ \forall i \\ l_1 + \cdots + l_k = n}} \binom{n}{l_1, \ldots, l_k} \prod_{i=1}^{k} \gamma_i(\alpha_s)^{l_i}$$

where the γ_i's are any set of generators for Π. In particular, this vanishes if $k > r$, where $\sum_{j=1}^{r} 2^{s_j}$ is the binary expansion of n.

Remark 9.4.

1. **Warning.** One should not confuse the h_i's here with elements of $\mathfrak{h}$; indeed, $h_i \in A$, so they really are in $\mathfrak{h}^*$.
2. The coefficient above is just the multinomial coefficient $n!/(\prod_i l_i!)$, which we also denote by $\binom{n}{l_1,\dots,l_{k-1}}$, just as $\binom{n}{k,n-k} = \binom{n}{k}$. The last line in the theorem follows because this coefficient is odd if and only if (r, s_j as above) we can partition $\{2^{s_j} : j\}$ into k nonempty subsets, and the l_i's are precisely the sums of the elements in the subsets. (This fails, for instance, if some two l_i's are equal, or $k > r$.) In turn, this fact follows (inductively) from the following easy-to-prove

   Lemma 9.5. *Suppose $p > 0$ is prime, $p^s \leq n < p^{s+1}$ for some $s \geq 0$, and $l_k \geq l_i\ \forall i$. If $l_k < p^s$ then p divides $\binom{n}{l_1,\dots,l_{k-1}}$. Otherwise p divides neither or both of $\binom{n}{l_1,\dots,l_{k-1}}$ and $\binom{n-p^s}{l_1,\dots,l_{k-1}}$.*

The rest of the subsection is devoted to the proof of the theorem. We once again mention a result crucial to the proof, then use it to prove the theorem, and conclude by proving the key claim.
Key claim. (char(R) arbitrary.) If W contains a Dynkin subgraph Ω of type A_2 or G_2, then both the simple roots in Ω are killed by all $\lambda \in \mathfrak{h}^W$. If Ω is of type B_2, then the long root in Ω is killed by all λ.

Proof modulo the key claim. We now show the theorem.

1. First suppose that char$(R) \neq 2$. If $\lambda \in \mathfrak{h}^W$, then $\lambda(\alpha) = \lambda(s_\alpha(\alpha)) = -\lambda(\alpha)$, whence $\lambda(\alpha) = 0$ for all $\alpha \in \Delta$, and $\mathfrak{h}^W = 0$.

   For the other claims, use the classification of simple Lie algebras in terms of Dynkin diagrams, as mentioned in [7, Chapter 3]. To show that a weight λ kills all of $\mathfrak{h}^*$, it suffices to show that $\lambda(\alpha) = 0\ \forall \alpha \in \Delta$, *i.e.*, that it kills each simple root or node of the corresponding Dynkin diagram.

   If the Dynkin diagram of a Lie algebra has (a sub-diagram of) type A_2 or G_2, then both nodes of that diagram (or both α_i's) are killed by all weights $\lambda \in \Gamma$, by the key claim above. This automatically eliminates all diagrams of type A_n for $n > 1$, as well as all D, E, F, G-type diagrams, leaving only type A_1 among these.

   Moreover, for types B, C, at most one simple root (the "last" one) is not killed by all λ's. If this root is long, then it is also killed by the key claim above (as a part of a B_2), and we are done.
2. First, $\lambda(s_\alpha^2) = \lambda(s_\alpha)^2 = 1$, whence $\lambda(s_\alpha) = \pm 1 = 1\ \forall \alpha \in \Delta$, if char$(R) = 2$. This implies that $\lambda(w) = 1 = \varepsilon(w)$ for all $w \in W, \lambda \in \Gamma$. Next, Theorem 7.8 above tells us that if Π has an element of order 4, then $\Sigma_\Pi(\mathbf{h}) = 0$. Finally, if some h_i has no "α_s-contribution", then it is killed by all λ, by the previous part, so $\lambda(\mathbf{h}) = 0\ \forall \lambda \in \Gamma$.
3. As we remarked after Theorem 7.10, $\Pi = (\mathbb{Z}/2\mathbb{Z})^k$ in characteristic 2, if Π does not have an element of order 4. (Reason: $\Gamma = \{\varepsilon\} \times \mathfrak{h}^W \cong (\mathfrak{h}^W, +)$ is a free R-module by the previous part, and $2\Gamma = 0$.)

   We now perform the computation. For this, suppose that $h_i - c_i\alpha_s$ is in the R-span of $\{\alpha \in \Delta : \alpha \neq \alpha_s\}$ (note that in the case of A_1, the condition $h_i \in R \cdot \alpha_s$ is automatic). Then $\Sigma_\Pi(\mathbf{h}) = (\prod_i c_i) \cdot \Sigma_\Pi(\alpha_s^n)$, so it suffices to compute $\Sigma_\Pi(\alpha_s^n)$.

   If $\{\gamma_i\}$ is any set of generators (or $\mathbb{Z}/2\mathbb{Z}$-basis) for Π, then the desired equation actually holds if we sum over all *nonnegative* tuples l_i that add up to k. Thus, the proof is similar to that of the key claim used to prove Theorem 7.10 above; simply note that if $I \subsetneq [k]$, then every $\prod_{j \in I} \gamma_j(\alpha_s^{|I_j|})$ occurs with an even coefficient.

Proof of the key claim. It helps to look at the pictures of these rank 2 root systems (drawn in [7, Chapter 3]). We use the W-invariance of $\lambda|_V \; \forall \lambda \in \Gamma$.

Consider the system A_2, with simple roots α, β. Given $\lambda \in \Gamma$, $\lambda(\alpha) = \lambda(\beta) = \lambda(\alpha+\beta)$, whence $\lambda(\alpha) = \lambda(\beta) = 0$.

The root system G_2 has two subsystems of type A_2, whence each λ must kill both subsystems.

Now consider B_2, with long root α and short root β. Clearly, $\beta + \alpha$ is another short root, whence $\lambda(\beta + \alpha) = \lambda(\beta)$, and we are done. □

9.4. The Reductive Case

We conclude by mentioning what happens in the reductive case. This uses the results proved in the simple case above. Recall also that the notation for this situation was set when we defined degenerate affine Hecke algebras with trivial parameter earlier. This notation will be used freely here, without recalling it from above.

Let V' be the direct sum of V_0 and the R-span of all the unique short simple roots $\alpha_{i,\, short}$ inside any of the simple components $V_i = \mathfrak{h}_i^*$ of the "correct" type (not G_2). Let the other simple roots in Δ span the R-submodule V''. Then $V = V' \oplus V''$, and each $\lambda \in \Gamma$ kills V''. There now are two cases.

Case 1. $\mathrm{char}(R) \neq 2$. Then λ in fact kills all $\alpha \in \Delta$, because $\lambda(\alpha) = \lambda(s_\alpha(\alpha)) = -\lambda(\alpha)$. This means that we are left with V_0, *i.e.*, if for all i, $h_i - v_{0,i} \in \oplus_{j>0} V_j$ for some $v_{0,i} \in V_0$, then $\Sigma_\Pi(\mathbf{h}) = \Sigma_\Pi(\prod_i v_{0,i})$.

Next, recall that $\Sigma_\Pi = \Gamma_W \times (V^*)^W$, so we are reduced to the case of every λ being represented (on V_0) by some element of $V_0^* = (V_0^*)^W$. We conclude this case by noting that some (partial) results on how to compute this were included in the previous section.

Case 2. $\mathrm{char}(R) = 2$. Then $\lambda(w) = 1$ for all w, λ, as seen above. Moreover, we are left only to consider the case of all $h_i \in V'$. Now, $\Gamma = \varepsilon_W \times (V')^*$, whence any finite subgroup $\Pi = (\mathbb{Z}/2\mathbb{Z})^k$ for some k (since it too is a $\mathbb{Z}/2\mathbb{Z}$-vector space). In this situation, Theorems 7.8 and 7.10 (and 9.3 as well) give us some information on how to compute $\Sigma_\Pi(\mathbf{h})$.

10. An Example that Attains Any Value

We conclude with examples where $\Sigma_\Pi(\mathbf{h})$ can take any value in R, if the h_i's are merely skew-primitive.

Example 12 (A skew-primitively generated algebra). By Proposition 5.5 above, if all h_i's are pseudo-primitive, then $\Sigma_\Pi(\mathbf{h}) = 0$ if $\mathrm{char}(R) \nmid |\Pi|$—whereas if $\mathrm{char}(R)$ divides $|\Pi|$, then this case was analyzed in Section 7 above.

One can ask if such results hold in general, *i.e.*, for products of skew-primitive elements. (Note by Theorem 5.6 that we need $\mathrm{char}(R) \nmid |\Pi|$.)

For the example that we now mention (for groups Π of *even* order), one needs to **assume** the following:

1. $\mathrm{char}(R) > 2^n$ and $\exp(\Pi)$, or $\mathrm{char}(R) = 0$ and $R \supset \mathbb{Q}$; and
2. If $d = \exp(\Pi)$ is the exponent, then d is even, and there exists a primitive dth root of unity in R, say z.

Beyond this, given n, Π (of even order), and $r \in R$, we will produce the desired Hopf algebra $\mathscr{H}$, a group of weights $\Pi \subset \Gamma_{\mathscr{H}}$, and skew-primitive $h_1, \ldots, h_n \in H$, such that $\Sigma_\Pi(\mathbf{h}) = r \in R$.

Given Π, use the Structure Theorem for finite Abelian groups to write $\Pi = \bigoplus_{i=1}^k (\mathbb{Z}/d_i\mathbb{Z})$, with $d_1|d_2|\ldots|d_k = \exp(\Pi)$. Then d_k is even, since Π has even order. Now define $\mathscr{H}$ to be the commutative R-algebra freely generated as: $\mathscr{H} = R[R^n] \otimes R[\mathbb{Z}^k]$. In other words, R is generated by $h_1, \ldots, h_n, g_1^{\pm 1}, \ldots, g_k^{\pm 1}$, with the relation that they all commute (and that the g_i's are invertible).

Now define the g_i's to be grouplike and $\Delta(h_j) = g_k \otimes h_j + h_j \otimes 1$. Also define (for all i, j):

$$\varepsilon(g_i) = 1,\; S(g_i) = g_i^{-1},\; \varepsilon(h_j) = 0,\; S(h_j) = -g_k^{-1} h_j$$

Since $\mathscr{H}$ is freely generated, the set of weights of $\mathscr{H}$ is $R^n \times (R^\times)^k$. Since (it can be checked that) $\mathscr{H}$ is also a Hopf algebra, the group operation is:

$$\begin{aligned} & (a_1, \ldots, a_n, z_1, \ldots, z_k) * (a_1', \ldots, a_n', z_1', \ldots, z_k') \\ = \; & (a_1 + z_k a_1', a_2 + z_k a_2', \ldots, a_n + z_k a_n', z_1 z_1', \ldots, z_n, z_n') \end{aligned}$$

We now produce the desired example. Define $\gamma_j \in \Gamma_{\mathscr{H}}$ on generators by: $\gamma_j(g_i) = z^{\delta_{ij} d_k / d_j}$, and $\gamma_j(h_i) = 0$ unless $j = k$. Moreover, $\gamma_k(h_i) = 1$ for $i < k$, and $\gamma_k(h_k) = (1-z)^n |\Pi|^{-1} r'$ for some $r' \in R$ (which we define later, and which depends on n).

It is now easy to check that each γ_j is of order d_j, and the γ_j's generate a subgroup of $\Gamma_{\mathscr{H}}$ isomorphic to Π. Moreover, $\gamma_1, \ldots, \gamma_{k-1}$ all kill $h_1, h_2, \ldots, h_n$. Adopting the notation of Proposition 4.2, $\gamma_j \in \Gamma_{h_i}$ for $j < k$ and all i. Since Γ_{h_i} is a subgroup of $\Gamma_{\mathscr{H}}$, hence $\Pi_1 := \langle \gamma_1, \ldots, \gamma_{k-1} \rangle \subset \Gamma_{h_i}$ for all i; by Proposition 4.2, $\Pi_1 \subset \Gamma_{\mathbf{h}}$.

Now use Lemma 4.4 (noting that Π_1 is normal in the Abelian group Π); then $\Sigma_\Pi(\mathbf{h}) = |\Pi/\Pi_1| \Sigma_{\Pi_1}(\mathbf{h})$. Use Theorem 6.3: $f_i = \frac{\gamma_k(h_i)}{\gamma_k(g_k) - 1}$, so

$$(-1)^n \prod_i f_i = (-1)^n \frac{(1-z)^n |\Pi|^{-1} r'}{\prod_{i=1}^n (z-1)} = \frac{r'}{|\Pi|}$$

Moreover, $S = \{I \subset \{1, 2, \ldots, n\} : g_k^{|I|} \text{ is fixed by } \Pi_1\}$, *i.e.*, all subsets I such that $d_k \mid |I|$. Since d_k is even, this means that $(-1)^{|I|} = 1 \; \forall I \in S$, whence

$$\Sigma_\Pi(\mathbf{h}) = |\Pi/\Pi_1| \Sigma_{\Pi_1}(\mathbf{h}) = \frac{|\Pi|}{|\Pi_1|} \cdot |\Pi_1| \cdot \frac{r'}{|\Pi|} \cdot \sum_{m \geq 0} \binom{n}{m d_k} = r' \sum_{m \geq 0} \binom{n}{m d_k}$$

By assumption (on R), the summation is a unit in R, so choosing r' suitably, one obtains any $r \in R$ as our answer. □

Also note (*e.g.*, by [17, Exercise 38, §1.2.6]), that the summation equals $\frac{1}{d_k} \sum_{l=0}^{d_k - 1} (1 + z^l)^n$.

Acknowledgments

I thank Susan Montgomery and Nicolás Andruskiewitsch for their comments and suggestions after reading a preliminary draft of this manuscript. I also thank the referees for their feedback, which helped improve the exposition of this paper.

References

1. Radford, D.E. The order of the antipode of a finite-dimensional Hopf algebra is finite. *Am. J. Math.* **1976**, *98*, 333–355.
2. Andruskiewitsch, N.; Schneider, H.-J. On the classification of finite-dimensional pointed Hopf algebras. *Ann. Math.* **2010**, *171*, 375–417.
3. Montgomery, S. *Hopf Algebras and Their Actions on Rings*; American Mathematical Society: Providence, RI, USA, 1993.
4. Abe, E. *Hopf Algebras*; Cambridge University Press: London, UK; New York, NY, USA, 1977.
5. Dascalescu, S.; Nastasescu, C.; Raianu, S. *Hopf Algebras: An Introduction*; Marcel-Dekker: New York, NY, USA, 2001.
6. Sweedler, M.E. *Hopf Algebras*; W.A. Benjamin: New York, NY, USA, 1969.
7. Humphreys, J.E. *Introduction to Lie Algebras and Representation Theory*; Springer-Verlag: New York, NY, USA, 1972.
8. Jantzen, J.C. *Lectures on Quantum Groups*; American Mathematical Society: Providence, RI, USA, 1995.
9. Majid, S. *A Quantum Groups Primer*; Cambridge University Press: New York, NY, USA, 2002.
10. Hu, N. Quantum group structure associated to the quantum affine space. *Algebra Colloq.* **2004**, *11*, 483–492.
11. Hu, N. Quantum group structure of the q-deformed virasoro algebra. *Lett. Math. Phys.* **1998**, *44*, 99–103.

12. Hadfield, T.; Krähmer, U. On the Hochschild homology of quantum $SL(N)$. *Comptes Rendus Math.* **2006**, *343*, 9–13.
13. Khare, A. Functoriality of the BGG category $\mathcal{O}$. *Commun. Algebra* **2009**, *37*, 4431–4475.
14. Dascalescu, S.; Iovanov, M.C.; Nastasescu, C. Quiver algebras, path coalgebras, and co-reflexivity. *Pacific J. Math.* arXiv:1208.4410.
15. Alperin, J.L.; Bell, R.B. *Groups and Representations*; Springer-Verlag: New York, NY, USA, 1995.
16. Cherednik, I. Integration of quantum many-body problems by affine knizhnik-zamolodchikov equations. *Adv. Math.* **1994**, *106*, 65–95.
17. Knuth, D.E. *The Art of Computer Programming, Volume 1: Fundamental Algorithms*; Addison-Wesley: Boston, MA, USA, 1997.

13

From Coalgebra to Bialgebra for the Six-Vertex Model: The Star-Triangle Relation as a Necessary Condition for Commuting Transfer Matrices

Jeffrey R. Schmidt

Departments of Physics and Mathematics, University of Wisconsin-Parkside 900 Wood Road, Kenosha, WI 53141, USA; E-Mail: jeff@rustam.uwp.edu

Abstract: Using the most elementary methods and considerations, the solution of the star-triangle condition $\frac{a^2+b^2-c^2}{2ab} = \frac{(a')^2+(b')^2-(c')^2}{2a'b'}$ is shown to be a necessary condition for the extension of the operator coalgebra of the six-vertex model to a bialgebra. A portion of the bialgebra acts as a spectrum-generating algebra for the algebraic Bethe ansatz, with which higher-dimensional representations of the bialgebra can be constructed. The star-triangle relation is proved to be necessary for the commutativity of the transfer matrices $T(a,b,c)$ and $T(a',b',c')$.

Keywords: vertex model; bialgebra; coalgebra; Bethe ansatz

---

1. Introduction

In two-dimensional lattice vertex models in which the state of a lattice point is specified by the states of the four links to its neighboring points, the matrix M of local Boltzmann weights can be used to construct the row-to-row transfer matrix by a matrix coproduct operation. In the six-vertex model a link between two lattice points has only two states, call them $0, 1$ or $\uparrow, \downarrow$, subject to the "ice-rule" (two arrows point into a vertex, two point out). The matrix M of local Boltzmann weights for the six-vertex model has two types of indices, "vertical" and "horizontal" [1],

$$M^{h_2,v_2}_{h_1,v_1}(\lambda) = \left(\begin{array}{cc|cc} M^{00}_{00} & M^{01}_{00} & M^{10}_{00} & M^{11}_{00} \\ M^{00}_{01} & M^{01}_{01} & M^{10}_{01} & M^{11}_{01} \\ \hline M^{00}_{10} & M^{01}_{10} & M^{10}_{10} & M^{11}_{10} \\ M^{00}_{11} & M^{01}_{11} & M^{10}_{11} & M^{11}_{11} \end{array}\right) = \left(\begin{array}{c|c} A(\lambda) & B(\lambda) \\ \hline C(\lambda) & D(\lambda) \end{array}\right) \tag{1}$$

in which

$$A(\lambda) = \begin{pmatrix} a(\lambda) & 0 \\ 0 & b(\lambda) \end{pmatrix}, \quad D(\lambda) = \begin{pmatrix} b(\lambda) & 0 \\ 0 & a(\lambda) \end{pmatrix}, \quad B(\lambda) = \begin{pmatrix} 0 & 0 \\ c(\lambda) & 0 \end{pmatrix}, \quad C(\lambda) = \begin{pmatrix} 0 & c(\lambda) \\ 0 & 0 \end{pmatrix} \tag{2}$$

have only vertical indices (for example $A^{v_2}_{v_1}$). All four matrices A, B, C, D typically depend on a set of parameters λ. The matrix C has a non-zero kernel $\uparrow = \begin{pmatrix} 1 \\ 0 \end{pmatrix}$ which is not annihilated by B, and A and D are invertible. This will be a principal ingredient of the solution of the model by the Bethe ansatz.

The matrix coproduct is a mapping $\Delta : \mathfrak{A} \to \mathfrak{A} \times \mathfrak{A}$ (a tensor product on vertical indices, a dot or matrix product on horizontal)

$$\begin{aligned}\Delta(M) = M \underset{\bullet}{\otimes} M &= \left(\begin{array}{c|c} A & B \\ \hline C & D \end{array}\right) \underset{\bullet}{\otimes} \left(\begin{array}{c|c} A & B \\ \hline C & D \end{array}\right) \\ &= \left(\begin{array}{c|c} A \otimes A + B \otimes C & A \otimes B + B \otimes D \\ \hline C \otimes A + D \otimes C & C \otimes B + D \otimes D \end{array}\right) \\ &= \left(\begin{array}{c|c} \Delta(A) & \Delta(B) \\ \hline \Delta(C) & \Delta(D) \end{array}\right)\end{aligned} \tag{3}$$

which in matrix index language would be

$$\left(M \underset{\bullet}{\otimes} M\right)^{h_2,v_2,v_2'}_{h_1,v_1,v_1'} = \sum_h M^{h,v_2}_{h_1,v_1} M^{h_2,v_2'}_{h,v_1'} \tag{4}$$

Because the $\otimes_\bullet$ product is fundamentally a matrix product (which is associative), Δ is coassociative, so Δ and a compatible counit map ϵ (see for example [2,3]) define a coalgebra structure on the set $\mathfrak{A}$.

If $\mathfrak{A}$ is an algebra and the coproduct and counit of the coalgebra structure on $\mathfrak{A}$ are compatible with the product and unit of the algebra (Δ and ϵ are algebra homomorphisms), then $\mathfrak{A}$ is a bialgebra (simultaneous coalgebra and algebra). The compatibility with the coalgebra structure is a very strong constraint on the algebraic product rules. The existence of an antipode map extends the bialgebra to a Hopf algebra [3].

For a lattice vertex model with n sites per row, we construct the L-matrix of Boltzmann weights whose horizontal components are higher dimensional representations of the elements $\mathcal{A}(\lambda), \mathcal{B}(\lambda), \mathcal{C}(\lambda), \cdots$ of $\mathfrak{A}$

$$L^{h_2,\mathbf{v}'}_{h_1,\mathbf{v}}(\lambda) = \left(M(\lambda) \underset{\bullet}{\otimes} M(\lambda) \underset{\bullet}{\otimes} \cdots \underset{\bullet}{\otimes} M(\lambda)\right)^{h_2,\mathbf{v}'}_{h_1,\mathbf{v}}, \qquad \left(\begin{array}{c|c} L^{0,\mathbf{v}'}_{0,\mathbf{v}} & L^{1,\mathbf{v}'}_{0,\mathbf{v}} \\ \hline L^{0,\mathbf{v}'}_{1,\mathbf{v}} & L^{1,\mathbf{v}'}_{1,\mathbf{v}} \end{array}\right) = \left(\begin{array}{c|c} \mathcal{A}^{\mathbf{v}'}_{\mathbf{v}} & \mathcal{B}^{\mathbf{v}'}_{\mathbf{v}} \\ \hline \mathcal{C}^{\mathbf{v}'}_{\mathbf{v}} & \mathcal{D}^{\mathbf{v}'}_{\mathbf{v}} \end{array}\right) \tag{5}$$

The transfer matrix for a lattice model with such a matrix M of Boltzmann weights is obtained from an iterated coproduct,

$$\begin{aligned}\Delta(L) &= L \underset{\bullet}{\otimes} M = \left(\begin{array}{c|c} \mathcal{A} \otimes A + \mathcal{B} \otimes C & \mathcal{A} \otimes B + \mathcal{B} \otimes D \\ \hline \mathcal{C} \otimes A + \mathcal{D} \otimes C & \mathcal{C} \otimes B + \mathcal{D} \otimes D \end{array}\right) \\ &= \left(\begin{array}{c|c} \Delta(\mathcal{A}) & \Delta(\mathcal{B}) \\ \hline \Delta(\mathcal{C}) & \Delta(\mathcal{D}) \end{array}\right)\end{aligned} \tag{6}$$

by summing over the last set of horizontal indices

$$\begin{aligned}T^{\mathbf{v}'}_{\mathbf{v}}(\lambda) = \left(Tr_h L\right)^{\mathbf{v}'}_{\mathbf{v}}(\lambda) &= \sum_h \left(M(\lambda) \underset{\bullet}{\otimes} M(\lambda) \underset{\bullet}{\otimes} \cdots \underset{\bullet}{\otimes} M(\lambda)\right)^{h,v_2,v_2',\cdots}_{h,v_1,v_1',\cdots}, \\ &= \left(Tr_h\, M(\lambda) \underset{\bullet}{\otimes} M(\lambda) \underset{\bullet}{\otimes} \cdots \underset{\bullet}{\otimes} M(\lambda)\right)^{\mathbf{v}'}_{\mathbf{v}}\end{aligned} \tag{7}$$

This is the trace of the L-matrix over its horizontal components

$$T(\lambda) = Tr_h\, L(\lambda) = \mathcal{A}(\lambda) + \mathcal{D}(\lambda) \tag{8}$$

Define the complementary product in which the tensor is on the horizontal indices, and matrix product on the vertical (note that the horizontal space sub-matrices depend on different parameter sets, carefully compare with Equation 3)

$$M \underset{\otimes}{\bullet} M = \left(\begin{array}{c|c} A & B \\ \hline C & D \end{array}\right) \underset{\otimes}{\bullet} \left(\begin{array}{c|c} A' & B' \\ \hline C' & D' \end{array}\right) = \left(\begin{array}{c|c|c|c} A\cdot A' & A\cdot B' & B\cdot A' & B\cdot B' \\ A\cdot C' & A\cdot D' & B\cdot C' & B\cdot D' \\ \hline C\cdot A' & C\cdot B' & D\cdot A' & D\cdot B' \\ C\cdot C' & C\cdot D' & D\cdot C' & D\cdot D' \end{array}\right) \tag{9}$$

The matrices A, B, C, D and $\mathcal{A}, \mathcal{B}, \mathcal{C}$ and $\mathcal{D}$ are functions of the parameter-sets λ (un-primed depend on λ, primed are functions of λ'), which for the six-vertex model is a collection of three energies upon which a, b, c depend.

For commutativity of the transfer matrix $[T(\lambda), T(\lambda')] = 0$, which makes the model integrable if a sufficiently large set of mutually commuting matrices can be found, it is sufficient that there exists an invertible matrix R such that [1]

$$R(L(\lambda) \underset{\otimes}{\bullet} L(\lambda')) = (L(\lambda') \underset{\otimes}{\bullet} L(\lambda))R, \qquad R^{bb'}_{aa'}(R^{-1})^{cc'}_{bb'} = \delta^c_a \delta^{c'}_{a'} \tag{10}$$

since if we write out the horizontal components

$$R\left(\begin{array}{c|c|c|c} \mathcal{A}\cdot\mathcal{A}' & \mathcal{A}\cdot\mathcal{B}' & \mathcal{B}\cdot\mathcal{A}' & \mathcal{B}\cdot\mathcal{B}' \\ \mathcal{A}\cdot\mathcal{C}' & \mathcal{A}\cdot\mathcal{D}' & \mathcal{B}\cdot\mathcal{C}' & \mathcal{B}\cdot\mathcal{D}' \\ \hline \mathcal{C}\cdot\mathcal{A}' & \mathcal{C}\cdot\mathcal{B}' & \mathcal{D}\cdot\mathcal{A}' & \mathcal{D}\cdot\mathcal{B}' \\ \mathcal{C}\cdot\mathcal{C}' & \mathcal{C}\cdot\mathcal{D}' & \mathcal{D}\cdot\mathcal{C}' & \mathcal{D}\cdot\mathcal{D}' \end{array}\right) R^{-1} = \left(\begin{array}{c|c|c|c} \mathcal{A}'\cdot\mathcal{A} & \mathcal{A}'\cdot\mathcal{B} & \mathcal{B}'\cdot\mathcal{A} & \mathcal{B}'\cdot\mathcal{B} \\ \mathcal{A}'\cdot\mathcal{C} & \mathcal{A}'\cdot\mathcal{D} & \mathcal{B}'\cdot\mathcal{C} & \mathcal{B}'\cdot\mathcal{D} \\ \hline \mathcal{C}'\cdot\mathcal{A} & \mathcal{C}'\cdot\mathcal{B} & \mathcal{D}'\cdot\mathcal{A} & \mathcal{D}'\cdot\mathcal{B} \\ \mathcal{C}'\cdot\mathcal{C} & \mathcal{C}'\cdot\mathcal{D} & \mathcal{D}'\cdot\mathcal{C} & \mathcal{D}'\cdot\mathcal{D} \end{array}\right) \tag{11}$$

(explicitly using invertibility of R) and take traces we obtain the desired commutativity

$$(\mathcal{A} + \mathcal{D})(\mathcal{A}' + \mathcal{D}') = (\mathcal{A}' + \mathcal{D}')(\mathcal{A} + \mathcal{D}), \quad \text{or} \quad [T(\lambda), T(\lambda')] = 0 \tag{12}$$

A sufficient condition that makes Equation 10 true is [1]

$$R(M(\lambda) \underset{\otimes}{\bullet} M(\lambda')) = (M(\lambda') \underset{\otimes}{\bullet} M(\lambda))R \tag{13}$$

Equation 13 becomes the Yang–Baxter equation (with a spectral parameter) if $R = M(\mu)$ for some choice of parameter μ, and this is not possible if M is not invertible.

If Equation 13 is written out in detail for the six-vertex model, with $\lambda = (a, b, c)$, $\lambda' = (a', b', c')$ and $\mu = (a'', b'', c'')$ and μ is eliminated from the resulting system of cubic equations, one obtains a much simpler constraint under which Equation 12 is true, which is the solution of the star-triangle relation

$$\frac{a^2 + b^2 - c^2}{2ab} = \frac{(a')^2 + (b')^2 - (c')^2}{2a'b'} \tag{14}$$

Suppose that one begins with a matrix M of local Boltzmann weights (possibly not invertible), which we think of as a low-dimensional representation of some operator coalgebra, and uses it to construct a transfer matrix for a lattice model by the matrix coproduct construction. Under what conditions can the coalgebra be extended to a bialgebra, or perhaps even a Hopf algebra?

The purpose of this paper is to show that Equation 14 is a necessary condition for the matrix coalgebra of the spectrum-generating operators $\mathcal{A}, \mathcal{B}, \mathcal{C}, \mathcal{D}$ with lowest dimensional representation given in Equation 2 (the six-vertex model) to extend to a bialgebra with matrix coproduct. A bialgebra need not be a Hopf algebra, and a Hopf algebra need not have an R-matrix. We show that, at least in the case of the six-vertex model, commuting transfer matrices are one of the bialgebra product relations. The star-triangle relation is therefore necessary for integrability. Whether or not the star-triangle equation is necessary for $[T(\lambda), T(\lambda')] = 0$ for any given model is still an open problem (see for example

P. 418 of [4]), one which seems to have been abundantly acknowledged but largely unaddressed in the literature.

We will prove that Equation 14 is necessary for commutativity of the transfer matrices for the six-vertex model by constructing a complete closed set of quadratic operator products that annihilate the entire vector space basis of the physical states. The operator products are found by exploiting the recursive nature of the coalgebra, and the requirement that the coproduct be an algebra homomorphism. If these products annihilate the lowest dimensional (single-site) state space, the recursions guarantee that they annihilate the state space for the model with arbitrarily long rows. In other words these operators are identically zero in any physical representation. We refer to these products as quadratic "zero-operators" [5], since they evaluate to zero on the basis of physical states. These relations establish the unique bialgebra structure compatible with the matrix coalgebra, and the given lowest dimensional representation of the coalgebra, namely the M-matrix, and are the familiar algebraic relations used in the algebraic Bethe ansatz. The algebraic Bethe ansatz begins with establishing a "vacuum" state Φ^0

$$\mathcal{C}\Phi^0 = 0 \tag{15}$$

and from there building a collection of states

$$\Psi(\lambda_1, \lambda_2, \cdots, \lambda_r) = \Big(\prod_{i=1}^{r} \mathcal{B}(\lambda_i)\Big)\,\Phi^0 \tag{16}$$

that are eigenvectors of the transfer matrix

$$T(\lambda)\,\Psi(\lambda_1, \lambda_2, \cdots, \lambda_r) = \Lambda(\lambda_1, \lambda_2, \cdots, \lambda_r)\,\Psi(\lambda_1, \lambda_2, \cdots, \lambda_r) \tag{17}$$

The set of parameters $\{\lambda_1, \lambda_2, \cdots, \lambda_r\}$ for which this can be done is determined by eliminating "unwanted" vectors using the zero-operator product rules.

For the eight-vertex model the $\mathcal{C}$ operator is non-singular, and one must perform local gauge transformations $\{\mathcal{A}, \cdots, \mathcal{D}\} \to \{\mathcal{A}', \cdots, \mathcal{D}'\}$ so that $\mathcal{C}'$ has a zero-eigenvector in order to finish the program of quadrature by the algebraic Bethe ansatz [6]. For the simple dimer model, one can find two vacuum states (both $\mathcal{B}$ and $\mathcal{C}$ have kernels [7]) and modify the Bethe ansatz accordingly. It may be very difficult or impossible to select parameters that eliminate unwanted terms, but that is a separate issue.

The plan of the article is to first establish those quadratic zero-operators that are compatible with the matrix coalgebra structure, and to show that this will require Equation 14 for closure, making it a necessary condition. The next step is to show that $[\mathcal{A} + \mathcal{D}, \mathcal{A}' + \mathcal{D}']$ is also a quadratic zero-operator if Equation 14 holds true, so that it is a necessary condition for commutativity of the transfer matrix.

The motivation behind this work is the desire to have a way of attacking lattice models that may not satisfy the Yang–Baxter equation. Perhaps M is not invertible, or a suitable R matrix cannot be found. It may still be possible to apply the algebraic Bethe ansatz if the operator coalgebra can be extended to o bialgebra by the means described here.

2. Operator Bialgebra Relations Compatible with the Coalgebra

We seek out binary product relations of matrix representations ($2^n \times 2^n$ matrices) of the operators $\mathcal{A}, \mathcal{B}, \mathcal{C}, \mathcal{D}$, which we denote by $\mathcal{A}_n, \mathcal{B}_n, \mathcal{C}_n, \mathcal{D}_n$ of the coalgebra that are invariant under the coproduct operation of Equation 6 (Δ is an algebra homomorphism) by which higher-dimensional representations are constructed from the lower. We work in the vector space of physical states. For the lowest dimensional representation, the 2×2, these four matrices are $\mathcal{A}_1 = A$, $\mathcal{B}_1 = B$, $\mathcal{C}_1 = C$, and $\mathcal{D}_1 = D$, and the basis for the vector space of physical states is

$$\uparrow = \begin{pmatrix} 1 \\ 0 \end{pmatrix}, \quad \downarrow = \begin{pmatrix} 0 \\ 1 \end{pmatrix} \tag{18}$$

We will use notation $A = A(\lambda)$, $A' = A(\lambda')$ with λ, λ' being different parameter sets $(a, b, c), (a', b', c')$. The action of simple associative matrix products of these operators on the basis of physical states is

$$\begin{aligned} A \cdot A' \cdot \uparrow &= aa' \uparrow, & A \cdot A' \cdot \downarrow &= bb' \downarrow \\ A \cdot B' \cdot \uparrow &= bc' \downarrow, & A \cdot B' \cdot \downarrow &= 0 \\ B \cdot A' \cdot \uparrow &= ca' \downarrow & B \cdot A' \cdot \downarrow &= 0 \\ B \cdot B' \cdot \uparrow &= 0, & B \cdot B' \cdot \downarrow &= 0 \end{aligned} \tag{19}$$

and so forth.

The coproducts (Equation 6)

$$\mathcal{A}_{n+1} = \mathcal{A}_n \otimes A + \mathcal{B}_n \otimes C, \qquad \mathcal{B}_{n+1} = \mathcal{A}_n \otimes B + \mathcal{B}_n \otimes D \tag{20}$$

give us another (the next higher dimensional) representation of the same coalgebra, the vector space of the representation is 2^{n+1}-dimensional, and we can decompose a product of operators as

$$\begin{aligned} \mathcal{A}_{n+1}\mathcal{B}'_{n+1} &= \left(\mathcal{A}_n \otimes A + \mathcal{B}_n \otimes C\right)\left(\mathcal{A}'_n \otimes B' + \mathcal{B}'_n \otimes D'\right) \\ &= \mathcal{A}_n\mathcal{A}'_n \otimes AB' + \mathcal{A}_n\mathcal{B}'_n \otimes AD' + \mathcal{B}_n\mathcal{A}'_n \otimes CB' + \mathcal{B}_n\mathcal{B}'_n \otimes CD' \end{aligned} \tag{21}$$

Let $\mathbf{v}$ be any basis vector of the vector space of the 2^n-dimensional representation Φ_n, upon which $\mathcal{A}_n, \mathcal{B}_n, \mathcal{C}_n, \mathcal{D}_n$ act. Then the basis of the 2^{n+1}-dimensional representation space Φ_{n+1} upon which $\mathcal{A}_{n+1}, \mathcal{B}_{n+1}, \mathcal{C}_{n+1}, \mathcal{D}_{n+1}$ act is the set of all

$$\{\mathbf{v}\otimes \uparrow, \quad \mathbf{v}\otimes \downarrow \quad | \quad \mathbf{v} \in \Phi_n\} \tag{22}$$

Apply the operator products Equation 21 to the basis of the vector space $\{\mathbf{v}\otimes \downarrow, \mathbf{v}\otimes \uparrow\}$ of states for the model with $n+1$ sites per row, in which $\mathbf{v} \in \Phi_n$ is any n-site vector

$$\begin{aligned} \mathcal{A}_{n+1}\mathcal{B}'_{n+1}(\mathbf{v}\otimes \uparrow) &= (\mathcal{A}_n\mathcal{A}'_n\mathbf{v}) \otimes (c'b \downarrow) + (\mathcal{A}_n\mathcal{B}'_n\mathbf{v}) \otimes (ab' \uparrow) + (\mathcal{B}_n\mathcal{A}'_n\mathbf{v}) \otimes (cc' \uparrow) \\ \mathcal{A}_{n+1}\mathcal{B}'_{n+1}(\mathbf{v}\otimes \downarrow) &= (\mathcal{A}_n\mathcal{B}'_n\mathbf{v}) \otimes (ba' \downarrow) + (\mathcal{B}_n\mathcal{B}'_n\mathbf{v}) \otimes (ca' \uparrow) \end{aligned} \tag{23}$$

$$\begin{aligned} \mathcal{B}_{n+1}\mathcal{A}'_{n+1}(\mathbf{v}\otimes \uparrow) &= (\mathcal{A}_n\mathcal{A}'_n\mathbf{v}) \otimes (a'c \downarrow) + (\mathcal{B}_n\mathcal{A}'_n\mathbf{v}) \otimes (a'b \uparrow) \\ \mathcal{B}_{n+1}\mathcal{A}'_{n+1}(\mathbf{v}\otimes \downarrow) &= (\mathcal{A}_n\mathcal{B}'_n\mathbf{v}) \otimes (cc' \downarrow) + (\mathcal{B}_n\mathcal{A}'_n\mathbf{v}) \otimes (ab' \downarrow) + (\mathcal{B}_n\mathcal{B}'_n\mathbf{v}) \otimes (bc' \uparrow) \end{aligned} \tag{24}$$

$$\begin{aligned} \mathcal{A}_{n+1}\mathcal{A}'_{n+1}(\mathbf{v}\otimes \uparrow) &= (\mathcal{A}_n\mathcal{A}'_n\mathbf{v}) \otimes (aa' \uparrow) \\ \mathcal{A}_{n+1}\mathcal{A}'_{n+1}(\mathbf{v}\otimes \downarrow) &= (\mathcal{A}_n\mathcal{A}'_n\mathbf{v}) \otimes (bb' \downarrow) + (\mathcal{A}_n\mathcal{B}'_n\mathbf{v}) \otimes (ac' \uparrow) + (\mathcal{B}_n\mathcal{A}'_n\mathbf{v}) \otimes (cb' \uparrow) \end{aligned} \tag{25}$$

$$\begin{aligned} \mathcal{B}_{n+1}\mathcal{B}'_{n+1}(\mathbf{v}\otimes \downarrow) &= (\mathcal{B}_n\mathcal{B}'_n\mathbf{v}) \otimes (aa' \downarrow) \\ \mathcal{B}_{n+1}\mathcal{B}'_{n+1}(\mathbf{v}\otimes \uparrow) &= (\mathcal{A}_n\mathcal{B}'_n\mathbf{v}) \otimes (cb' \downarrow) + (\mathcal{B}_n\mathcal{A}'_n\mathbf{v}) \otimes (ac' \downarrow) + (\mathcal{B}_n\mathcal{B}'_n\mathbf{v}) \otimes (bb' \uparrow) \end{aligned} \tag{26}$$

We can see that these binary products close in the sense that exactly the same set of binary products appear on both sides of each equation, a powerful constraint on the form of algebraic relations compatible with the coproduct.

The relations of Equations 25 and 26 can be written as

$$\begin{aligned}
(OP_1)_{n+1}(\mathbf{v}\otimes\uparrow) &= aa'\Big((OP_1)_n(\mathbf{v})\Big)\otimes\uparrow \\
(OP_1)_{n+1}(\mathbf{v}\otimes\downarrow) &= bb'\Big((OP_1)_n(\mathbf{v})\Big)\otimes\downarrow+\Big((OP_4)_n(\mathbf{v})\Big)\otimes\uparrow \\
(OP_2)_{n+1}(\mathbf{v}\otimes\downarrow) &= aa'\Big((OP_2)_n(\mathbf{v})\Big)\otimes\downarrow \\
(OP_2)_{n+1}(\mathbf{v}\otimes\uparrow) &= bb'\Big((OP_2)_n(\mathbf{v})\Big)\otimes\downarrow+\Big((OP_3)_n(\mathbf{v})\Big)\otimes\uparrow
\end{aligned} \tag{27}$$

if the operator products $OP_1,\cdots,OP_4$ are given by

$$\begin{aligned}
(OP_1) &= [\mathcal{A},\,\mathcal{A}'] \\
(OP_2) &= [\mathcal{B},\,\mathcal{B}'] \\
(OP_3) &= cb'\,\mathcal{A}\mathcal{B}'+ac'\,\mathcal{B}\mathcal{A}'-c'b\,\mathcal{A}'\mathcal{B}-a'c\,\mathcal{B}'\mathcal{A} \\
(OP_4) &= ac'\,\mathcal{A}\mathcal{B}'+cb'\,\mathcal{B}\mathcal{A}'-a'c\,\mathcal{A}'\mathcal{B}-c'b\,\mathcal{B}'\mathcal{A}
\end{aligned} \tag{28}$$

The operators given in Equation 28 annihilate the entire vector space of states for the $n=1$ representation, $\{\uparrow,\downarrow\}$. This is easy to verify using Equation 19. Therefore Equation 27 have all zeros on the right-hand side, and the operator products (OP_1) and (OP_2) annihilate the entire $n=2$ vector space of states.

In the next section we will show that Equations 23 and 24 can also be written with right-hand sides expressed entirely in terms of (OP_i) for $i=1,2,3,4$, and (OP_3) and (OP_4) also have recursion relations of the same form, making a complete set of recursions for the entire AB-subalgebra;

$$\begin{aligned}
(OP_3)_{n+1}(\mathbf{v}\otimes\uparrow) &= cc'(bb'+aa')\Big((OP_1)_n\mathbf{v}\Big)\otimes\downarrow+x\Big((OP_3)_n\mathbf{v}\Big)\otimes\uparrow+y\Big((OP_4)_n\mathbf{v}\Big)\otimes\uparrow \\
(OP_3)_{n+1}(\mathbf{v}\otimes\downarrow) &= (cca'b'+c'c'ab)\Big((OP_2)_n\mathbf{v}\Big)\otimes\uparrow+x''\Big((OP_3)_n\mathbf{v}\Big)\otimes\downarrow+y''\Big((OP_4)_n\mathbf{v}\Big)\otimes\downarrow \\
(OP_4)_{n+1}(\mathbf{v}\otimes\uparrow) &= (cca'b'+c'c'ab)\Big((OP_1)_n\mathbf{v}\Big)\otimes\downarrow+x'\Big((OP_3)_n\mathbf{v}\Big)\otimes\uparrow+y'\Big((OP_4)_n\mathbf{v}\Big)\otimes\uparrow \\
(OP_4)_{n+1}(\mathbf{v}\otimes\downarrow) &= cc'(aa'+bb')\Big((OP_2)_n\mathbf{v}\Big)\otimes\uparrow+x'''\Big((OP_3)_n\mathbf{v}\Big)\otimes\downarrow+y'''\Big((OP_4)_n\mathbf{v}\Big)\otimes\downarrow
\end{aligned} \tag{29}$$

the action of $(OP_3)_{n+1}$ on a basis of the $n+1$-vector space can be written entirely in terms of the actions of $(OP_i)_n$, $i=1,2,3,4$ on the basis of the n-vector space provided a,b,c and a',b',c' obey the star triangle equation, and the same for (OP_4). This is the core of the method; in the matrix coalgebra structure, sets of operator products (here the AA', AB', BA', BB') have recursion relations for their actions on basis vectors that close within the set. A complete set of such recursion relations mean that these binary operator products annihilate the entire vector space for any n, since the recursions are single-step, and they annihilate the $n=1$ vector space. Matrices that annihilate the entire basis of a vector space are identically zero. These operator combinations are the zero matrix on the spaces of physical states, and we have constructed those binary product laws (algebraic relations between $\mathcal{A},\mathcal{B}$ preserved by the coproduct). From these we establish the product relations for the subalgebra spanned by $\mathcal{A}$ and $\mathcal{B}$ compatible with the coalgebra structure.

Equation 29 follow directly from Equation 28 by the fact that the coproduct is an algebra homomorphism in the bialgebra (for instance $\Delta(\mathcal{A}\mathcal{A}')=\Delta(\mathcal{A})\Delta(\mathcal{A}')$), and the only question is closure—is the coproduct of (OP_n) a linear combination of only $(OP_1),\cdots,(OP_4)$ for $n=1,2,3,4$. If so, then if each operator annihilates the physical basis, so does its coproduct, and the operator is zero on the basis of all physical representations. The coproduct is an algebra homomorphism requirement of compatibility of coalgebra and algebra structure in the bialgebra; it is what guarantees the existence of our basis-annihilating operator products.

We will also show that commutativity of the transfer matrix is one of the algebraic relations (a binary product that identically annihilates the entire vector space of states) if and only if this same constraint is imposed. The constraint is the star-triangle relation.

2.1. The AB Subalgebra and the Star-Triangle Relations

We prove that (OP_3) and (OP_4) have recursion relations of the form Equation 27, so that the operator products $(OP_i)_{n+1}$ for $i = 1,2,3,4$ acting on any basis vector of the $n+1$-vector space can be expressed entirely in terms of the operator products $(OP_i)_n$ for $i = 1,2,3,4$ acting on any basis vector of the n-vector space (the set is closed with a recursive action).

Let us explicitly calculate the action of $(OP_3)_{n+1}$ on the basis of the $n+1$-site vector space. Let $\mathbf{v}$ be any basis vector of the n-site vector space Φ_n, then

$$\begin{aligned}(OP_3)(\mathbf{v}\otimes\uparrow) &= cc'(bb'+aa')\Big((OP_1)\mathbf{v}\Big)\otimes\downarrow+\Big(cab'b'\,\mathcal{A}\mathcal{B}' \\ &+ (cb'cc'+ac'a'b)\mathcal{B}\mathcal{A}'-c'a'bb\,\mathcal{A}'\mathcal{B}-(c'bcc'+a'cab')\mathcal{B}'\mathcal{A}\Big)\mathbf{v}\otimes\uparrow \end{aligned} \tag{30}$$

Force closure of the set of relations $\{OP_1, OP_2, OP_3, OP_4\}$; is there an x and y such that

$$(OP_3)(\mathbf{v}\otimes\uparrow) = cc'(bb'+aa')\Big((OP_1)\mathbf{v}\Big)\otimes\downarrow+x\Big((OP_3)\mathbf{v}\Big)\otimes\uparrow+y\Big((OP_4)\mathbf{v}\Big)\otimes\uparrow \tag{31}$$

This requires that there be a solution to the equations

$$\begin{aligned} c^2b'c'+aba'c' &= x\,ac'+y\,cb' \\ c'c'bc+a'b'ac &= x\,a'c+y\,c'b \\ cab'b' &= x\,cb'+y\,ac' \\ c'a'bb &= x\,c'b+y\,a'c \end{aligned} \tag{32}$$

(and of course there might not be a solution) the first and third of these giving

$$x = \frac{ac^2b'(c')^2+a^2ba'(c')^2-ac^2(b')^3}{a^2(c')^2-c^2(b')^2}, \qquad y = cc'\Big(\frac{c^2(b')^2+aa'bb'-a^2(b')^2}{c^2(b')^2-a^2(c')^2}\Big) \tag{33}$$

Note that the proposed zero-operators Equation 27 are unchanged by $(a,b,c) \rightleftharpoons (a',b',c')$, so the left-hand side of Equation 31 has this symmetry. Solution of the second and fourth requiring that for x to be symmetric under $(a,b,c) \rightleftharpoons (a',b',c')$ leads to

$$\frac{ac^2b'(c')^2+a^2ba'(c')^2-ac^2(b')^3}{a^2(c')^2-c^2(b')^2} = \frac{a'(c')^2bc^2+(a')^2b'ac^2-a'(c')^2b^3}{(a')^2c^2-(c')^2b^2} \tag{34}$$

which can be factored into

$$a'b'(aa'+bb')\Big(c^2-a^2-b^2\Big) = ab(aa'+bb')\Big((c')^2-(a')^2-(b')^2\Big) \tag{35}$$

or

$$\frac{a^2+b^2-c^2}{2ab} = \frac{(a')^2+(b')^2-(c')^2}{2a'b'} \tag{36}$$

the familiar solution of the star-triangle relation for the six-vertex model. The $(a,b,c) \rightleftharpoons (a',b',c')$ symmetry of y results in the same constraint

$$a'b'\Big(a^2+b^2-c^2\Big)\Big(a'b'c^2+ab(c')^2\Big) = ab\Big((a')^2+(b')^2-(c')^2\Big)\Big(ab(c')^2+a'b'c^2\Big) \tag{37}$$

The second of Equation 29 results in

$$\begin{aligned} cbb'a' + c'c'ac &= x''\,cb' + y''\,ac' \\ cca'c' + bb'ac' &= x''\,bc' + y''\,a'c \\ aab'c' &= x''\,ac' + y''\,cb' \\ a'a'bc &= x''\,a'c + y''\,c'b \end{aligned} \tag{38}$$

which are identical to Equation under interchange of a and b, and of a' with b', and also lead to Equation 35, and so to Equation 36.

Following the same set of steps, (which we do not repeat here) we discover that exactly the same condition gives us the desired recursion relation for the action of $(OP_4)_{n+1}$ on a basis of the $n+1$ vector space decomposing into actions of $(OP_1)_n, (OP_2)_n, (OP_3)_n$ and $(OP_4)_n$ on a basis of the n-vector space. Since all four operator products annihilate the basis of the $n = 1$ space, the closure of these recursions forces the operator products to annihilate the bases of their vector spaces in their higher-dimensional representations. The product relations are universal (not representation dependent).

The star-triangle relations are a necessary condition for the operators Equation 28 to annihilate the entire vector space of physical states, a full set of recursions for the four zero-operators Equation 28 can be constructed in this way.

Combining OP_3 and OP_4 we obtain one of two important spectrum-generating relations used in the algebraic Bethe ansatz

$$\mathcal{A}\mathcal{B}' = \Big(\frac{(a')^2c^2 - (c')^2b^2}{c^2a'b' - (c')^2ab}\Big)\,\mathcal{B}'\mathcal{A} - cc'\Big(\frac{aa' - bb'}{c^2a'b' - (c')^2ab}\Big)\mathcal{B}\mathcal{A}' \tag{39}$$

2.2. The BD Subalgebra and the Star-Triangle Relations

The other portion of the full bialgebra needed by the algebraic Bethe ansatz is the DB subalgebra, which we obtain by decomposing the coproducts as

$$\mathcal{D}_{n+1}\mathcal{D}'_{n+1} = \Big(C \otimes \mathcal{B}_n + D \otimes \mathcal{D}_n\Big)\Big(C' \otimes \mathcal{B}'_n + D' \otimes \mathcal{D}'_n\Big) \tag{40}$$

Let $\mathbf{v}$ be any basis of the n-site vector space, and apply the four possible binary products involving $\mathcal{B}$ and $\mathcal{D}$ to a basis of the $n+1$-site vector space

$$\begin{aligned} \mathcal{D}_{n+1}\mathcal{D}'_{n+1}(\uparrow \otimes \mathbf{v}) &= bb' \uparrow \otimes (\mathcal{D}_n\mathcal{D}'_n\mathbf{v}) \\ \mathcal{D}_{n+1}\mathcal{D}'_{n+1}(\downarrow \otimes \mathbf{v}) &= ca' \uparrow \otimes (\mathcal{B}_n\mathcal{D}'_n\mathbf{v}) + bc' \uparrow \otimes (\mathcal{D}_n\mathcal{B}'_n\mathbf{v}) + aa' \downarrow \otimes (\mathcal{D}_n\mathcal{D}'_n\mathbf{v}) \end{aligned} \tag{41}$$

$$\begin{aligned} \mathcal{B}_{n+1}\mathcal{B}'_{n+1}(\uparrow \otimes \mathbf{v}) &= aa' \uparrow \otimes (\mathcal{B}_n\mathcal{B}'_n\mathbf{v}) + bc' \downarrow \otimes (\mathcal{B}_n\mathcal{D}'_n\mathbf{v}) + ca' \downarrow \otimes (\mathcal{D}_n\mathcal{B}'_n\mathbf{v}) \\ \mathcal{B}_{n+1}\mathcal{B}'_{n+1}(\downarrow \otimes \mathbf{v}) &= bb' \downarrow \otimes (\mathcal{B}_n\mathcal{B}'_n\mathbf{v}) \end{aligned} \tag{42}$$

$$\begin{aligned} \mathcal{B}_{n+1}\mathcal{D}'_{n+1}(\uparrow \otimes \mathbf{v}) &= ab' \uparrow \otimes (\mathcal{B}_n\mathcal{D}'_n\mathbf{v}) + cb' \downarrow \otimes (\mathcal{D}_n\mathcal{D}'_n\mathbf{v}) \\ \mathcal{B}_{n+1}\mathcal{D}'_{n+1}(\downarrow \otimes \mathbf{v}) &= ac' \uparrow \otimes (\mathcal{B}_n\mathcal{B}'_n\mathbf{v}) + ba' \downarrow \otimes (\mathcal{B}_n\mathcal{D}'_n\mathbf{v}) + cc' \downarrow \otimes (\mathcal{D}_n\mathcal{B}'_n\mathbf{v}) \end{aligned} \tag{43}$$

and finally

$$\begin{aligned} \mathcal{D}_{n+1}\mathcal{B}'_{n+1}(\uparrow \otimes \mathbf{v}) &= cc' \uparrow \otimes (\mathcal{B}_n\mathcal{D}'_n\mathbf{v}) + ba' \uparrow \otimes (\mathcal{D}_n\mathcal{B}'_n\mathbf{v}) + ac' \downarrow \otimes (\mathcal{D}_n\mathcal{D}'_n\mathbf{v}) \\ \mathcal{D}_{n+1}\mathcal{B}'_{n+1}(\downarrow \otimes \mathbf{v}) &= cb' \uparrow \otimes (\mathcal{B}_n\mathcal{B}'_n\mathbf{v}) + ab' \downarrow \otimes (\mathcal{D}_n\mathcal{B}'_n\mathbf{v}) \end{aligned} \tag{44}$$

The first two sets of these identities suggest examining the combinations

$$
\begin{aligned}
(OP_5) &= [\mathcal{D},\mathcal{D}'] \\
(OP_6) &= (OP_2) = [\mathcal{B},\mathcal{B}'] \\
(OP_7) &= ca'\,\mathcal{B}\mathcal{D}' + bc'\,\mathcal{D}\mathcal{B}' - c'a\,\mathcal{B}'\mathcal{D} - b'c\,\mathcal{D}'\mathcal{B} \\
(OP_8) &= bc'\,\mathcal{B}\mathcal{D}' + ca'\,\mathcal{D}\mathcal{B}' - b'c\,\mathcal{B}'\mathcal{D} - c'a\,\mathcal{D}'\mathcal{B}
\end{aligned}
\tag{45}
$$

all of which annihilate the $n = 1$ vector space basis, and after following the same construction used to find the AB subalgebra relations, we discover that (OP_i) for $i = 5,6,7,8$ all have a set of one-step recursion relations that are closed among themselves if Equation 14 is true. Therefore each of these products annihilates the entire state-vector basis for any n, and are zero-operators. They constitute the algebraic relations of the BD subalgebra compatible with (preserved by) the matrix coalgebra structure. This pattern is repeated for all of the other subalgebras.

The second product law needed to complete the algebraic Bethe ansatz is made by combining (OP_7) with (OP_8)

$$
\mathcal{D}\mathcal{B}' = \left(\frac{(c')^2a^2 - c^2(b')^2}{(c')^2ab - c^2a'b'}\right)\mathcal{B}'\mathcal{D} - cc'\left(\frac{aa'-bb'}{(c')^2ab - c^2a'b'}\right)\mathcal{B}\mathcal{D}' \tag{46}
$$

Relation Equation 14 imposed in the standard way by re-parameterization $\lambda = (a,b,c) = (a(\theta,\gamma),b(\theta,\gamma),c(\theta,\gamma))$ with

$$
a = \sin(\gamma-\theta), \qquad b = \sin\theta, \qquad c = \sin\gamma \tag{47}
$$

simplifies the coefficients of Equations 39 and 46

$$
\begin{aligned}
(a')^2 - b^2 &= \sin(\gamma-(\theta'-\theta))\,\sin(\gamma-(\theta'+\theta)) = a(\theta'-\theta)\,a(\theta'+\theta) \\
a'b' - ab &= \sin(\gamma-(\theta'+\theta))\,\sin(\theta'-\theta) = a(\theta'+\theta)\,b(\theta'-\theta) \\
aa' - bb' &= \sin\gamma\,\sin(\gamma-(\theta'+\theta)) = c\,a(\theta'+\theta)
\end{aligned}
\tag{48}
$$

to the standard forms seen elsewhere

$$
\begin{aligned}
A(\theta)B(\theta') &= \frac{a(\theta'-\theta)}{b(\theta'-\theta)}\,B(\theta')A(\theta) - \frac{c(\theta'-\theta)}{b(\theta'-\theta)}\,B(\theta)A(\theta') \\
D(\theta)B(\theta') &= \frac{a(\theta-\theta')}{b(\theta-\theta')}\,B(\theta')D(\theta) - \frac{c(\theta-\theta')}{b(\theta-\theta')}\,B(\theta)D(\theta')
\end{aligned}
\tag{49}
$$

The analogous integrability condition for the five-vertex model found by constructing the zero-operators is much simpler [5,8]

$$
\frac{b}{a} = \frac{b'}{a'} = q, \qquad \frac{c}{a} = q\,e^{i\theta}, \quad \theta = \frac{2\pi u}{n}, \qquad u = 0,1,\cdots,n-1 \tag{50}
$$

which is not a re-parameterization of the six-vertex model; the matrix of Boltzmann weights is singular;

$$
A = \begin{pmatrix} c & 0 \\ 0 & b \end{pmatrix}, \quad D = \begin{pmatrix} a & 0 \\ 0 & 0 \end{pmatrix}, \quad B = \begin{pmatrix} 0 & 0 \\ \sqrt{ab} & 0 \end{pmatrix}, \quad C = \begin{pmatrix} 0 & \sqrt{ab} \\ 0 & 0 \end{pmatrix} \tag{51}
$$

and so no invertible R-matrix can be constructed from it.

2.3. Commuting Transfer Matrices. The AD and BC Subalgebras

By building the binary product relations for each subalgebra that exhibits this closure inherent in the matrix coalgebra structure, we can complete the entire set of algebraic relations that make the operator coalgebra into a bialgebra. This is tedious, but each step is simply a repetition of what we have done for the AB and BD subalgebras. The AD and BC subalgebras have recursion relations that close among themselves, and one of these relations is the transfer matrix commutator.

From the relations

$$\begin{aligned} \mathcal{A}_{n+1}\mathcal{D}'_{n+1}(\mathbf{v}\otimes\downarrow) &= (\mathcal{A}_n\mathcal{D}'_n\mathbf{v})\otimes(ba'\downarrow)+(\mathcal{B}_n\mathcal{D}'_n\mathbf{v})\otimes(ca'\uparrow) \\ \mathcal{D}_{n+1}\mathcal{A}'_{n+1}(\mathbf{v}\otimes\downarrow) &= (\mathcal{C}_n\mathcal{B}'_n\mathbf{v})\otimes(cc'\downarrow)+(\mathcal{D}_n\mathcal{A}'_n\mathbf{v})\otimes(ab'\downarrow)+(\mathcal{D}_n\mathcal{B}'_n\mathbf{v})\otimes(bc'\uparrow) \\ \mathcal{B}_{n+1}\mathcal{C}'_{n+1}(\mathbf{v}\otimes\downarrow) &= (\mathcal{A}_n\mathcal{D}'_n\mathbf{v})\otimes(cc'\downarrow)+(\mathcal{B}_n\mathcal{C}'_n\mathbf{v})\otimes(ab'\downarrow)+(\mathcal{B}_n\mathcal{D}'_n\mathbf{v})\otimes(bc'\uparrow) \\ \mathcal{C}_{n+1}\mathcal{B}'_{n+1}(\mathbf{v}\otimes\downarrow) &= (\mathcal{C}_n\mathcal{B}'_n\mathbf{v})\otimes(ba'\downarrow)+(\mathcal{D}_n\mathcal{B}'_n\mathbf{v})\otimes(ca'\uparrow) \end{aligned} \tag{52}$$

we find that (doing away with the subscripts; we know what these matrices act upon)

$$\begin{aligned} \Big([\mathcal{A},\mathcal{D}']-[\mathcal{A}',\mathcal{D}]\Big)(\mathbf{v}\otimes\downarrow) &= \Big(\Big(a'b[\mathcal{A},\mathcal{D}']-ab'[\mathcal{A}',\mathcal{D}]+cc'(\mathcal{C}\mathcal{B}'-\mathcal{C}'\mathcal{B})\Big)\mathbf{v}\Big)\otimes\downarrow \\ &+ \Big(\Big(ca'\mathcal{B}\mathcal{D}'-c'a\mathcal{B}'\mathcal{D}+bc'\mathcal{D}\mathcal{B}'-b'c\mathcal{D}'\mathcal{B}\Big)\mathbf{v}\Big)\otimes\uparrow \end{aligned} \tag{53}$$

Calling

$$\begin{aligned} (OP_9) &= \Big(a'b[\mathcal{A},\mathcal{D}']-ab'[\mathcal{A}',\mathcal{D}]+cc'(\mathcal{C}\mathcal{B}'-\mathcal{C}'\mathcal{B})\Big) \\ (OP_{10}) &= [\mathcal{A},\mathcal{D}']-[\mathcal{A}',\mathcal{D}] \end{aligned} \tag{54}$$

(OP_9) and (OP_{10}) could be zero-operators if for some x and y

$$\begin{aligned} (OP_{10})(\mathbf{v}\otimes\downarrow) &= \Big((OP_9)\mathbf{v}\Big)\otimes\downarrow+\Big((OP_7)\mathbf{v}\Big)\otimes\uparrow \\ (OP_9)(\mathbf{v}\otimes\downarrow) &= (ab'+a'b)\Big((OP_{10})\mathbf{v}\Big)\otimes\downarrow+\Big(\Big(a'bca'\mathcal{B}\mathcal{D}'-a'bb'c\mathcal{D}'\mathcal{B} \\ &- ab'c'a\mathcal{B}'\mathcal{D}+abb'c'\mathcal{D}\mathcal{B}'+c^2c'a'\mathcal{D}\mathcal{B}'-c(c')^2a\mathcal{D}'\mathcal{B}\Big)\mathbf{v}\Big)\otimes\uparrow \end{aligned} \tag{55}$$

$$= (ab'+a'b)\Big((OP_{10})\mathbf{v}\Big)\otimes\downarrow+x\Big((OP_7)\mathbf{v}\Big)\otimes\uparrow+y\Big((OP_8)\mathbf{v}\Big)\otimes\uparrow \tag{56}$$

which requires that x and y be symmetric under $(a,b,c)\rightleftharpoons(a',b',c')$

$$(a')^2bc = x\,ca'+y\,bc', \qquad abb'c'+c^2c'a' = x\,bc'+y\,ca' \tag{57}$$

or

$$x = \frac{bc^2(a')^3-ab^2b'(c')^2-c^2(c')^2a'b}{c^2(a')^2-b^2(c')^2} \tag{58}$$

which is identical to Equation 33 with $a\rightleftharpoons b$, $a'\rightleftharpoons b'$. We conclude that in order for the condition of commuting transfer matrices ($OP_{10}=0$) to hold, it is necessary for (a,b,c) and (a',b',c') to be related by the star-triangle relations Equation 14. Analysis of y in Equation 57 leads to the same conclusion. One still needs to show that (OP_9) and (OP_{10}) annihilate the other half of the basis, namely $\mathbf{v}\otimes\uparrow$, and a short calculation verifies that they do. The same conclusion is drawn for the five-vertex model, (OP_{10}) is a zero-operator and transfer matrices with different spectral parameters commute [7].

2.4. Generality of the Method

Despite the lack of elegance that other constructions of bialgebras utilizing a solution R of the Yang–Baxter equation may have (such as the RTT construction [3]), the methods used here appear to be quite general. Application to Sweedler's Hopf algebra [9] generated by x, g with coproduct $\Delta(g) = g \otimes g$ and $\Delta(x) = x \otimes g + 1 \otimes x$ leads uniquely to the algebraic relations $g \cdot g = 1$ and $x \cdot x = 0$ making the coalgebra a bialgebra (they are preserved by the coproduct), starting with a two-dimensional representation [8]. The less trivial example of the Hopf algebra $U_q(\mathfrak{b}_+)$ with generators K and X (the quantum deformation of the universal enveloping algebra of the upper Borel subalgebra of $\mathfrak{sl}(2)$) whose coproducts are

$$\Delta(K) = K \otimes K, \qquad \Delta(X) = X \otimes K^{-1} + K \otimes X \tag{59}$$

uniquely produces $qX \cdot K = K \cdot X$ from such a representation [8]. The two-dimensional representation is the set of local Boltzmann weights, quantities that the model-builder would possess.
Starting with a two-dimensional representation of the Yangian Hopf algebra $Y(\mathfrak{gl}(2))$

$$\begin{aligned} A(u) &= \begin{pmatrix} 1+u^{-1} & 0 \\ 0 & 1 \end{pmatrix}, & B(u) &= \begin{pmatrix} 0 & u^{-1} \\ 0 & 0 \end{pmatrix} \\ C(u) &= \begin{pmatrix} 0 & 0 \\ u^{-1} & 0 \end{pmatrix}, & D(u) &= \begin{pmatrix} 1 & 0 \\ 0 & 1+u^{-1} \end{pmatrix} \end{aligned} \tag{60}$$

with matrix coproduct, analytical construction of the zero-operators leads to the well-known product rules [8]

$$\begin{aligned} [A(u), B(v)] &= \frac{A(u)B(v) - A(v)B(u)}{u-v}, & [A(u), D(v)] &= \frac{C(u)B(v) - C(v)B(u)}{u-v} \\ [A(u), C(v)] &= \frac{C(u)A(v) - C(v)A(u)}{u-v}, & [B(u), C(v)] &= \frac{D(u)A(v) - D(v)A(u)}{u-v} \\ [A(u), A(v)] &= 0, & [B(u), B(v)] &= 0 \end{aligned} \tag{61}$$

and finally when applied to the five-vertex or hexagonal lattice dimer model with spectral parameters a, b and c (using notation $\mathcal{X} = \mathcal{X}(a, c)$, $\mathcal{X}' = \mathcal{X}(a', c')$) it results in a bialgebra structure

$$\begin{aligned} \mathcal{A}'\mathcal{B} = \mathcal{A}\mathcal{B}' &= q\Big(\frac{\mathcal{B}'\mathcal{A} - \mathcal{B}\mathcal{A}'}{\frac{c}{a} - \frac{c'}{a'}}\Big), & \mathcal{D}'\mathcal{B} = \mathcal{D}\mathcal{B}' &= -q\Big(\frac{\mathcal{B}'\mathcal{D} - \mathcal{B}\mathcal{D}'}{\frac{c}{a} - \frac{c'}{a'}}\Big) \\ \mathcal{C}\mathcal{A}' = \mathcal{C}'\mathcal{A} &= q\Big(\frac{\mathcal{A}'\mathcal{C} - \mathcal{A}\mathcal{C}'}{\frac{c}{a} - \frac{c'}{a'}}\Big), & \mathcal{C}'\mathcal{D} = \mathcal{C}\mathcal{D}' &= -q\Big(\frac{\mathcal{D}'\mathcal{C} - \mathcal{D}\mathcal{C}'}{\frac{c}{a} - \frac{c'}{a'}}\Big) \\ [\mathcal{A}, \mathcal{D}'] &= \frac{q}{\frac{c'}{a'} - \frac{c}{a}}\Big(\mathcal{B}'\mathcal{C} - \mathcal{B}\mathcal{C}'\Big), & \mathcal{C}'\mathcal{B} = \mathcal{C}\mathcal{B}' &= \frac{q}{\frac{c}{a} - \frac{c'}{a'}}\Big(\mathcal{D}'\mathcal{A} - \mathcal{D}\mathcal{A}'\Big) \end{aligned} \tag{62}$$

closely related to the Yangian, but which has no R-matrix or antipode [5,8]. It was for the purposes of studying such models that the technique was developed in the first place. Note that Sweedler's algebra and the Yangian both possess R-matrices, but $U_q(\mathfrak{b}_+)$ and the five-vertex model do not, yet the methods of this article lead uniquely to their bialgebra structure equations.

3. The Algebraic Bethe Ansatz

Once the spectrum-generating relations Equation 49 have been established and a pseudo-vacuum state Φ_n^0 with $\mathcal{C}\Phi_n^0 = 0$ has been found, the eigenvalues of the transfer matrix can be found by the algebraic Bethe ansatz.

The vector

$$\Phi_n^0 = \uparrow \otimes \uparrow \otimes \uparrow \otimes \cdots \otimes \uparrow \otimes \uparrow, \qquad n - \text{factors} \tag{63}$$

satisfies the requirements of the Bethe ground state (which we show inductively)

$$\begin{aligned}
\mathcal{C}_n\Phi_n^0 &= \left(\mathcal{C}_{n-1}\Phi_{n-1}^0\right)\otimes\left(A\uparrow\right)+\left(\mathcal{D}_{n-1}\Phi_{n-1}^0\right)\otimes\left(C\uparrow\right)\\
&= a\left(\mathcal{C}_{n-1}\Phi_{n-1}^0\right)\otimes\uparrow\\
&= \cdots = a^{n-1}(C\uparrow)\otimes\uparrow\otimes\cdots\otimes\uparrow\otimes\uparrow=0 \quad \text{by iteration}\\
\mathcal{A}_n\Phi_n^0 &= \left(\mathcal{A}_{n-1}\Phi_{n-1}^0\right)\otimes\left(A\uparrow\right)+\left(\mathcal{B}_{n-1}\Phi_{n-1}^0\right)\otimes\left(C\uparrow\right)\\
&= a\left(\mathcal{A}_{n-1}\Phi_{n-1}^0\right)\otimes\uparrow=a^n\Phi_n^0\\
\mathcal{D}_n\Phi_n^0 &= \left(\mathcal{C}_{n-1}\Phi_{n-1}^0\right)\otimes\left(B\uparrow\right)+\left(\mathcal{D}_{n-1}\Phi_{n-1}^0\right)\otimes\left(D\uparrow\right)\\
&= b\left(\mathcal{D}_{n-1}\Phi_{n-1}^0\right)\otimes\uparrow=b^n\Phi_n^0
\end{aligned}\tag{64}$$

This state is an eigenstate of the transfer matrix

$$T(\theta)\Phi_n^0=\left(\mathcal{A}_n+\mathcal{D}_n\right)\Phi_n^0=\left(a^n(\theta)+b^n(\theta)\right)\Phi_n^0 \tag{65}$$

Products of the $\mathcal{B}$ operators applied to the vacuum state will produce vectors that are eigenvalues of the transfer matrix if the unwanted vectors in the expansion of $T(\theta)\Phi$ have zero coefficients

$$\Phi=\mathcal{B}(\theta_1)\cdots\mathcal{B}(\theta_r)\,\Phi_n^0,\qquad [\mathcal{B}(\theta_i),\,\mathcal{B}(\theta_j)]=0,\qquad T(\theta)\,\Phi=\Lambda\,\Phi+\text{unwanted vectors} \tag{66}$$

The eigenvalues are

$$\Lambda = \left(a^n(\theta)\prod_i^r\frac{a(\theta_i-\theta)}{b(\theta_i-\theta)}+b^n(\theta)\prod_i^r\frac{a(\theta-\theta_i)}{b(\theta-\theta_i)}\right) \tag{67}$$

provided the set $\{\theta_1,\theta_2,\cdots,\theta_r\}$ is chosen such that

$$\left(\frac{b(\theta_j)}{a(\theta_j)}\right)^n=-\prod_{i\neq j}^r\frac{b(\theta_j-\theta_i)\,a(\theta_i-\theta_j)}{a(\theta_j-\theta_i)\,b(\theta_i-\theta_j)} \tag{68}$$

which eliminates the unwanted terms in the expansion of $T(\theta)\,\Phi$ [6].

The conditions for elimination of the unwanted terms in the five-vertex model (hexagonal lattice dimer model) is far simpler [5,8], from

$$\mathcal{A}\mathcal{B}'=\frac{bc^{-1}}{1-\frac{c'b}{cb'}}\left(\mathcal{B}'\mathcal{A}-\sqrt{\frac{ab'}{a'b}}\mathcal{B}\mathcal{A}'\right) \tag{69}$$

with a similar relation involving $\mathcal{B}$ and $\mathcal{D}$, together with $[\mathcal{B},\mathcal{B}']=0$

$$T\left(\mathcal{B}'\Phi^0\right)=\frac{bc^{-1}}{1-\frac{c'b}{cb'}}\left(\left(c^n-a^n\right)\mathcal{B}'\Phi^0-\left(\sqrt{\frac{ab'}{a'b}}\,c'^n-\sqrt{\frac{a'b}{ab'}}\,a'^n\right)\mathcal{B}\Phi^0\right) \tag{70}$$

one is led to Equation 50, and this easily extends to the higher excitations [5,8].

4. Conclusions

The method of constructing zero-operators, quadratic operator products that annihilate the entire state-space which are preserved by the coproduct, can be used to deduce the conditions (on the model parameters) under which a lattice model has commuting transfer matrices. Coassociativity of the coproduct operation, which is the operation by which the transfer matrix is built up from local Boltzmann weights, is used to obtain recursion relations for a set of operators that in their lowest dimensional representation annihilate the state space. The recursions guarantee that they will annihilate all higher-dimensional state spaces. This promotes the operator coalgebra to a bialgebra. If the bialgebra can be shown to possess an antipode it may actually be a Hopf algebra, and the existence of an R-matrix would make it quasi-triangular, but these steps are not needed for a complete quadrature. Only two sub-algebras, the AB and DB, are required to construct Bethe vectors and perform the algebraic Bethe ansatz, assuming that the resulting equations for the elimination of unwanted terms can be solved. Simple possession of these subalgebras is no guarantee that this can be done.

In this article we have shown that a necessary condition for the zero-operator recursions to exist (for the six vertex model) is that the Boltzmann weights are constrained by

$$\frac{a^2+b^2-c^2}{2ab} = \frac{(a')^2+(b')^2-(c')^2}{2a'b'}$$

the star-triangle relations. One of these zero-operators is in fact $[T, T']$, $T = \mathcal{A} + \mathcal{D}$, and therefore a necessary condition for commutativity of transfer matrices with different Boltzmann weights is the star-triangle condition. This condition is well-known to be sufficient [1].

The zero-operator method does not make use of the Yang–Baxter equation or require the existence of an R-matrix. It is a direct determination of the spectrum-generating bialgebra of a lattice model, requiring only the lowest dimensional non-trivial representation of the bialgebra (the M matrix) and has been used to perform the algebraic Bethe ansatz for models with a singular M-matrix [5,7,8], in which the bialgebra is closest in structure to a Yangian quantum group, but possesses no antipode. This approach for constructing a bialgebra is complementary to the RTT method [3,10] which begins with a solution to the Yang–Baxter equation (an invertible R-matrix constructed from M) and produces a bialgebra with matrix coalgebra.

The zero-operators themselves are linear combinations of products of the $\mathcal{A}, \mathcal{B}, \mathcal{C}$ and $\mathcal{D}$ operators for pairs of spectral parameters. The commutator of the transfer matrices is just such an object, and so the method is particularly well-suited to the problem of establishing necessary spectral conditions under which the transfer matrices will commute.

The methods used here are particularly amenable to the use of computer algebra systems. All of the calculations here were facilitated, and verified, by the use of REDUCE [11].

References

1. Baxter, R.J. *Exactly Solved Models in Statistical Mechanics*; Academic Press: Waltham, MA, USA, 1982.
2. Chari, V.; Pressley, A. *A Guide to Quantum Groups*; Cambridge University Press: Cambridge, UK, 1994.
3. Kauffman, L.H. *Knots and Physics*; World Scientific: Singapore, 1991.
4. McCoy, B.M. *Advanced Statistical Mechanics*; Oxford University Press: Oxford, UK, 2010.
5. Schmidt, J.R. The algebraic Bethe Ansatz without the Yang–Baxter equation. *Can. J. Phys.* **2008**, *86*, 1177–1193.
6. Takhtadzhan, L.A.; Faddeev, L.D. The quantum method of the inverse problem and the Heisenberg XYZ model. *Uspekhi Mat. Nauk* **1979**, *34*, 13–63.
7. Schmidt, J.R. A modified Bethe Ansatz for the two-dimensional dimer problem. *Can. J. Phys.* **2007**, *85*, 745–762.
8. Schmidt, J.R. Spectrum-Generating bialgebra of the hexagonal-lattice dimer model. *Can. J. Phys.* **2009**, *87*, 1099–1125.
9. Sweedler, M.E. *Hopf Algebras*; W.A. Benjamin, Inc.: New York, NY, USA, 1969.
10. Faddeev, L.D.; Reshetikhin, N.Y.; Takhtajan, L.A. Quantization of Lie Groups and Lie Algebras. *Leningrad Math. J.* **1990**, *1*, 193–225.
11. REDUCE. Available online: http://reduce-algebra.sourceforge.net/ (accessed on 18 December 2008).

Permissions

All chapters in this book were first published by MDPI; hereby published with permission under the Creative Commons Attribution License or equivalent. Every chapter published in this book has been scrutinized by our experts. Their significance has been extensively debated. The topics covered herein carry significant findings which will fuel the growth of the discipline. They may even be implemented as practical applications or may be referred to as a beginning point for another development.

The contributors of this book come from diverse backgrounds, making this book a truly international effort. This book will bring forth new frontiers with its revolutionizing research information and detailed analysis of the nascent developments around the world.

We would like to thank all the contributing authors for lending their expertise to make the book truly unique. They have played a crucial role in the development of this book. Without their invaluable contributions this book wouldn't have been possible. They have made vital efforts to compile up to date information on the varied aspects of this subject to make this book a valuable addition to the collection of many professionals and students.

This book was conceptualized with the vision of imparting up-to-date information and advanced data in this field. To ensure the same, a matchless editorial board was set up. Every individual on the board went through rigorous rounds of assessment to prove their worth. After which they invested a large part of their time researching and compiling the most relevant data for our readers.

The editorial board has been involved in producing this book since its inception. They have spent rigorous hours researching and exploring the diverse topics which have resulted in the successful publishing of this book. They have passed on their knowledge of decades through this book. To expedite this challenging task, the publisher supported the team at every step. A small team of assistant editors was also appointed to further simplify the editing procedure and attain best results for the readers.

Apart from the editorial board, the designing team has also invested a significant amount of their time in understanding the subject and creating the most relevant covers. They scrutinized every image to scout for the most suitable representation of the subject and create an appropriate cover for the book.

The publishing team has been an ardent support to the editorial, designing and production team. Their endless efforts to recruit the best for this project, has resulted in the accomplishment of this book. They are a veteran in the field of academics and their pool of knowledge is as vast as their experience in printing. Their expertise and guidance has proved useful at every step. Their uncompromising quality standards have made this book an exceptional effort. Their encouragement from time to time has been an inspiration for everyone.

The publisher and the editorial board hope that this book will prove to be a valuable piece of knowledge for researchers, students, practitioners and scholars across the globe.

List of Contributors

Tahsin Oner and Ibrahim Senturk Tugce Katican
Department of Mathematics, Ege University, ˙Izmir 35100, Turkey

Gulsah Oner
Multidisciplinary Vocational School Technical Programme, Dokuz Eylul University, ˙Izmir 35210, Turkey

Florin F. Nichita
Institute of Mathematics of the Romanian Academy, 21 Calea Grivitei Street, 010702 Bucharest, Romania

Solomon Marcus
Institute of Mathematics of the Romanian Academy, 21 Calea Grivitei Street, 010702 Bucharest, Romania
Romanian Academy, 125 Calea Victoriei Street, 010071 Bucharest, Romania

Alexander E. Hoffnung
Department of Mathematics, Temple University, 1805 N. Broad Street, Philadelphia, PA 19122, USA

Joel Lemay
Department of Mathematics and Statistics, University of Ottawa, Ottawa, K1N 6N5, Canada

Kenichi Shimizu
Graduate School of Mathematics, Nagoya University, Furo-cho, Chikusa-ku, Nagoya 464-8602, Japan

Michiel Hazewinkel
Burg. 's Jacob Laan 18, NL-1401BR BUSSUM, The Netherlands

Jon Links
Centre for Mathematical Physics, School of Mathematics and Physics, The University of Queensland, Brisbane 4072, Australia

Robert G. Underwood
Department of Mathematics/Informatics Institute, Auburn University Montgomery, Montgomery, AL 36124, USA

Bartosz Zieliński
Department of Theoretical Physics and Computer Science, University of Łód´z, Pomorska 149/153, 90-236, Łód´z, Poland

Tomasz Brzeziński and Simon A. Fairfax
Department of Mathematics, Swansea University, Singleton Park, Swansea SA2 8PP, UK

Apoorva Khare
Department of Mathematics and Department of Statistics, Stanford University, 390 Serra Mall, Stanford, CA 94305, USA

Jeffrey R. Schmidt
Departments of Physics and Mathematics, University of Wisconsin-Parkside 900 Wood Road, Kenosha, WI 53141, USA

Index

A

Algebraic Structures, 2, 5, 133

Algebras, 1-3, 5-7, 9-12, 14, 17, 22, 26-31, 33-34, 36, 46, 50-51, 53, 56-60, 65, 67-69, 71-72, 75-77, 83, 86-93, 99, 104-116, 165, 170-173, 176-178, 180-181, 183, 187, 192-196, 198-199, 201-202, 204, 214-218, 220, 233

Antipode, 95, 112, 120-121, 125, 128, 146, 173, 176-178, 195-196, 231, 233

Anyonic Pairing, 133-134, 141

Augmentation Ideal, 214

Automorphism, 59-61, 70-72, 76, 78, 83-85, 90, 92, 103, 106, 113, 122, 124

Axiomatic System, 1-2, 9

Axioms, 1-3, 5, 9-10, 22, 28-29, 32, 37, 43-45, 55, 159

B

Bialgebra, 144, 146, 149-151, 154-158, 196, 204, 221, 224, 226, 228, 230-231, 233

Bilinear Form, 80, 88, 91, 104, 108, 112, 114, 119, 121, 124, 199

Binary Operation, 2-3, 10, 156

Binary Relation, 7, 14, 54

Boolean Algebra, 7-8, 14-16, 28

C

Categorification, 30-36, 38, 43, 45, 50-51, 53, 56, 59

Coaction, 115-116, 118, 120, 172, 176-179, 181, 183-188, 190-192

Coalgebra, 93-94, 112-113, 115-116, 118-120, 124, 131, 144-149, 155, 160, 185, 196-197, 204-205, 221, 224-226, 230-231, 233

Cocontinuous Functor, 33, 47, 49

Coinvariant Subalgebra, 177, 180, 190

Commutativity, 2, 71, 98, 176, 221, 224, 227, 233

Comultiplication Map, 146, 148, 150

Corings, 160-164, 166, 170-171

Counit Map, 146-147, 149-150, 157

Coupling Parameters, 133-134, 141-142

D

Dihedral Group, 133, 142

Duality, 91-99, 102-103, 105-107, 113, 118, 120, 124, 148, 160-161, 164, 170-171

E

Eigenvalues, 97, 134, 231-232

Enriched Bicategories, 30-33, 38, 42-43, 47

Equivalent Set, 1-3

G

Groupoidification, 30-32, 34, 38-39, 56

Groupoids, 30-34, 38-43, 45-48, 50, 57

H

Haar Measure, 91, 93, 120, 177, 179

Hasse-schmidt Derivation, 127-131

Hecke Algebras, 30-31, 33-34, 36, 46, 50-51, 53, 56, 196, 201, 215-216, 218

Higher Derivation, 127

Homomorphism), 65, 196, 224

Hopf Algebra, 29, 91-93, 95, 99-100, 103-104, 106, 108, 112-115, 119-122, 124, 127-134, 140, 142, 146, 172-173, 176-179, 181-182, 184, 188, 194-197, 199, 201-202, 204-206, 214-215, 218, 231, 233

Hyperbolic Functions, 23-24, 27-28

I

Idempotent, 7-8, 67, 93, 110-112, 114, 119

Incidence Geometries, 30-31, 33, 55

Indecomposable Representation, 00, 02-05

Invariant, 10, 54, 83-85, 90-91, 93, 98-99, 109, 125, 133, 141, 143, 179, 185, 190, 224

Isomorphism, 32-33, 37-42, 45-47, 54-55, 58-59, 65-67, 72, 75, 80, 82-87, 95-97, 99, 102-104, 107-110, 113-114, 118-121, 131, 151, 157, 161-163, 166, 177, 181-185, 190, 201

L

Lemma, 6-8, 11-12, 14-21, 49, 57, 60, 63, 65, 69, 72, 76, 81-83, 86, 95-96, 98, 100, 102, 104-106, 109, 111, 114, 150, 162, 167, 169, 194, 196-199, 201-204, 206-208, 210-211, 214-215, 217

Lie Algebra, 58-60, 81-82, 86-88, 128, 130, 133, 194, 199, 201, 214-217

Linear Operator, 10, 32, 35, 42, 47, 49

M

Matrix Coproduct, 221, 231

Monoidal Bicategories, 33, 42-44, 47-48

Monoidal Category, 32, 43, 92-93, 95-96, 98-99, 105-106, 113

Monoidal Functor, 33, 105

Morphism, 38, 42-45, 48-50, 59-61, 63-66, 69-70, 75, 78, 84, 95-96, 107, 109, 113, 130, 162-166, 168-169, 176

Myhill-nerode Bialgebra, 144, 158

Myhill-nerode Theorem, 144, 155-156

N

Natural Bijection, 94-96, 103

O

Operator Coalgebra, 221, 224, 230, 233

P

Permutation, 30-33, 35-36, 38, 43, 45-46, 51, 67, 80, 122, 125, 201

Physical States, 224-226, 228

Primitive Element, 210

Principal Comodule Algebra, 172-173, 178-180, 183-184, 187-188, 191-192

Q

Quantum Groups, 2, 9-10, 22, 26, 58-59, 89-91, 93, 126, 142, 171, 194, 196, 198-199, 201, 233

Quasitriangular Structure, 144, 151, 156-158

Quiver, 58-61, 63-73, 75-78, 80-87, 89-90, 196, 204, 220

R

Representation Theory, 5, 30-31, 33-34, 36-37, 55, 58-60, 67, 78, 80, 83, 89-90, 126, 201

Ring Extension, 160, 162-165

Ringel-hall Algebra, 58-60, 86-89

Root System, 58, 78, 82, 218

S

Scalar Multiples, 91, 104, 108, 114, 187

Skew-primitive, 194-197, 202-204, 206-208, 211-212, 218

Subalgebra, 86, 133, 145, 172, 176-177, 180-181, 184-185, 190-191, 198-199, 201, 205, 207, 215-216, 226-228, 230-231

Symmetric Euler, 80-81, 88

Symmetry Algebra, 142

T

Tensor Algebra, 58-59, 68-69, 71-72, 75

Tensor Product, 42-43, 45, 47, 88, 94, 97, 105, 135, 139, 146, 151, 161

Topos Theory, 31, 47, 57

Toposes, 33, 47-49, 57

Transcendence, 23-25, 28-29

Transcendental Numbers, 22-26, 28

Transfer Matrix, 135, 139-142, 221, 224, 227, 230-232

Trivial Path, 67, 70

U

Unary Operation, 2-3, 10

V

Valued Graph, 58-59, 82-83

Vertex Model, 133-134, 142, 221, 224, 227, 230-233

W

Wajsberg-algebra, 10

Y

Yang-baxter Equation, 1-2, 5-13, 15-17, 19, 21-24, 26-29, 53, 55, 59, 89, 133, 135, 142, 152, 159-160, 164, 166, 171, 224, 230, 233

www.ingramcontent.com/pod-product-compliance
Lightning Source LLC
LaVergne TN
LVHW060940220926
848485LV00015B/1148